HANDBOOK OF NEUROPSYCHOLOGY, 2nd Edition

VOLUME 7

THE FRONTAL LOBES

HANDBOOK OF NEUROPSYCHOLOGY, 2nd Edition

Series Editors

FRANÇOIS BOLLER

INSERM, Unité 324, Centre Paul Broca, 75014 Paris, France

and

JORDAN GRAFMAN

National Institute of Neurological Disorders and Stroke, National Institutes of Health, Bethesda, MD 20892, USA

Volume 7

THE FRONTAL LOBES

Editor

JORDAN GRAFMAN

National Institute of Neurological Disorders and Stroke, National Institutes of Health, Bethesda, MD 20892, USA

ELSEVIER

Amsterdam – London – New York – Oxford – Paris – Shannon – Tokyo

2002

ELSEVIER SCIENCE B.V.
Sara Burgerhartstraat 25
P.O. Box 211, 1000 AE Amsterdam, The Netherlands

First edition 1993
Second edition, first printing 2002

Library of Congress Cataloging in Publication Data
A catalog record from the Library of Congress has been applied for.

British Library Cataloguing in Publication Data
A catalogue record from the British Library has been applied for.

ISBN series 0 444 50376 5 (HB)
ISBN series 0 444 50377 3 (PB)
ISBN this edition (hardbound) 0 444 50365 X
ISBN this edition (paperback) 0 444 50374 9

∞ The paper used in this publication meets the requirements of ANSI/NISO Z39.48-1992 (Permanence of Paper).
Printed in The Netherlands

Preface

We are pleased to present the seventh volume of the second edition of the Handbook of Neuropsychology. As already demonstrated by the first edition and by the first six volumes of the second edition of the Handbook, neuropsychology is a field of science that has undergone extraordinary growth and changes in recent years. Planning for the first edition of the Handbook started over sixteen years ago and even though the more recent volumes of the first edition were designed to reflect some of the changes that have taken place in the field of neuropsychology, we have decided, with the encouragement of the publisher, that it would be worthwhile to prepare a new edition. As the series co-editors, we intend to ensure that the new edition of the Handbook of Neuropsychology remains the principal reference source in the field, continuing to provide comprehensive and current coverage of both experimental and clinical aspects of neuropsychology. To this end, we have asked the authoritative chapter authors to produce new in-depth reviews that go beyond a summary of their results and point of view. Each chapter is up-to-date, covering the latest developments in methodology and theory. Discussion of 'bedside' evaluations, laboratory techniques, as well as extensive discussions of theoretical models are all to be found in the Handbook. The Handbook also presents the latest findings and methodologies of functional neuroimaging techniques such as PET, fMRI, and transcranial magnetic stimulation (TMS).

We are confident that the Handbook will continue to be an essential reference source for clinicians such as neuropsychologists, neurologists, psychiatrists, and psychologists, as well as for all scientists engaged in research in the cognitive neurosciences. This second all-new edition is designed to update chapters covering research domains where considerable developments have occurred. In addition, there has been an in-depth reorganization of content areas that have spawned new approaches since the first edition. All the chapters included in this new edition will provide the most recent data and references for further studies and research.

The first volume included an introduction section that focused on practical and theoretical issues of general interest. Two chapters covered, in a novel and comprehensive fashion, clinical evaluation and neuropsychological assessment, with an emphasis not so much on the description of tests, but on their rationale. One of the features of neuropsychology in recent years has been the spectacular comeback of single-case studies and that is why the chapter on statistical approaches compared statistical procedures appropriate for groups to those of single cases. Hemispheric specialization remains an important topic, which was examined in the introduction under two different points of view. One chapter summarized the contribution to neuropsychology provided by the commissurotomy ('split brain') model, while another chapter reviewed experimental assessments of hemispheric specialization in normal individuals. An introduction to current neuroimaging techniques used language disorders as an illustration. Neurophysiological techniques were reviewed with an emphasis on evoked potentials (ERP). Several chapters dealt with the application of theoretical models to neuropsychology including a discussion of the lesion method and of computer modeling.

In the first volume, the Introduction was followed by the section on Attention edited by Giacomo Rizzolatti. It included four chapters. Two of them concerned selective attention.

The first was essentially devoted to visuo/spatial attentional phenomena, while the focus of the second was on the temporal aspects of attention. The phenomenon of failure to orient, neglect and neglect-related phenomena was dealt with in the third chapter. This chapter included a large section devoted to the anatomical localization of lesions producing neglect in humans. Finally, the last chapter reviewed the anatomy and the neurophysiological properties of the circuits whose lesion produces neglect deficits in primates. In that chapter, the various theories of neglect were reviewed and their validity discussed from a neurophysiological perspective.

The second volume of the Handbook was a special 'in memoriam' volume. It was dedicated to the memory of Laird Cermak who, despite his grave illness, vigorously and resolutely took on the task of preparing a new section on memory and its disorders. The volume included chapters on animal models and neuropsychological assessment. Memory was discussed from the anatomical and clinical viewpoints, and memory disorders resulting from specific diseases such as Herpes simplex encephalitis and Alzheimer's disease were considered in detail. The section further provided a cognitive neuropsychological analysis of various forms of memory, including explicit memory, remote memory and semantic memory. Chapters on confabulation and on functional amnesia were also included. We were all deeply saddened by the news of Laird's untimely death. We are very much indebted to Mieke Verfaellie for supervising the final stages of the preparation of this volume for the revised handbook.

The third volume, edited by Rita Sloan Berndt, covered traditional approaches to, as well as new techniques for, investigating language disorders by leading researchers in the field of language and aphasia research. The volume is divided into four parts. The first part, entitled the study of aphasia, includes a chapter on historical developments, a discussion of the relationship between neuroanatomy and language, one on sign language, one on cross-lingual studies and a review of aphasia in bilinguals and polyglots. The second section ('understanding the symptoms of aphasia') discussed and analyzed the symptoms of aphasia and related disorders including reading and writing. The next section went further into related disorders with chapters that discuss the relationship between language and memory and attention as well as disorders of skilled movements, of body representation and of number processing. The volume concluded with a review of emerging methods for the study of language and aphasia including neuroimaging, ERP, TMS and studies of 'split brain' subjects.

Volume 4, edited by Marlene Behrmann, covered disorders of visual behavior. The volume began with a chapter reviewing the neurophysiology of spatial vision with special emphasis on single unit recordings in non-human primates. The next three chapters reviewed the recent work on recognition deficits for faces (prosopagnosia), objects (visual object agnosia) and words (peripheral dyslexias). Disorders of spatial representation, of color processing and of mental imagery were presented next including a detailed discussion of the neuropsychological behavior as well as the underlying neural substrate of these disorders. Additional chapters dealt with Balint's syndrome, with blindsight and with visuospatial or constructional disorders. Finally, the relationship between eye movements and brain damage has been described in detail.

The fifth volume edited by Guido Gainotti covered emotional behavior and its disorders. It included introductory chapters dealing with basic theoretical and anatomical issues in the neuropsychological study of emotions. A central part of this volume addressed the problem of hemispheric asymmetries in emotional representation and a final group of chapters examined the neural mechanisms of the stress response and reviewed the main emotional disorders. In the introductory chapters, an effort was made to present both neurobiologically oriented and cognitively oriented theories of emotion. Both the detailed anatomo-clinical and theoretical aspects of the anatomical substrates of emotions were covered in depth. In the central part of the volume, the claims for right hemisphere dominance for emotions and emotional communication were contrasted with those assuming a different hemispheric specialization for positive versus negative emotions and with models assuming asymmetric cortico-limbic control of human emotion. Finally, in the last chapters of the volume, individual differences in the hemispheric control of the stress response were discussed and the neural mechanisms of affective/emotional disturbances were approached with neuropsychological methods and with functional neuroimaging techniques.

The sixth volume of the Handbook, edited by François Boller and Stefano Cappa, was devoted to topics related to Aging and Dementia. The volume began with two chapters dealing with age-related cognitive and neurobiological alterations in animals, including a detailed review of data obtained with transgenic and knockout mouse technology. The next chapter reviewed the cognitive changes associated with normal aging. The gamut of symptoms that occur in Alzheimer's disease (AD) were then described and analyzed. They include effects on attention, language, memory, non-verbal functions with emphasis on spatial abilities, olfaction and the motor system. The discussion of dementia syndromes was presented in two sections. The first concerned AD, discussed from the points of view of epidemiology, neuropathology and neurochemistry, concluding with a review of current and future treatments. The other section dealt with non-AD dementias including Fronto-temporal and Lewy body dementias and specific conditions such as Parkinson's and Huntington's disease, as well as HIV infection. The volume included a review of brain imaging and cerebral metabolism findings in aging and dementia. The final chapters reviewed the relations between culture and dementia and the special syndrome of severe dementia.

The volume you now have in your hands, Volume 7 of this new edition of the Handbook, comprises a completely revamped section on the Frontal Lobes edited by Jordan Grafman. A chapter on the neural architecture/anatomy of the prefrontal cortex leads off this volume. Animal research has contributed greatly to our understanding of the special capability of the frontal lobes to respond to a variety of input from 'lower order' sensory and posterior association cortex and this and other observations are reported in this volume. Functions dependent on the frontal lobes emerge late in ontogeny and appear to decline early in normal aging. These findings are reviewed and their implications for neuropsychology discussed. Over the last 15 years, the functioning of the frontal lobes has become associated with the term working memory. In this volume the concept of working memory is discussed in relationship to both functional neuroimaging and patient studies. Gross distinctions in the functioning of the prefrontal cortex have divided it topographically into dorsolateral and ventromedial sectors. This volume offers chapters highlighting the role of each sector from both neuroimaging and lesion perspectives. Many of the views of the prefrontal cortex characterize it as involved in 'processing.' A chapter in this volume takes a slightly different perspective and attempts to characterize the nature of knowledge representation within the frontal lobes. Many theories of the frontal lobes suggest that they are concerned with maintaining information across time. A chapter reviews the role of the frontal lobes in temporal processing. Since the frontal lobes are affected by many neuropsychiatric disorders, a chapter in this volume places their dysfunction in the context of the parallel fronto-subcortical networks and suggests that each network, when damaged, contributes specific symptoms to the patient's clinical presentation. Finally, computational modeling has taken center-stage in cognitive neuroscience and its usefulness in testing different theoretical stances about the role of the frontal lobes in information processing is presented in a concluding chapter.

The revised edition of the Handbook will continue with two volumes on Child Neuropsychology edited by Sid Segalowitz and Isabelle Rapin. In childhood, the impact of acquired and degenerative disorders is strongly colored by the immaturity of the brain, so that what is said in other volumes of the Handbook rarely applies directly to children. This section starts with a consideration of aberrant brain development and the limits of plasticity of the immature brain when it is damaged. The epidemiology of the developmental disorders follows a discussion of left handedness and the emergence of cerebral dominance. The next seven chapters consider the strengths and limitations of the techniques of neuroimaging and electrophysiology as they apply to normal and handicapped children and of clinical neuropsychological testing in infancy, pre-school and schoolage children. The following six chapters are devoted to developmental disorders of motor and somatosensory perception, and to the functional impact of impaired vision and hearing. Inadequate language acquisition and language disorders are among the most frequent developmental disorders bringing children to neuropsychologists. Chapters on language development, developmental language disorders and acquired aphasias are followed by discussions of the ubiquitous academic problems, reading disability and dyscalculia. Learning disabilities and academic difficulties also arise from other cognitive losses, discussed in chapters devoted to deficits of memory in children, some of whom present with severe mental deficiency and autistic behaviors, and of attention, including criteria for its diagnosis and pharmacological management. Executive disorders have come to center stage in both adult and pediatric neuropsychology in the last decade, and their role in a variety of developmental disorders is explored. In addition to the pervasive focus on cognitive disorders, child neuropsychology needs to consider aberrant drive and affect and some of their potential correlates with major societal impact such as substance abuse and eating disorders. These disorders have their roots in childhood even though they may not appear until later in life. Epilepsy and autistic spectrum disorders have separate chapters because of their particularly high prevalence in children. Finally, while neuropsychological syndromes in children often directly inform us about normal brain–behavior relations, the neuropsychology of normal development is also based on data obtained from nonclinical paradigms. Such developments are described in separate chapters on normal cognitive and affective development.

The final volume on Rehabilitation, edited by Jordan Grafman and Ian Robertson contains topics not specifically covered in the previous edition of the Handbook. In particular, we were so impressed by the advances and current direction of studies on neuroplasticity and rehabilitation that we thought the time was right to include these topics in the new edition of the Handbook of Neuropsychology. Neuroplasticity is among the most exciting areas of research in cognitive neuroscience. In this volume, there are chapters on animal models of neuroplasticity, cortical map changes with practice and following brain damage, cross-modal reassignment of function, auditory system reassignment following learning and brain-damage, the effects of amputation on phantom limb and pain perceptions, the effects of age on plasticity, the ability of the non-damaged hemisphere to take over functions of the damaged hemisphere, and the effects of cognitive skill learning on neural organization and function. A section describing how basic science findings can be translated into practical rehabilitation of patients follows. Rehabilitation programs for neglect, language, executive functions and motor skills are described. The use of functional neuroimaging to provide a neural window on cognitive plasticity is discussed. The last section of the Handbook has traditionally been reserved for a discussion of the state-of-the-art of new technologies. Here we provide updates on MRI, fMRI, PET, transcranial magnetic stimulation, and other technologies that have burst upon the cognitive neuroscience scene in the last ten years.

As can be seen, there are important changes between the first edition and the present one in terms of organization and content. Other changes include timing, presentation and availability. The eleven volumes of the first edition appeared over a span of ten years because in most cases, volumes were written and edited consecutively.

In the second edition, several volumes have been concurrently planned and we estimate that the entire series will appear over a span of approximately two years. The first edition consisted of hardbound volumes that were followed by paperback volumes after an interval ranging from two to three years. In the current edition, the paperback edition will appear at the same time as the hardbound one and its price will make it more easily available to students and fellows.

Besides the printed second edition of the Handbook of Neuropsychology that you have in your hands, there are already plans for a web-based version. This will make it easier to adapt the Handbook to changes that will undoubtedly occur in the near future. For example, neuroimaging techniques are rapidly developing, opening new in-roads into the mapping of brain and behavior relationships. The rising interest in the aging population is also very likely to increase even further, as new advances will occur concerning the early detection of cognitive impairment and the maintenance of cognitive functions into old age. The same advances will undoubtedly also occur for the developing brain.

Many people have contributed to the successful preparation of the Handbook. We again wish to emphasize our appreciation for the commitment of the volume editors who have spent long hours in the planning stage and in the actual compiling of the various sections. Throughout the development and production of the Series, the editorial staff of Neurology and Neuroscience of Elsevier Science B.V. in Amsterdam has provided invaluable assistance.

François Boller   Jordan Grafman

List of contributors

Anderson, S.W.   Department of Neurology, Division of Behavioral Neurology and Cognitive Neuroscience, 2155 RCP, The University of Iowa Hospitals and Clinics, Iowa City, IA 52242, USA

Asaad, W.F.   Department of Brain and Cognitive Sciences, Massachusetts Institute of Technology, Cambridge, MA 02139, USA

Barbas, H.   Department of Health Sciences, Boston University, 635 Commonwealth Avenue, Room 431, Boston, MA 02215, USA

Bechara, A.   Department of Neurology, University of Iowa, Hospitals and Clinics, Iowa City, IA 52242, USA

Cohen, J.D.   Department of Psychology, Center for the Study of Brain, Mind and Behavior, Princeton University, Green Hall, Princeton, NJ 08544, USA

Cummings, J.L.   Departments of Neurology *and* Psychiatry and Biobehavioral Sciences, UCLA School of Medicine, Los Angeles, CA 90095, USA

Damasio, A.R.   Department of Neurology, University of Iowa, Hospitals and Clinics, Iowa City, IA 52242, USA

Della Sala, S.   Psychology Department, University of Aberdeen, Old Aberdeen, AB24 2UB, UK

Geva, A.   Department of Psychology, University of Michigan, 525 East University Avenue, Ann Arbor, MI 48109-1109, USA

Ghashghaei, H.T.   Department of Health Sciences, Boston University, 635 Commonwealth Avenue, Boston, MA 02215, USA

Grafman, J.   Cognitive Neuroscience Section, NINDS, NIH, Building 10, Room 5C205, 10 Center Drive, MSC 1440, Bethesda, MD 20892-1440, USA

MacPherson, S.E.S.   Psychology Department, University of Aberdeen, Old Aberdeen, AB24 2UB, UK

Marshuetz, C.   Department of Psychology, Yale University, 2 Hillhouse, P.O. Box 208205, New Haven, CT 06520-8205, USA

McPherson, S.   Departments of Neurology, Psychiatry and Biobehavioral Sciences, UCLA School of Medicine, Los Angeles, CA 90095, USA

Miller, E.K.　　Center for Learning and Memory, RIKEN-MIT, Neuroscience Research Center, Bldg. E25, Room 236, Cambridge, MA 02139, USA

Nichelli, P.　　Dipartimento di Patologia, Neuropsicosensoriale, Università di Modena e, Reggio Emilia, Via Del Pozzo 71, 41100 Modena, Italy

Phillips, L.H.　　Psychology Department, University of Aberdeen, Old Aberdeen, AB24 2UB, UK

Rempel-Clower, N.L.　　Department of Health Sciences, Boston University, 4th floor, 635 Commonwealth Avenue, Boston, MA 02215, USA

Smith, E.E.　　Department of Psychology, University of Michigan, 525 East University Avenue, Ann Arbor, MI 48109-1109, USA

Tranel, D.　　University of Iowa College of Medicine, Department of Neurology, University of Iowa Hospitals and Clinics, Iowa City, 52242 IA, USA

Xiao, D.　　Boston University, 635 Commonwealth Avenue, 4th floor, Boston, MA 02215, USA

Contents

Handbook of Neuropsychology, 2nd Edition, Vol. 7
J. Grafman (Ed)

CHAPTER 1

Anatomic basis of functional specialization in prefrontal cortices in primates

H. Barbas *, H.T. Ghashghaei, N.L. Rempel-Clower and D. Xiao

Department of Health Sciences, Boston University, 635 Commonwealth Ave., Room 431, Boston, MA 02215, USA

Overview

The prefrontal cortex in primates participates in complex cognitive processes, specific aspects of memory and emotions through an extensive network of connections with other cortical and subcortical structures (for reviews see Barbas, 1995; Fuster, 1989; Goldman-Rakic, 1988). Moreover, there is evidence of a certain degree of specialization and division of labor within the vast and heterogeneous extent of the prefrontal cortex in primates (Fig. 1). Lateral prefrontal areas, which stretch from the arcuate sulcus on the lateral surface to the frontal pole (Fig. 1B,E), have been associated with cognitive processes requiring selective attention, extraction of information, and monitoring of responses in order to perform tasks that rely on holding information temporarily in mind (for reviews see Fuster, 1989; Goldman-Rakic, 1988; Petrides, 1996a). Medial prefrontal (Fig. 1A,D) and orbitofrontal (Fig. 1C) areas are as expansive as the lateral, stretching from the rostral extent of the corpus callosum on the medial surface, and from the anterior border of the olfactory tubercle on the orbitofrontal surface to the medial and basal aspects of the frontal pole. Through their strong affiliation with the limbic system, medial and orbitofrontal areas have a role in emotional processes and long-term memory. The specialization of distinct sectors of the prefrontal cortex is achieved through a specific set of connections with cortical and subcortical structures and a distinct mode of anatomic communication. Moreover, the distinct sectors of the prefrontal cortex have differential vulnerability in psychiatric and neurologic diseases, which may be related to the diversity in their structure and neural interactions.

Detailed sensory information processed by lateral prefrontal areas versus a global overview of the sensory environment processed by orbitofrontal cortices

The prefrontal cortex in primates has a fundamental role in guiding behavior through targeted attention and selection of information for action. Achievement of these tasks presupposes extraction of the necessary information, and the prefrontal cortex has access to cortices associated with all sensory modalities. Sensory input to different prefrontal regions is organized topographically and by the attributes of the signals processed. Cortices representing the modalities of vision, audition and somatic sensation project to lateral prefrontal cortices. Cortices associated with each of these modalities also project to orbitofrontal areas, which, in addition, receive projections from primary olfactory areas (for reviews see Cavada, Company, Tejedor et al., 2000; Tagaki, 1986). In contrast to the lateral and orbital areas, medial prefrontal areas receive only sporadic information from most sensory association cortices, although they receive preferential projections from several auditory association cortices (Barbas, 1988; Barbas, Ghashghaei, Dombrowski and Rempel-Clower, 1999).

---

* Corresponding author. Tel.: +1 (617) 353–5036; Fax: +1 (617) 353–7567; E-mail: barbas@bu.edu

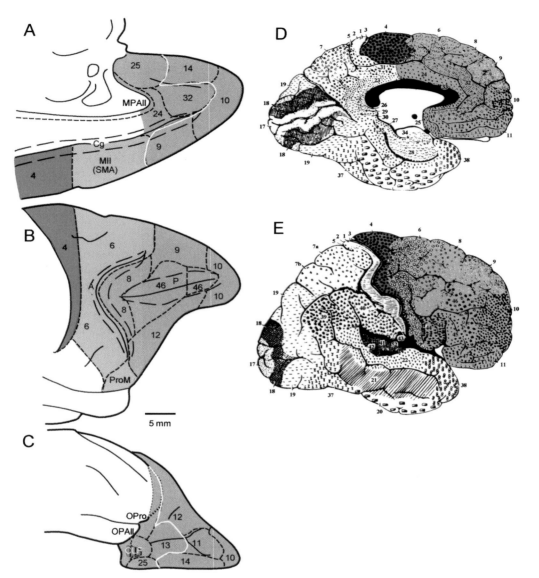

Fig. 1. The frontal cortex in the rhesus monkey (left panels) and in the human (right panels), which includes the prefrontal (blue), premotor (orange), and motor cortices (red). The extent of the prefrontal cortex on the medial (A), lateral (B) and orbital (C) surfaces of the rhesus monkey brain is superimposed on a map of the prefrontal cortex by Barbas and Pandya (1989), and in the human is shown on the medial (D) and lateral (E) surfaces on the map of Brodmann (Brodmann, 1909). Agranular and dysgranular (limbic) prefrontal areas of the rhesus monkey cortex include several posterior medial (A) and basal (C) areas delimited roughly by a white line. For a map of further premotor areas in the cingulate region see (Picard and Strick, 1996). Abbreviations in A–C: A, arcuate sulcus; MPAll, OPAll, medial and orbital periallocortex (agranular); MII (SMA), supplementary motor area; OLF, olfactory cortex. OPro, orbital periallocortex (dysgranular); P, principal sulcus; ProM, dysgranular premotor area.

Projections from sensory cortices reaching distinct sectors of the prefrontal cortex are likely to serve different functions, based on fundamental differences in topography and physiologic characteristics of neurons in the areas giving rise to the projections. For example, while area 8 in the arcuate concavity as well as caudal orbitofrontal areas receive visual input, the origin of the projections differs considerably. Visual input to caudal periarcuate areas originates from caudal visual areas including areas V2, V3, V4 and posterior inferior temporal cortex (Barbas, 1988; Barbas and Mesulam, 1981; Distler, Boussaoud, Desimone and Ungerleider, 1993; Schall, Morel, King and Bullier, 1995; Webster, Bachevalier and Ungerleider, 1994), which represent relatively early stages of visual processing. Thus, even though area 8 is situated at a considerable distance from the occipital cortices, its visual input is comparable to what unimodal visual areas receive (Fig. 2). This pattern of projection suggests that the periarcuate cortex receives relatively detailed information about specific aspects of the visual environment. In contrast, orbitofrontal cortices receive projections from anterior inferior temporal cortices, where the visual receptive fields are global, and from ventral temporal polar cortices that are polymodal (Barbas, 1988, 1993; Desimone and Gross, 1979; Gross, Bender and Rocha-Miranda, 1969).

Lateral and orbitofrontal areas also receive auditory input, albeit from different auditory cortices (Barbas, 1988, 1993; Barbas and Mesulam, 1981, 1985; Barbas et al., 1999; Chavis and Pandya, 1976; Müller-Preuss, Newman and Jürgens, 1980; Petrides and Pandya, 1988; Romanski, Bates and Goldman-Rakic, 1999; Vogt and Pandya, 1987). Periarcuate area 8 and the adjacent part of dorsal area 46 receive projections from caudal superior temporal auditory cortices that are close to the primary auditory area (Barbas, 1988; Barbas and Mesulam, 1981, 1985; Petrides and Pandya, 1988; Romanski et al., 1999). On the other hand, orbitofrontal areas receive projections from late processing auditory areas in the anterior superior temporal gyrus (Barbas, 1993), which have the structural and connectional characteristics of limbic cortices (for review see Pandya, Seltzer and Barbas, 1988). In addition to the above modalities, orbitofrontal areas receive projections from gustatory and olfactory cortices, enriching their access to the entire spectrum of sensory information (Fig. 3), and surpassing most other areas of the cerebral cortex in this respect (Barbas, 1993; Carmichael and Price, 1995b; Cavada et al., 2000; Morecraft, Geula and Mesulam, 1992). On the basis of the information received, the orbitofrontal cortex seems to get a global overview of the sensory environment, and may act as an environmental integrator.

Connectional differences as a basis of functional biases in distinct prefrontal regions

Involvement of lateral prefrontal areas in cognitive processes and short-term memory

The detailed sensory information to lateral prefrontal cortices from relatively early processing visual areas may be necessary for interpreting the sensory environment for action. Classic studies have associated the periarcuate cortex (area 8) with eye movements, on the basis of temporary disturbances of gaze after its damage, which provided a name for this area, the frontal eye fields (for review of early literature and references see Barbas and Mesulam, 1981). Although not essential for the initiation of eye movements, the frontal eye fields are essential in behavioral tasks requiring search of the environment, exemplified by a set of neurons with visual receptive fields that accelerate their firing activity when the monkey orients to a stimulus and uses it to guide behavior (Goldberg and Bushnell, 1981; Wurtz and Mohler, 1976). Moreover, there is a certain degree of specificity in this function within the frontal eye fields, so that large saccadic eye movements are elicited by electrical microstimulation of the rostral part (Robinson and Fuchs, 1969), an area that receives input from auditory cortices and from visual cortices representing peripheral visual fields (Barbas and Mesulam, 1981). On the other hand, small and medium size saccades are elicited by microstimulation of the caudal frontal eye field at the junction of the upper and lower limbs of the arcuate sulcus (Bruce, Goldberg, Bushnell and Stanton, 1985; Robinson and Fuchs, 1969), an area that receives strong projections from visual cortices, including substantial input from areas representing central visual fields (Barbas, 1988; Barbas and Mesulam, 1981). Based on their connections with visual and auditory cortices, the frontal eye fields seem to be well suited for scanning central parts of the visual field using small saccades, or orienting to peripheral visual and auditory stimuli using large saccades and head movement. These findings are consistent with the classic notion that the frontal eye fields have a

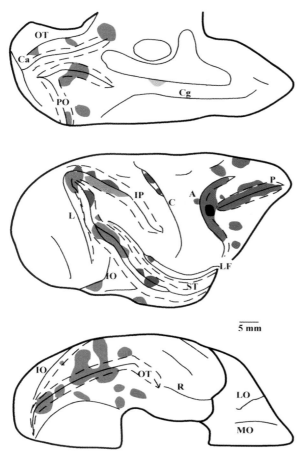

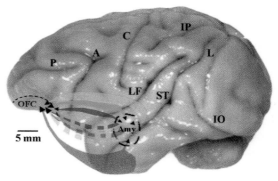

Fig. 3. Multimodal input to the orbitofrontal cortex and the amygdala. Lateral view of the rhesus monkey brain showing projections from sensory cortices (represented in solid color lines and black arrows), emanating from visual (red), auditory (yellow), somatosensory (green), olfactory (blue), and directed to the orbitofrontal cortex (OFC, which extends beneath the inferior convexity of the frontal lobe), and to the amygdala (Amy, buried in the temporal lobe). Robust projections from the sensory recipient parts of the amygdala are directed to orbitofrontal cortices (dotted lines). A, arcuate sulcus; Amy, amygdala; C, central sulcus; IO, inferior occipital sulcus; IP, intraparietal sulcus; L, lunate sulcus; LF, lateral fissure; OFC, orbitofrontal cortex; P, principal sulcus; ST, superior temporal sulcus.

Fig. 2. Visual input to the periarcuate cortex. An example of the distribution of the sites of all projection neurons in the cortex directed to the concavity of the arcuate sulcus (black area). Most projection neurons are distributed in visual cortices (red), on the medial (top), lateral (center) and basal (bottom) surfaces, and in the lateral intraparietal visuomotor region (IP, orange (center), and considerably fewer are seen in other areas (center: somatosensory, green; auditory, yellow; polymodal areas, purple; top: cingulate cortex, light green). Projections from visual areas to the arcuate concavity also surpass local projections from other prefrontal areas (blue). Abbreviations for sulci: A, arcuate; C, central sulcus; Ca, calarine fissure; Cg, cingulate; IO, inferior occipital; IP, intraparietal; L, lunate; LF, lateral fissure; LO, lateral orbital; MO, medial orbital; OT, occipitotemporal; P, principal; PO, parietooccipital; R, rhinal; ST, superior temporal.

role in searching the environment in behavioral tasks through accurate timing of eye movements (Schiller and Chou, 1998; for review see Schiller, 1998).

The specific involvement of the frontal eye fields and the adjacent caudal area 46 in oculomotor func-

tions in behavior is further substantiated by the fact that they are the only prefrontal areas that receive significant projections from visuomotor areas in the lateral bank of the caudal intraparietal sulcus (Andersen, Asanuma, Essick and Siegel, 1990; Barbas, 1988; Barbas and Mesulam, 1981, 1985; Cavada and Goldman-Rakic, 1989; Petrides and Pandya, 1984). The homologous frontal region in cats receives proprioceptive signals from muscles that move the head and the eyes, suggesting a role in coordinated eye and head movement (Barbas and Dubrovsky, 1981). Human and non-human primates with unilateral damage to the frontal eye fields ignore stimuli that appear on the side opposite the lesion (for reviews see Mesulam, 1981, 1999). The same deficit is observed after unilateral lesion of the parietal cortex. Moreover, the profound neglect of stimuli in the contralateral visual field is not restricted to the external sensory environment, but extends to the remembered environment as well (Bisiach, Capitani, Luzzatti and Perani, 1981).

Lateral prefrontal areas situated rostral to the frontal eye fields, at the mid-principalis level in macaque monkeys, have been associated with related, though distinct, functions. Within the domain of mnemonic processing, lateral prefrontal cortices have generally been associated with cognitive tasks that depend on the ability to keep information temporarily in mind and to monitor self-generated responses in order to perform a specific task (for reviews see Fuster, 1989, 1993; Goldman-Rakic, 1988). This set of cognitive functions has classically been associated with an area around the principal sulcus (area 46), whose damage in non-human primates perturbs the ability to remember, after a few seconds delay, where a food reward was hidden (Jacobsen, 1936).

In recent years, the role of dorsolateral prefrontal cortices in working memory tasks has been further refined with evidence that cognitive tasks with different requirements employ different prefrontal cortices (e.g., Funahashi, Bruce and Goldman-Rakic, 1989, 1993; Petrides, 1996b; Petrides, Alivisatos and Evans, 1995). For example, delayed response tasks, in general, engage area 46 (e.g., Funahashi et al., 1993). On the other hand, the prefrontal polar area 10 appears to have a specialization within the sphere of working memory, and is recruited when a main goal must be remembered while simultaneously juggling secondary tasks (Koechlin, Basso, Pietrini et al., 1999). Of particular interest is a robust projection from anterior auditory association cortices to frontal polar area 10 (Barbas and Mesulam, 1985; Barbas et al., 1999), suggesting that there may be a major auditory center within area 10 similar to a visual center in the frontal eye fields (Fig. 4C). Auditory input may be particularly relevant in managing multiple tasks within working memory. Thus, analogous to the frontal eye fields, area 10 may have a role in directing attention to behaviorally meaningful stimuli.

The relationship of prefrontal areas with sensory cortices is bidirectional, as they not only receive information but also issue projections to sensory cortices (Rempel-Clower and Barbas, 2000). Within the domain of the visual modality, lateral prefrontal cortices appear to influence attentional mechanisms, visual memory, and task-related activity of neurons in inferior temporal areas (Desimone, 1996; Fuster, Bauer and Jervey, 1985; Miller, 1999; Petrides, 1996b; Rainer, Asaad and Miller, 1998; Tomita, Ohbayashi, Nakahara et al., 1999). For example,

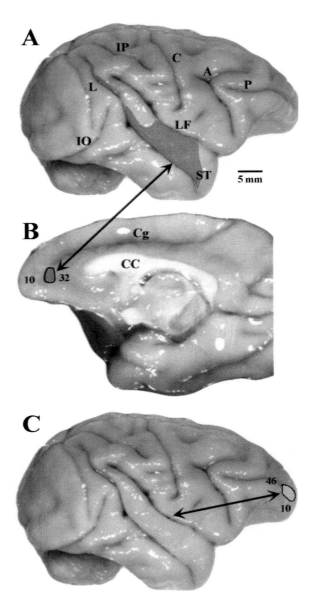

Fig. 4. Some connections between prefrontal areas and auditory association areas. Medial and frontal polar areas are preferentially connected with auditory association areas, as shown with an example of the connections of medial area 32 (B, red with black outline), and for the connections of frontal polar area 10 (C, yellow with black outline). The auditory connections in each case are shown in the superior temporal gyrus (A, red area; C, yellow area). A, arcuate sulcus; C, central sulcus; CC, corpus callosum; Cg, cingulate sulcus; IO, inferior occipital sulcus; IP, intraparietal sulcus; L, lunate sulcus; LF, lateral fissure; P, principal sulcus; ST, superior temporal sulcus. Numbers in B and C indicate architectonic areas.

when macaque monkeys are engaged in a variety of delayed response tasks, neurons in prefrontal area 46 increase their firing activity if the stimulus matches the sample, or they fire selectively to relevant locations, or to an anticipated target (Desimone, 1996; Miller, Erickson and Desimone, 1996; Rainer et al., 1998; Rainer, Rao and Miller, 1999). This evidence suggests that reciprocal projections from area 46 to the inferior temporal cortex may support active selection and comparison of stimuli held in short-term memory. Moreover, the selectivity of delay-related activity in neurons in the visual inferotemporal cortex is reduced after incapacitation of the dorsolateral prefrontal cortex by cooling (Fuster et al., 1985). In addition, stimulus selective activity in neurons of the inferior temporal cortex persists even after caudal commissurectomy deprives the inferior temporal cortex of one hemisphere from receiving bottom-up input from the opposite visual field (Tomita et al., 1999). In this case, information on the relevant stimulus apparently is issued through feedback pathways from the prefrontal cortex (Tomita et al., 1999). Recent evidence provides the anatomic basis for such feedback projections from lateral prefrontal areas, which terminate in the upper layers of the inferior temporal cortex (Rempel-Clower and Barbas, 2000).

The reciprocal projections from prefrontal areas to auditory areas (Barbas et al., 1999) appear to be essential in filtering relevant sensory information and suppressing irrelevant information. This idea emerged from evidence that humans with dorsolateral prefrontal lesions are impaired when irrelevant auditory stimuli are introduced in an auditory discrimination task (Chao and Knight, 1998). Moreover, their performance is correlated with decreased neural activity in dorsolateral prefrontal cortices and a concomitant increase of activity in auditory cortices. Similar behavioral/neurophysiological correlates have been reported for aged humans, suggesting that neural changes in prefrontal cortex impair the ability to ignore irrelevant sounds and to distinguish signal from noise (Chao and Knight, 1997).

Involvement of medial and orbitofrontal areas in long-term memory and emotions

Like orbitofrontal areas, the medial sector of the prefrontal cortex has a different set of connections with sensory and other areas than the lateral. In particular, the medial prefrontal sector has stronger connections with areas implicated in long-term memory. Moreover, the connections of the posterior and anterior medial prefrontal cortices are not uniform (Barbas et al., 1999). Among medial cortices, those situated caudally around the corpus callosum, are aligned through their connections with hippocampal and anterior medial temporal cortices, and their damage in humans, results in anterograde amnesia comparable to the classic amnesic syndrome seen after hippocampal lesions (Alexander and Freedman, 1984; Crowell and Morawetz, 1977; Talland, Sweet and Ballantine, 1967; for review see Barbas, 1997). Preferential projections from memory-related medial temporal cortices originate from the entorhinal (area 28) and perirhinal (areas 35 and 36) areas (Bachevalier, Meunier, Lu and Ungerleider, 1997; Barbas, 1988; Carmichael and Price, 1995a; Vogt and Pandya, 1987) and target most heavily caudal medial area 25, the adjacent area 24, and the posterior extent of area 32 (Barbas et al., 1999). The same caudal medial prefrontal areas receive robust projections from the hippocampal formation (Barbas and Blatt, 1995), limbic thalamic nuclei, including the midline, the magnocellular sector of the mediodorsal nucleus (MDmc), and the caudal part of parvicellular MD (MDpc) (Baleydier and Mauguiere, 1980; Barbas, Henion and Dermon, 1991; Dermon and Barbas, 1994; Goldman-Rakic and Porrino, 1985; Morecraft et al., 1992; Ray and Price, 1993). All of the above structures have been implicated in long-term memory (for reviews see Amaral, Insausti, Zola-Morgan et al., 1990; Markowitsch, 1982; Squire, 1992; Squire and Zola-Morgan, 1988; Zola-Morgan and Squire, 1993). The significance of the above connections in memory is revealed by visual recognition deficits after damage to caudal medial prefrontal areas in monkeys (Bachevalier and Mishkin, 1986; Voytko, 1985), and the amnesic syndrome in humans, manifested after vascular lesions of the anterior communicating artery, which supplies the ventromedial prefrontal areas (Crowell and Morawetz, 1977; D'Esposito, Alexander, Fischer et al., 1996) including Brodmann's area 25, 24 and 32, around the rostrum of the corpus callosum (Brodmann, 1909). In contrast, medial prefrontal areas situated anteriorly (rostral area 32, area 14, area 9 and area 10) can be differentiated from caudal medial prefrontal areas by the paucity of projections from memory-related structures (Barbas and Blatt, 1995; Barbas et al., 1999; Dermon and Barbas, 1994).

All medial prefrontal areas, however, appear to be preferential targets of auditory cortices (Fig. 4A,B), emanating mostly from anterior auditory association and temporal polar cortices (Barbas, 1988; Barbas et al., 1999; Petrides and Pandya, 1988; Vogt and Pandya, 1987). Neurons in these areas are broadly tuned (Kosaki, Hashikawa, He and Jones, 1997; Rauschecker, 1998; Rauschecker, Tian and Hauser, 1995) and respond best to complex species-specific vocalizations and emotional communication, such as distress calls emitted by infant monkeys when separated from their mothers (for review see Vogt and Barbas, 1988).

Imaging studies have provided evidence that caudal medial areas are involved in speech production and overlap with areas whose damage results in the akinetic mute syndrome in humans (e.g., Barris and Schuman, 1953; Buge, Escourolle, Rancurel and Poisson, 1975; Nielsen and Jacobs, 1951). Physiological studies in monkeys suggest a specific functional interaction between anterior cingulate areas and auditory cortices (Jürgens and Müller-Preuss, 1977; Müller-Preuss et al., 1980; Müller-Preuss and Ploog, 1981; Heffner and Heffner, 1986), manifested by decreases in auditory evoked activity in superior temporal auditory areas after electrical stimulation of the cingulate vocalization cortices (Müller-Preuss et al., 1980; Müller-Preuss and Ploog, 1981). This evidence suggests that areas involved in vocalization have a modulatory effect on auditory association areas during phonation. Parallel functional imaging studies in humans demonstrated that in normal, but not in schizophrenic subjects, cognitive tasks involving verbal fluency activate the anterior cingulate region and result in reduction of activity in superior temporal auditory cortices (Dolan, Fletcher, Frith et al., 1995; Frith and Dolan, 1996). In contrast, when subjects are asked to imagine sentences spoken in other peoples' voices, the anterior cingulate, the adjacent supplementary motor, and temporal auditory association cortices are robustly activated in normal subjects, but show decreased activity in schizophrenic patients who have auditory hallucinations (McGuire, Silbersweig, Wright et al., 1995,

1996). Moreover, whereas decreased activation in medial prefrontal areas is found in schizophrenic patients, in general, the decreased activity in auditory cortices during imagined speech is restricted to those schizophrenic patients who experience auditory hallucinations. Medial prefrontal areas thus may have a role in distinguishing external auditory stimuli from internal auditory representations. The altered pattern of activity in patients who hallucinate, may be due to a functional breakdown of a pathway from medial prefrontal to auditory cortices.

Recent studies have demonstrated that the anterior cingulate region in humans has a role in cognitive tasks involving conflict (Botvinick, Nystrom, Fissell et al., 1999; MacDonald, Cohen, Stenger and Carter, 2000). Moreover, the role of the cingulate cortex in cognitive tasks is dissociated from the engagement of dorsolateral areas 9 and 46 in executive functions. The anterior cingulate region in humans overlaps with caudal medial prefrontal areas 32, 25 and 24 in non-human primates (Fig. 1). As noted above, the latter areas are affiliated with other cortical and subcortical limbic structures, and their connections are distinct from those of dorsolateral prefrontal areas.

Posterior orbitofrontal areas receive projections from memory-related entorhinal and perirhinal cortices and the hippocampal formation as well, though to a lesser extent than the medial (Barbas, 1993; Barbas and Blatt, 1995; Barbas et al., 1999). On the other hand, caudal orbitofrontal cortices are distinguished by heavy connections with the amygdala, which also reach caudal medial areas (Amaral and Price, 1984; Barbas and De Olmos, 1990; Carmichael and Price, 1995a; Porrino, Crane and Goldman-Rakic, 1981).

The strong bidirectional interactions between orbitofrontal cortices and the amygdala suggest, that the two structures may share a role in emotions. In fact, there are several similarities between the orbitofrontal limbic cortices and the amygdala in primates. At the behavioral level, lesions of the orbitofrontal cortex or the amygdala impair monkeys in exhibiting appropriate emotional responses and render them incapable of engaging in normal social interactions (for review see Kling and Steklis, 1976). At the connectional level, both structures receive robust projections from all sensory cortices (Barbas,

1993; Morecraft et al., 1992; Turner, Mishkin and Knapp, 1980) (Fig. 3). In both structures, the sensory input originates in areas where the emphasis in processing appears to be in the significance of features of stimuli and their memory (for reviews see Barbas, 1992, 1995).

The question arises of whether the role of the amygdala and the orbitofrontal cortex in emotions can be distinguished at all. Previous studies have established a key role of the amygdala in emotions (Damasio, 1994; Davis, 1992; LeDoux, 1996; Nishijo, Ono and Nishino, 1988). Further studies indicate that a short subcortical loop connecting the amygdala with the thalamus can support fear conditioning in rats (Romanski and LeDoux, 1992). In humans, the amygdala increases its activity during viewing of masked fearful faces when there is no awareness of the event (Whalen, Rauch, Etcoff et al., 1998). This evidence suggests that processing of emotional information at the level of the amygdala is not necessarily a conscious event. The amygdala, therefore, may be likened to an elaborate central reflex structure for the internal milieu, whereby it receives signals concerned with the internal environment and issues efferent signals to brainstem autonomic structures. This raises the question of the neural structure(s) that may account for emotional awareness. Classic studies implicate the cortex as necessary for the conscious perception of emotion (Kennard, 1945), and the orbitofrontal cortex may have a pivotal role in this process, based on behavioral manifestations after its damage, and bolstered by its massive connections with the amygdala in macaque monkeys (Amaral and Price, 1984; Damasio, 1994; Porrino et al., 1981; for review see Barbas, 1995).

The orbitofrontal cortex has strong reciprocal connections with the entorhinal cortex (Carmichael and Price, 1995a; Morecraft et al., 1992; Rempel-Clower and Barbas, 2000; Van Hoesen, Pandya and Butters, 1975), which is the primary conduit through which information from the cortex reaches the hippocampus (Leonard, Amaral, Squire and Zola-Morgan, 1995; Nakamura and Kubota, 1995; Suzuki, Miller and Desimone, 1997; Witter, Van Hoesen and Amaral, 1989; for reviews see Rosene and Van Hoesen, 1987; Squire and Zola, 1996). Through its connections with the amygdala and widespread sensory ar-

eas, the orbitofrontal cortex may process information about the emotional significance of stimuli, and the association of stimuli with reward (for reviews see Rolls, 1996; Watanabe, 1998). Orbitofrontal neurons respond to stimuli that predict reward (Tremblay and Schultz, 1999), which requires memory of previous experience with the same stimuli. The emotional significance of events is an important factor in whether they are encoded as long-term memories (for review see Cahill and McGaugh, 1998), and thus the orbitofrontal cortex is in a position to provide information to the entorhinal cortex that could influence the formation of long-term memories. Recent studies in humans suggest a role for the orbitofrontal cortex in distinguishing between mental representations of the current situation and irrelevant memories (Schnider and Ptak, 1999; Schnider, Treyer and Buck, 2000). It is particularly interesting that this function appears analogous to the functions of the frontal eye fields in directing visual attention and medial area 10 in directing auditory attention to currently relevant stimuli.

In addition to the amygdala, several other structures may be involved in emotions. In Papez's classic circuit for emotions, the anterior thalamic nuclei occupy a central position in pathways that also include the cingulate cortex, the hippocampus and the mammillary body (Papez, 1937). The amygdala was not mentioned as part of the original circuit by Papez. Since the publication of Papez's classic paper, evidence has been provided that the anterior thalamic nuclei, the hippocampus and the mammillary body have a role in memory as well (Braak and Braak, 1991; Kaitz and Robertson, 1981; Gaffan, Aggleton, Gaffan and Shaw, 1990; Parker and Gaffan, 1997; Thomas and Gash, 1985). Moreover, the anterior nuclei have an additional link with the prefrontal cortices (Barbas et al., 1991; Dermon and Barbas, 1994). The extent of interaction between the anterior nuclei and the prefrontal cortex was recently shown with the use of combined approaches to study prefrontal axonal terminations in the anterior nuclei, as well as the sources of projection neurons to the anterior nuclei (Xiao and Barbas, 2001). Both medial prefrontal and orbitofrontal limbic cortices are linked with the anterior nuclei to a much greater extent than lateral prefrontal cortices. Moreover, the hippocampal formation, including the subicular complex as well as the adjoining entorhinal, perirhinal, and parahippocampal cortices project to the anterior nuclei as well. Finally, a massive projection from the ipsilateral mammillary body reaches the anterior nuclei. These findings suggest that the anterior nuclei are a common link for distinct orbitofrontal and medial prefrontal areas, medial temporal cortices, and the hypothalamic mammillary body, in pathways processing both emotional and mnemonic information (Xiao and Barbas, 2001).

Differences in the pattern of connections of lateral, orbitofrontal and medial prefrontal cortices

Restricted thalamic projections to lateral prefrontal areas and widespread projections to orbitofrontal and medial prefrontal areas

The above discussion suggests that lateral, medial, and orbitofrontal cortices have a distinct set of connections with cortical and subcortical structures in circuits that may underlie their role in cognition, memory and emotion. The observed differences are not restricted to topography, but extend to the pattern of connections. With respect to the relationship of prefrontal cortices with the thalamus, classic studies have associated the magnocellular division of the mediodorsal (MD) thalamic nucleus with orbitofrontal areas, and the parvicellular and multiform divisions with lateral prefrontal areas (for reviews see Jones, 1985; Steriade et al., 1997). While classic studies had provided evidence of the participation of other nuclei, besides MD, in the projection to prefrontal cortices, it was not until the advent of neural tracing procedures that the extent of this projection system was fully realized (e.g., Kievit and Kuypers, 1977). Moreover, studies employing quantitative procedures indicated that the vast majority of thalamic neurons directed to lateral prefrontal cortices originate in MD and only a few are found in other nuclei. In contrast, orbitofrontal and medial prefrontal cortices are targeted by MDmc, and by numerous other limbic thalamic nuclei including the midline and anterior (Barbas et al., 1991; Dermon and Barbas, 1994). In fact, MD does not even include the majority of afferent neurons that project to some medial prefrontal cortices, including areas

25 and 14. The latter, as well as area 32, the olfactory cortex, and posterior orbitofrontal areas are targeted heavily by midline thalamic nuclei, which also project to other medial areas, including area 9. In contrast, the midline thalamus has no significant connections with lateral prefrontal cortices (Dermon and Barbas, 1994). The intralaminar thalamic nuclei also provide a significant projection to all medial and posterior orbitofrontal areas, and to a lesser extent to rostral orbital and lateral prefrontal cortices (Dermon and Barbas, 1994).

Another major distinction among different sectors of the prefrontal cortex pertains to their connections with the contralateral thalamus. While most projection neurons directed to the prefrontal areas are found in the ipsilateral thalamus, the contralateral thalamus issues projections to prefrontal cortices as well, albeit to a lesser extent (Andersen, Asanuma and Cowan, 1985; Preuss and Goldman-Rakic, 1987; Tigges, Tigges, Cross et al., 1982). Moreover, neurons from the contralateral thalamus project preferentially to caudal orbitofrontal and medial prefrontal cortices but not to lateral prefrontal areas (Dermon and Barbas, 1994). Nuclei such as MDmc that project to caudal orbitofrontal and caudal medial (limbic) and to the rostrally situated medial and orbital areas issue bilateral projections to limbic areas, but issue strictly ipsilateral projections to eulaminate prefrontal areas (Dermon and Barbas, 1994).

Contralateral thalamic projections have been considered transient in development in rats (Laemle and Sharma, 1986; Takada, Fishell, Li et al., 1987; Minciacchi and Granato, 1989), but the above evidence suggests the persistence of significant contralateral projections in adult monkeys. Moreover, their continued presence appears to be largely dependent on the target area, as they are observed mostly for orbitofrontal and medial limbic areas in adult primates. This evidence suggests that the maturation of thalamocortical interactions may differ temporally for distinct sectors of the prefrontal cortex.

Cortical structure as a basis for the pattern of corticocortical connections

The above discussion suggests that medial and orbitofrontal limbic cortices can be distinguished from lateral prefrontal cortices by their connections. Prefrontal cortices also differ fundamentally by structure (Barbas and Pandya, 1989). Caudal medial and orbitofrontal areas belong to the cortical component of the limbic system, and are identified structurally as areas that have fewer than 6 layers and either lack, or have a rudimentary granular layer 4. On the other hand, lateral prefrontal areas situated in the periarcuate region and at the caudal extent of the principal sulcus are eulaminate, with six distinct layers. Areas situated rostrally on the lateral surface, or on the anterior medial and anterior orbitofrontal surfaces are also eulaminate, but the definition between layers is less distinct, particularly as they abut dysgranular (limbic) areas. We recently showed that agranular and dysgranular (limbic) prefrontal cortices have a lower cell density than the eulaminate, and differ by the prevalence of the calcium binding proteins calbindin and parvalbumin (Dombrowski and Barbas, 1998; Dombrowski, Hilgetag and Barbas, 2001).

Consideration of cortical structure is important because it appears to be the best indicator of the pattern of connections between cortices (Barbas, 1986; Barbas and Rempel-Clower, 1997; Rempel-Clower and Barbas, 2000). Moreover, the relationship of structure to connections is based on global structure rather than local architecture. Global structure refers to the number of layers and the distinction between layers in a given area, and can be described as cortical types, such as agranular, dysgranular or eulaminate. Each cortical type, in turn, includes several architectonic areas. Consistent patterns of connections are revealed when cortical areas are grouped by cortical type, suggesting that broad structural attributes underlie common patterns in corticocortical connections. On the basis of such analysis, it was possible to determine that projection neurons from eulaminate areas with distinct layers originate mostly in the upper layers and their axons terminate predominantly in the deep layers (4–6) of areas with fewer layers or less laminar definition (Fig. 5). Connections of areas with the reverse structural relationship originate mostly in the deep layers and their axons terminate predominantly in the upper layers (1–3). Moreover, according to the structural model for connections (summarized in Fig. 5), the proportion of projection neurons or axonal terminals in the upper to the deep layers varies as a function of the relative difference in structure between the connected areas (Barbas and Rempel-Clower, 1997). Thus the pattern is exaggerated when structurally dissimilar areas are connected (e.g., caudal orbital with lateral areas) than when structurally similar areas are interconnected (e.g., caudal orbitofrontal with rostral orbitofrontal areas).

The above patterns of connections, described for connections between pairs of prefrontal areas (Barbas and Rempel-Clower, 1997), apply to other connections as well, because parcelling by cortical type can be accomplished in all cortical systems, since it does not rely on local architectonic particularity. We recently showed that structure can be used to describe the connections between prefrontal areas and temporal areas (Rempel-Clower and Barbas, 2000). Thus, projections arising from the same origin terminated in different laminar patterns in structurally distinct target areas. Conversely, axonal projections arising from several structurally different areas terminated in distinct laminar patterns within a single target area (Fig. 6). The above patterns in connections between prefrontal and temporal cortices are exemplified in the projections from dysgranular orbitofrontal areas, which terminated predominantly in the deep layers of agranular temporal area 28, in all layers of dysgranular area 36, and mostly in the upper layers of granular area TE (Fig. 6A).

Differences are also seen in the terminations of medial prefrontal areas with temporal association and auditory areas (Barbas et al., 1999). Axons from medial prefrontal cortices terminated most densely, though not exclusively, in the upper layers of auditory cortices, resembling a 'feedback' pattern, by analogy with sensory areas. In contrast, axons from medial prefrontal cortices terminated mostly in layer 4 and the deep layers of dorsal temporal polar cortices, resembling a 'feedforward' projection, by analogy with sensory areas. These specific interactions of medial prefrontal cortices with dorsal superior temporal polar cortices may have a role in encoding the emotional significance of auditory stimuli.

The above evidence suggests that a given prefrontal area does not have a single pattern of connection with temporal areas, but rather issues projections and interacts with elements in different layers of a target area in a pattern that depends on the structural relationship of the interconnected areas (Fig. 6). The significance of identifying patterns of connections

A. Large differences in laminar definition   B. Moderate differences in laminar definition

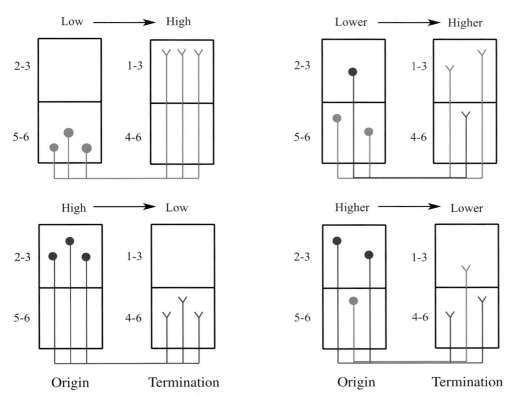

Fig. 5. The pattern of connections predicted by the structural model (A, top). Projections between areas with large differences in laminar definition originate predominantly in the deep layers of areas with low laminar definition and their axons terminate mostly in the upper layers of areas with high laminar definition (A, top). The opposite pattern is seen for reciprocal projections (A, bottom). A less extreme version of the above pattern is predicted in the interconnections of areas with moderate differences in laminar definition, for example, when a cortex with a lower laminar definition projects to a cortex with higher laminar definition; or (B, top) when a cortex with higher laminar definition projects to a cortex with lower laminar definition (B, bottom). Adapted from Barbas and Rempel-Clower (1997).

as a function of cortical structure is based on the potential to predict patterns of connections in the human cortex, where invasive procedures are precluded (Barbas and Rempel-Clower, 1997). In a general sense, the pattern of connections has functional implications, since axons terminating in the upper layers are likely to influence different populations of neurons and processes than axons terminating in the deep layers. Differences in morphology, receptors, and neurochemical properties in different cortical layers have been described widely in the cortex (e.g., De Lima, Voigt and Morrison, 1990; Hof, Nimchinsky and Morrison, 1995; Gabbott and Bacon, 1996; Goldman-Rakic, Lidow and Gallager, 1990).

An important variable within the microenvironment of connections in different layers is the level of inhibitory control. In the cortex, inhibition is effected, in part, through inhibitory interneurons, which have been classified by their expression of specific markers, including calcium binding proteins (CBP). One of these groups is positive for the CBP parvalbumin, expressed in basket and chandelier cells (DeFelipe, Hendry and Jones, 1989; Kawaguchi and Kubota, 1997). Parvalbumin-positive neurons are most densely distributed in the middle layers of the cortex, and synapse with pyramidal cell bodies, proximal dendrites and axon initial segments (DeFelipe et al., 1989; Shao and Burkhalter, 1999). Another group

11

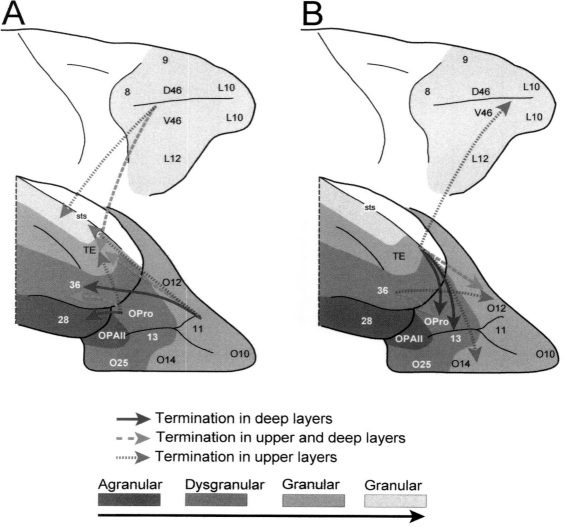

Fig. 6. Application of the structural model in the connections between prefrontal and temporal areas. The pattern of termination of efferent fibers (A) from prefrontal to temporal, and (B) from temporal to prefrontal areas, drawn on lateral (top) and basal (bottom) views of the rhesus monkey brain. Shades of gray indicate cortical type with laminar definition increasing in the direction of the arrow. Two shades of light gray are used to indicate granular type cortex: the darker shade is used for areas with less laminar definition and a narrower layer 4. Termination mostly into the deep layers is illustrated by solid blue arrows. Termination predominantly into the upper layers is illustrated by dotted red arrows. Termination distributed relatively equally between the upper and deep layers is illustrated with dashed green arrows. Letters before numbers for architectonic areas designate: D, dorsal; L, lateral; O, orbital; V, ventral. From Rempel-Clower and Barbas (2000).

of interneurons is positive for the CBP calbindin, expressed in inhibitory double bouquet cells in the cortex, which are most prevalent in cortical layers 2 and 3, and innervate distal dendrites and spines of other neurons (e.g., Peters and Sethares, 1997).

The laminar bias in the distribution of parvalbumin- and calbindin-positive neurons suggests that axonal terminations in the middle layers, as opposed to the superficial layers, are likely to interact within a qualitatively different microenvironment. Physiologic evidence supporting the functional implications of different patterns of connections is provided in studies in sensory systems, where stimulation of 'forward' pathways from LGn to V1, which terminate in and around layer 4, or forward pathways in the cortex, leads to monosynaptic excitation followed by disynaptic inhibition (Douglas, Martin and Whitteridge, 1991; Shao and Burkhalter, 1999). In contrast, in pathways that terminate in layer 1 (feedback), excitatory influences predominate (Sandell and Schiller, 1982; Shao and Burkhalter, 1999). The laminar pattern of terminations, therefore, may critically affect the balance of excitatory and inhibitory influences exerted by the prefrontal cortex on temporal and other cortices.

In prefrontal agranular and dysgranular cortices (which are considered limbic), as well as in eulaminate cortices, neurons positive for CBP are more prevalent in the upper layers (Dombrowski et al., 2001). However, as summarized above, prefrontal limbic and eulaminate areas differ markedly in the pattern of their connections (Barbas, 1986; Barbas and Rempel-Clower, 1997; Rempel-Clower and Barbas, 2000). In limbic areas the deep layers are the principal sources and targets of corticocortical connections, whereas in eulaminate areas it is the upper layers that primarily issue and receive cortical connections (Barbas and Rempel-Clower, 1997). This evidence suggests that the focus of connections and the prevalence of CBP are matched in eulaminate cortices, but mismatched in limbic areas. The mismatch in limbic areas may have functional consequences, since neurons with CBP are inhibitory and have the capacity to buffer and sequester calcium (for reviews see Baimbridge, Celio and Rogers, 1992; Heizmann, 1992). These molecular/connectional features may provide a clue as to why limbic areas have a predilection for epileptiform activity (Penfield and Jasper, 1954).

The recruitment of bidirectional pathways in behavior

The above discussion suggests that the prefrontal cortex is enriched in bidirectional connections with a variety of cortical and subcortical structures. This pattern is not restricted to the connections of prefrontal cortices, but applies to other cortices as well.

In fact, reciprocity of neural pathways is a cardinal feature of communication in the nervous system, and underlies processes ranging from elementary sensory perception to complex processes of learning, memory and emotion (for review see Barbas, 2000a). Reciprocal pathways, as described for the cortex, do not generally involve the same neurons. Similarly, thalamocortical projections terminate in the middle layers of the cortex, whereas corticothalamic projection neurons originate primarily in the deep layers (for reviews see Jones, 1985; Steriade et al., 1997).

An important class of connections in neural communication are the so-called 'feedback' projections from the cortex to the thalamus. In the sensory systems, these projections have been called feedback because they originate from later processing areas (the cortex) and terminate in an earlier processing area (the thalamus). An important feature of feedback pathways is their expansiveness in comparison with feedforward pathways, not only between the cortex and the thalamus, but in corticocortical pathways as well. The extensive feedback projections originating from the deep layers of all cortices to the thalamus (for reviews see Jones, 1984, 1985) appear to have multiple but always critical roles in neural function. For example, it has been suggested that in the visual system activation of a feedback pathway between the primary cortex and the lateral geniculate increases the efficacy of feature specific input to the visual cortex (Sillito, Jones, Gerstein and West, 1994).

In association areas that are not primarily sensory, it is not clear what function is served by major feedback systems from the cortex to the thalamus. The thalamic MD nucleus, which projects to all prefrontal areas, has a role in working memory as well as long-term memory (for review see Markowitsch, 1982). In view of this evidence, and the distinct role of lateral prefrontal cortices in working memory and limbic cortices in long-term memory, it is possible that anatomically distinct feedback projection systems from each of these cortices have specific roles in memory. A pathway from MD to lateral prefrontal cortices, and feedback projections from lateral cortices to MD may act as a positive loop to hold events in working memory. In contrast, bidirectional connections between caudal orbitofrontal and caudal medial (limbic) prefrontal cortices and

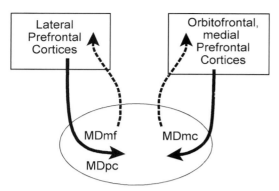

Fig. 7. Bidirectional connections linking the prefrontal areas with subdivisions of the mediodorsal (MD) nucleus of the thalamus. Lateral prefrontal cortices are connected with the multiform (MDmf) and parvicellular (MDpc) subdivisions of MD. Orbitofrontal and medial prefrontal cortices are connected with the magnocellular (MDmc) subdivision of MD. Activation of each set of connections may have a role in the distinct mnemonic functions of lateral prefrontal areas in working memory, and orbitofrontal and medial prefrontal areas in long-term memory.

MD may form a loop involved in long-term memory, according to the model in Fig. 7.

The above idea is supported by physiologic and behavioral studies. For example, in the lateral prefrontal cortex and in MD neurons are active when animals must hold information temporarily in memory in delayed response tasks (Fuster and Alexander, 1973). When the lateral prefrontal cortex is incapacitated by cooling, neuronal responses in MD associated with working memory are disrupted (Alexander and Fuster, 1973). This evidence implicates feedback projections from the prefrontal cortex for activation of MD neurons in a specific task that requires mnemonic processing, a function associated with both of these structures (Alexander and Fuster, 1973; Kubota, Niki and Goto, 1972; Kubota, Tonoike and Mikami, 1980; Wilson, Scalaidhe and Goldman-Rakic, 1993).

Another example is provided in cases with Alzheimer's disease, this time disrupting long-term memory. In the prefrontal cortex brains from patients with Alzheimer's disease show degeneration preferentially in limbic cortices involving mostly the deep layers (Chu, Tranel, Damasio and Van Hoesen, 1997), which issue feedback projections. As noted above, limbic prefrontal areas are affiliated with structures associated with long-term memory.

In addition to the magnocellular part of MD, the above thalamocortical loop may involve the anterior nuclei, the midline nuclei and the intralaminar nuclei as well, which have particularly robust connections with prefrontal limbic areas (Barbas et al., 1991; Dermon and Barbas, 1994; Xiao and Barbas, 2001). Thus, separate sets of bidirectional systems involving distinct sectors of the prefrontal cortex and subdivisions of MD may participate in different aspects of memory.

Another strongly interconnected system involves the orbitofrontal cortex and the amygdala. Activation of bidirectional pathways linking these structures may be essential for the conscious appreciation of the emotional significance of events. These pathways may also involve limbic thalamic nuclei included in the original Papez circle for emotions, such as the anterior nuclei. In addition, the amygdala, the thalamic MD, and midline thalamic nuclei are robustly connected with each other (Aggleton, Burton and Passingham, 1980; Aggleton and Mishkin, 1984; Russchen, Amaral and Price, 1987) and may have, to some extent, shared roles in memory and emotion (Oyoshi, Nishijo, Asakura et al., 1996).

Complementary roles of pathways from lateral, orbitofrontal and medial prefrontal cortices in behavior

The distinct differences in the pattern of connections between lateral prefrontal areas, on one hand, and orbitofrontal and medial prefrontal areas, on the other hand, suggest that they may have specific roles in cognitive processes, memory and emotions. However, because these areas are connected with structures associated with different aspects of neural processing, they must function in concert even in simple aspects of behavior. Take for instance, an ambiguous stimulus in the environment, such as a loud noise that elicits an alerting fear response that likely activates the amygdala. Activation of a pathway from the amygdala to prefrontal limbic cortices may mediate the conscious perception of fear, as suggested above. In turn, a descending projection from the limbic prefrontal cortices to the amygdala may reinforce a loop on the emotional significance of the event. The role of lateral prefrontal cortices in this event is likely to be in directing attention to the relevant stimulus, and to determine if the noise signifies danger or not. The robust connections of lateral prefrontal cortices with early sensory cortices (for review see Barbas, 1995), as well as their strong connections with premotor cortices in macaque monkeys (Arikuni, Watanabe and Kubota, 1988; Barbas and Pandya, 1987; Matelli, Camarda, Glickstein and Rizzolatti, 1986; McGuire, Bates and Goldman-Rakic, 1991) provide the anatomic basis for the discriminatory component as well as the executive response for this process. Intrinsic connections linking lateral prefrontal and limbic prefrontal cortices (e.g., Barbas and Pandya, 1989; Barbas et al., 1999), provide the anatomic basis of coordinated activation.

The executive functions of the prefrontal cortices are further supported through their privileged reception (shared only by their neighboring premotor areas) of the output from the basal ganglia through the thalamic MD nucleus (for review see Alexander, Delong and Strick, 1986). The ventral anterior and ventral lateral nuclei of the thalamus receive output from the basal ganglia as well, and project to all prefrontal areas (Barbas et al., 1991; Dermon and Barbas, 1994). Pathways through the basal ganglia have been implicated in several cognitive processes, sequential planning, and monitoring of behavior (Koechlin, Corrado, Pietrini and Grafman, 2000; MacDonald et al., 2000; Middleton and Strick, 2000). As in the case of the preserved topographic relationships between the subdivisions of the thalamic MD nucleus and different regions of the prefrontal cortex, the pathways through the basal ganglia form several parallel circuits, which retain a certain degree of topographic organization at each level of the circuit in the striatum, pallidum, distinct subdivisions of MD and the ventral anterior and ventral lateral thalamic nuclei (Alexander et al., 1986; Groenewegen, Berendse, Wolters and Lohman, 1990). However, there is a certain degree of interaction in these pathways, suggested by the connections as well as the multiple neuropathologies associated with the basal ganglia, manifested in multiple symptoms in motor performance and in cognitive and emotional dysfunctions (Alexander et al., 1986; Delong and Georgopoulos, 1981; Groenewegen et al., 1990; Groenewegen, Wright and Uylings, 1997; Joel and Weiner, 1994; Koechlin et al., 2000).

Differential interface of distinct prefrontal areas with extrathalamic systems

Connections of prefrontal cortices with the basal forebrain and the hypothalamus

The discussion above suggests that distinct sectors of the prefrontal cortex are connected topographically and according to a set of rules with other cortical and subcortical structures which may be accessed during specific aspects of behavior. Like other cortical systems, the prefrontal cortex is targeted by ascending modulatory systems, including the basal forebrain, the locus coeruleus, the substantia nigra/ventral tegmental area, and the raphe nuclei (for reviews see Aston-Jones, Rajkowski, Kubiak et al., 1996; Foote and Morrison, 1987; Mesulam, 1995; Steriade, 1995, 1996). There is strong evidence that these systems play a key role in the activities of the prefrontal cortex. For example, fluctuation in the activity of neurons in the locus coeruleus affects attentional and cognitive processes (Usher, Cohen, Servan-Schreiber et al., 1999). In addition, the level of available dopamine, supplied through projections from the ventral tegmental area, and noradrenaline, supplied from projection neurons in the locus coeruleus, has marked consequences on cognitive function in the prefrontal cortex (for review see Arnsten, 1997).

The anatomic interactions of these systems with the prefrontal cortex show a higher degree of organization than previously thought. For example, among prefrontal cortices, the orbitofrontal and medial areas appear to have strong anatomic interactions with the basal forebrain in monkeys (Irle and Markowitsch, 1986; Kievit and Kuypers, 1975; Mesulam and Mufson, 1984; Mesulam, Mufson and Wainer, 1986a; Mesulam, Mufson, Wainer and Levey, 1983b; Mesulam and Van Hoesen, 1976; Russchen, Amaral and Price, 1985) and rats (Carlsen, Zaborszky and Heimer, 1985; Gaykema, Luiten, Nyakas and Traber, 1990; Mesulam et al., 1983b; Rye, Wainer, Mesulam et al., 1984; Zaborszky, Pang, Somogyi et al., 1999). While there is considerable overlap in the origin of basal forebrain projections to orbitofrontal, medial and lateral prefrontal cortices in rats and primates (Gaykema et al., 1990; Grove, 1988a,b; Luiten, Spencer, Traber and Gaykema,

1985; McKinney, Coyle and Hedreen, 1983; Mesulam et al., 1983a,b; Mesulam, Volicer, Marquis et al., 1986b; Pearson, Gatter, Brodal and Powell, 1983; Russchen et al., 1985; Selden, Gitelman, Salamon-Murayama et al., 1998; Stewart, MacFabe and Leung, 1985), there are notable biases in these connections as well. The posterior part of the basal forebrain preferentially projects to lateral prefrontal cortices, whereas anterior and ventrally located diagonal band nuclei project to medial and orbitofrontal cortices (Ghashghaei and Barbas, 2001).

Moreover, the posterior part of the nucleus basalis not only projects to lateral prefrontal cortices but also to auditory association and temporal polar areas (Mesulam et al., 1983a,b), which, in turn, are interconnected through corticocortical pathways (for reviews see Barbas, 1992, 1995; Pandya et al., 1988). This evidence suggests that through a set of common connections, the posterior part of the basal forebrain may temporally coordinate activation of prefrontal and sensory cortices that are linked through corticocortical connections (Ghashghaei and Barbas, 2001). Such activation may facilitate the recruitment of signals necessary for cognitive processes that rely on lateral prefrontal cortices.

Similarly, the anteromedial parts of the diagonal band nuclei issue preferential projections to medial prefrontal areas and to the hippocampal formation, parahippocampal, rhinal and perirhinal areas, and midline thalamic nuclei (Grove, 1988a; Gaykema et al., 1990; Hreib, Rosene and Moss, 1988; McKinney et al., 1983; Mesulam et al., 1983a,b; Russchen et al., 1985), which also project preferentially to medial prefrontal areas (for reviews see Barbas, 1997, 2000b). In addition, the ventral aspect of the diagonal band nuclei projects to orbitofrontal cortices (Ghashghaei and Barbas, 2001), and to the olfactory bulb (Mesulam et al., 1983a,b; Rosene, Heimer and Van Hoesen, 1978), matching an equally robust projection from olfactory areas to the posterior orbitofrontal cortices (Barbas, 1993; Carmichael, Clugnet and Price, 1994; Morecraft et al., 1992; Potter and Nauta, 1979). Finally, there is evidence that the projections from the basal forebrain to the amygdala originate predominantly from the lateral part of the nucleus basalis (Mesulam et al., 1983a,b), which also projects robustly to orbitofrontal cortices.

The above evidence suggests that even though the projections of the basal forebrain to different prefrontal regions overlap, a set of basal forebrain nuclei issues projections to interconnected cortices forming a more elaborate but unique network (Ghashghaei and Barbas, 2001) (Fig. 8). Thus, within a seemingly diffuse system, specificity may be afforded by concomitant activation by the basal forebrain of interconnected neural structures. This pattern of innervation suggests that the arousal and attentional functions of the basal forebrain (Robbins, Granon, Muir et al., 1998; Sarter and Bruno, 2000; Voytko, Olton, Richardson et al., 1994) may be effected through activation of circuits that recruit functionally distinct prefrontal cortices and the areas they are connected with (Ghashghaei and Barbas, 2001; Grove, 1988a; Pearson et al., 1983; Zaborszky et al., 1999).

Another system with widespread connections to all sectors of the prefrontal cortex is the hypothalamus (Jacobson, Butters and Tovsky, 1978; Kievit and Kuypers, 1975; Mizuno, Uemura-Sumi, Yasui et al., 1982; Morecraft et al., 1992; Potter and Nauta, 1979; Rempel-Clower and Barbas, 1998; Tigges, Walker and Tigges, 1983). However, even though the hypothalamus is generally regarded as a diffuse projection system to the cortex, there is a certain degree of topographic organization in this projection system to different prefrontal regions. Thus, lateral prefrontal areas receive projections mostly from the posterior hypothalamus, whereas projections to caudal orbitofrontal and medial prefrontal cortices originate along the entire caudal to rostral extent of the hypothalamus, within its posterior, tuberal and anterior divisions; the latter contribute only a minor projection to lateral prefrontal cortices (Rempel-Clower and Barbas, 1998).

Selective output from medial and orbitofrontal cortices to the basal forebrain and hypothalamus

The most striking difference in the connections of different sectors of the prefrontal cortex with extrathalamic systems is the restricted descending projection from medial and orbitofrontal cortices to the basal forebrain and to the hypothalamus. In the basal forebrain axons from medial and orbitofrontal cortices terminate rostrally within the nucleus

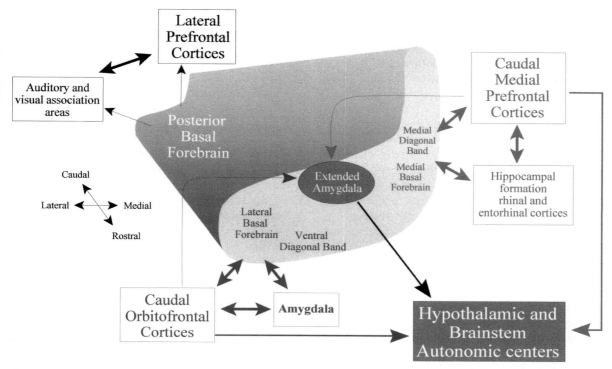

Fig. 8. Summary of the connections of the basal forebrain with the prefrontal cortex. Lateral prefrontal areas receive projections preferentially (though not exclusively) from the posterior basal forebrain, but do not project to the basal forebrain. Caudal orbitofrontal areas are preferentially (though not exclusively) connected through reciprocal pathways with rostral parts of the basal forebrain, and medial prefrontal areas are preferentially connected with the medial part of the basal forebrain and the diagonal band nuclei, and project to the motor-related system of the extended amygdala; the latter projects to hypothalamic and brainstem autonomic centers. The diagram also shows the preferential connections of basal forebrain regions with other cortices and structures described in the literature, which may form elaborate interconnected networks recruited in behavior (for discussion and references see Ghashghaei and Barbas, 2001).

basalis and the diagonal band nuclei (Gaykema, Vanweeghel, Hersh and Luiten, 1991; Ghashghaei and Barbas, 2001; Mesulam and Mufson, 1984).

The basal forebrain is a heterogeneous region, composed of several systems, including cholinergic and non-cholinergic neurons (Alheid and Heimer, 1996; Heimer, Harlan, Alheid et al., 1997; Mesulam et al., 1983a,b; Rye et al., 1984; Zaborszky et al., 1999). Of particular interest are two non-cholinergic systems that have motor-related functions, the ventral striatopallidum, which occupies roughly the top half of the basal forebrain, and the extended amygdala, composed of punctuated groups of neurons extending from the central nucleus of the amygdala through the basal forebrain to the bed nucleus of the stria terminalis (for reviews see De Olmos, 1990; De Olmos and Heimer, 1999; Heimer et al., 1997). We

recently deciphered these systems in rhesus monkeys with the aid of neurochemical markers, and determined that descending projections from some orbitofrontal, and from medial area 32, reach the extended amygdala and the ventral striatopallidal region (Ghashghaei and Barbas, 2001), as shown on the basis of connections in several other species (Cassell and Wright, 1986; Kapp, Schwaber and Driscoll, 1985; Room, Russchen, Groenewegen and Lohman, 1985; Yasui, Breder, Saper and Cechetto, 1991). This evidence indicates that the connections of the basal forebrain with orbitofrontal and medial prefrontal cortices are bidirectional.

In marked contrast, the linkage of lateral prefrontal areas is unidirectional, composed of an ascending limb but lacking an efferent projection to the basal forebrain (Ghashghaei and Barbas, 2001;

17

Mesulam and Mufson, 1984) (Fig. 8). This system is, therefore, an open-loop system, which may be related to 'on-line' processing in cognitive tasks supported by lateral prefrontal cortices and their interconnected visual, auditory and other sensory cortices, which may be recruited in temporal synchrony to retrieve information to solve the problem at hand.

The tightly linked orbitofrontal, medial prefrontal, and basal forebrain regions, have common connections with diencephalic structures and the amygdala (for reviews see Alheid and Heimer, 1996; Barbas, 1997, 2000b; De Olmos, 1990; McGaugh, Cahill and Roozendaal, 1996; Zola-Morgan and Squire, 1993). Components of this elaborate network have been associated with processing novelty and the behavioral significance of sensory stimuli (Gallagher, McMahan and Schoenbaum, 1999; Lipton, Alvarez and Eichenbaum, 1999; Schoenbaum, Chiba and Gallagher, 1998, 1999, 2000; Rolls, Critchley, Browning et al., 1999; Tremblay and Schultz, 1999; Wilson and Rolls, 1990a,b). For example, neuronal activity in the amygdala and orbitofrontal areas is modified as the salience of cues in a behavioral task changes (Gallagher et al., 1999; Schoenbaum et al., 1998; Schoenbaum et al., 2000), a process that may depend on descending projections from the orbitofrontal cortex and the amygdala (Ghashghaei and Barbas, 2001). This elaborate network may be activated by the basal forebrain in the process of monitoring the motivational significance of associated events and their encoding into long-term memory.

The selective descending pathway from restricted medial and orbitofrontal areas in the territory of the extended amygdala and the ventral striatopallidal region provides a link to the output of the amygdala through hypothalamic and brainstem autonomic structures (for reviews see Alheid and Heimer, 1996; Heimer et al., 1997; Holstege, 1991; Price, Russchen and Amaral, 1987). The selective axonal projections from caudal orbitofrontal and medial prefrontal areas to the basal forebrain are complemented by another selective descending pathway from the same prefrontal cortices to hypothalamic autonomic centers (Rempel-Clower and Barbas, 1998). This evidence suggests that there are parallel pathways from medial prefrontal and orbitofrontal cortices to hypothalamic and brainstem autonomic centers (Leichnetz and Astruc, 1976; Neafsey, 1990; Ongur,

An and Price, 1998; Rempel-Clower and Barbas, 1998; Saper, Loewy, Swanson and Cowan, 1976; Wouterlood, Steinbusch, Luiten and Bol, 1987), as well as spinal autonomic centers (Rempel-Clower, Ghashghaei and Barbas, 1999), the final common pathway innervating peripheral autonomic organs.

Anatomic basis of theories for emotions

As discussed above, caudal orbitofrontal and medial prefrontal cortices not only receive projections from the amygdala, but also issue robust descending projections to autonomic structures, and thus can change cardiac and respiratory responses in emotional situations. Because the cortex appears to be essential for cognitive awareness of the emotional significance of events, activation of autonomic responses that accompany complex emotional stimuli likely involve orbitofrontal pathways. The inability of patients with orbitofrontal lesions to respond autonomically in emotional situations (Bechara, Tranel, Damasio and Damasio, 1996; Damasio, Tranel and Damasio, 1990) suggests disconnection of a prefrontal cortical system from hypothalamic, brainstem and spinal autonomic centers, the motor link of the emotional system. Although patients with orbitofrontal lesions retain cognitive function, they make poor decisions, suggesting that cognitive processes become disconnected from emotionally driven autonomic responses.

Classic theories of emotion have addressed the temporal interaction of autonomic activation and the conscious perception of emotion (for review see LeDoux, 1996). The James–Lange theory postulates that the autonomic arousal occurring in the presence of an emotional stimulus precedes the experience of emotion. In contrast, the Cannon–Bard theory holds that the cortex plays a critical role in generating the conscious experience of emotion, and that this occurs along with (but does not rely on) the sensation of autonomic activation.

Neural circuitry suggests that both theories may be correct but for different types of situations. For example, a sudden loud sound of a siren can activate the amygdala and trigger autonomic arousal through brainstem structures apart from the conscious experience of fear. However, when driving a car, our visual system takes a bit longer to provide information to determine whether the patrol car with the siren is following us. The assessment of this complex visual scene is likely to depend on projections from visual areas to the lateral prefrontal cortex, and then to orbitofrontal and medial prefrontal areas. In turn, orbitofrontal and medial prefrontal areas issue direct projections to autonomic areas in the hypothalamus and brainstem, which innervate visceral organs (Rempel-Clower and Barbas, 1998). These prefrontal areas are interconnected with the amygdala, which also has direct access to brainstem autonomic centers, as reviewed above. If we determine that we are about to receive a ticket for speeding, our conscious emotional experience as well as our level of autonomic activity is quite different from the situation in which the patrol car passes us.

Conclusion

The distinct patterns of connections noted in lateral, orbitofrontal and medial prefrontal cortices suggest that they have specific but complementary roles in cognition, memory and emotion. In all systems connections between neural structures are generally reciprocal, suggesting closed-loop control through feedforward and feedback communication. Moreover, structures that are robustly interconnected, such as the orbitofrontal cortex and the thalamic MDmc, are also connected with lateral prefrontal areas, albeit more weakly. In turn, lateral prefrontal cortices are connected with relatively early-processing sensory areas, forming a more intricate network. The connections between cortices with diverse functions suggest that cognitive processes and emotional processes are inexorably linked in the nervous system (Barbas, 1995, 2000b). Subcortical systems such as the basal forebrain, the locus coeruleus and the substantia nigra once thought to have diffuse projections to the cortex appear to engage interconnected cortical systems in distinct aspects of behavior.

Functional compromise of neural structures in several neurologic and psychiatric diseases is likely to disrupt parts of interdependent networks recruited in complex behavior. For example, the engagement of the orbitofrontal cortex in obsessive compulsive disorder (Rapoport and Fiske, 1998) may involve persistent activation of bidirectional pathways between the orbitofrontal cortex and the amygdala. It

is also interesting that subclinical cases of obsessive 'checkers' show enhanced performance on long-term memory, but have a somewhat reduced working memory function (Roth and Baribeau, 1996). The enhanced long-term memory function may be mediated through repetitive activation of a loop linking orbitofrontal cortices with MDmc, whereas reduced working memory may reflect a functional compromise of the pathway linking lateral prefrontal cortices with MDpc. This evidence suggests that distinct mnemonic mechanisms can be affected in opposite directions in this disorder. Anatomic pathways suggest that there are qualitatively different feedback systems that may subserve specific aspects of mnemonic and emotional processes.

Conversely, disruption of bidirectional pathways between the amygdala and the orbitofrontal cortex is likely to affect a circuit associated with attaching the appropriate emotional significance to an event, evidenced in several psychiatric diseases, such as sociopathic personality, or autistic behavior. This idea is consistent with observations that activity in prefrontal limbic cortices is reduced in some of these disorders, and is exemplified by inappropriate affect when orbitofrontal cortices are damaged (e.g., Damasio, 1994; Damasio et al., 1990).

The identification of limbic cortices, as primarily feedback systems, on the basis of the pattern of their connections, is important from a clinical perspective because limbic cortices are preferentially vulnerable in several neurologic and psychiatric diseases, including Alzheimer's disease, epilepsy, schizophrenia, obsessive compulsive disorder, and Tourette's syndrome (Abbruzzese, Bellodi, Ferri and Scarone, 1995; Breiter, Rauch, Kwong et al., 1996; Weeks, Turjanski and Brooks, 1996; for reviews see Rapoport and Fiske, 1998; Weinberger, 1988). Limbic cortices are enriched in connections, so that their pathology is likely to disrupt a massive feedback system to cortical and subcortical structures and upset intricate neural rhythms and behavior.

Acknowledgements

This work was supported by Grants NS57414, MH11151 (NIMH) and NS24760 (NINDS).

References

Abbruzzese M, Bellodi L, Ferri S, Scarone S: Frontal lobe dysfunction in schizophrenia and obsessive–compulsive disorder: a neuropsychological study. Brain and Cognition: 27; 202–212, 1995.

Aggleton JP, Burton MJ, Passingham RE: Cortical and subcortical afferents to the amygdala of the rhesus monkey (*Macaca mulatta*). Brain Research: 190; 347–368, 1980.

Aggleton JP, Mishkin M: Projections of the amygdala to the thalamus in the cynomolgus monkey. Journal of Comparative Neurology: 222; 56–68, 1984.

Alexander GE, Delong MR, Strick PL: Parallel organization of functionally segregated circuits linking basal ganglia and cortex. Annual Review of Neuroscience: 9; 357–381, 1986.

Alexander GE, Fuster JM: Effects of cooling prefrontal cortex on cell firing in the nucleus medialis dorsalis. Brain Research: 61; 93–105, 1973.

Alexander MP, Freedman M: Amnesia after anterior communicating artery aneurysm rupture. Neurology: 34; 752–757, 1984.

Alheid GF, Heimer L: Theories of basal forebrain organization and the 'emotional motor system'. Progress in Brain Research: 107; 461–484, 1996.

Amaral DG, Insausti R, Zola-Morgan S, Squire LR, Suzuki WA: The perirhinal and parahippocampal cortices and medial temporal lobe memory function. In Vision, Memory, and the Temporal Lobe. New York: Elsevier Science, pp. 149–161, 1990.

Amaral DG, Price JL: Amygdalo-cortical projections in the monkey (*Macaca fascicularis*). Journal of Comparative Neurology: 230; 465–496, 1984.

Andersen RA, Asanuma C, Cowan WM: Callosal and prefrontal associational projecting cell populations in area 7a of the macaque monkey: a study using retrogradely transported fluorescent dyes. Journal of Comparative Neurology: 232; 443–455, 1985.

Andersen RA, Asanuma C, Essick G, Siegel RM: Corticocortical connections of anatomically and physiologically defined subdivisions within the inferior parietal lobe. Journal of Comparative Neurology: 296; 65–113, 1990.

Arikuni T, Watanabe K, Kubota K: Connections of area 8 with area 6 in the brain of the macaque monkey. Journal of Comparative Neurology: 277; 21–40, 1988.

Arnsten AFT: Catecholamine regulation of prefrontal cortex. J. Psychopharmacology: 11; 151–162, 1997.

Aston-Jones G, Rajkowski J, Kubiak P, Valentino RJ, Shipley MT: Role of the locus coeruleus in emotional activation. Progress in Brain Research: 107; 379–402, 1996.

Bachevalier J, Meunier M, Lu MX, Ungerleider LG: Thalamic and temporal cortex input to medial prefrontal cortex in rhesus monkeys. Experimental Brain Research: 115; 430–444, 1997.

Bachevalier J, Mishkin M: Visual recognition impairment follows ventromedial but not dorsolateral prefrontal lesions in monkeys. Behavioral Brain Research: 20; 249–261, 1986.

Baimbridge KG, Celio MR, Rogers JH: Calcium-binding proteins in the nervous system. Trends in Neuroscience: 15; 303–308, 1992.

Baleydier C, Mauguiere F: The duality of the cingulate gyrus in monkey. Neuroanatomical study and functional hypothesis. Brain: 103; 525–554, 1980.

Barbas H: Pattern in the laminar origin of corticocortical connections. Journal of Comparative Neurology: 252; 415–422, 1986.

Barbas H: Anatomic organization of basoventral and mediodorsal visual recipient prefrontal regions in the rhesus monkey. Journal of Comparative Neurology: 276; 313–342, 1988.

Barbas H: Architecture and cortical connections of the prefrontal cortex in the rhesus monkey. In Chauvel P (Ed), Advances in Neurology, Vol. 57. New York: Raven Press, pp. 91–115, 1992.

Barbas H: Organization of cortical afferent input to orbitofrontal areas in the rhesus monkey. Neuroscience: 56; 841–864, 1993.

Barbas H: Anatomic basis of cognitive–emotional interactions in the primate prefrontal cortex. Neuroscience and Biobehavioral Reviews: 19; 499–510, 1995.

Barbas H: Two prefrontal limbic systems: Their common and unique features. In Sakata H, Mikami A, Fuster JM (Eds), The Association Cortex: Structure and Function. Amsterdam: Harwood Academic, pp. 99–115, 1997.

Barbas H: Complementary role of prefrontal cortical regions in cognition, memory and emotion in primates. Advances in Neurology: 84; 87–110, 2000a.

Barbas H: Connections underlying the synthesis of cognition, memory, and emotion in primate prefrontal cortices. Brain Research Bulletin: 52; 319–330, 2000b.

Barbas H, Blatt GJ: Topographically specific hippocampal projections target functionally distinct prefrontal areas in the rhesus monkey. Hippocampus: 5; 511–533, 1995.

Barbas H, De Olmos J: Projections from the amygdala to basoventral and mediodorsal prefrontal regions in the rhesus monkey. Journal of Comparative Neurology: 301; 1–23, 1990.

Barbas H, Dubrovsky B: Central and peripheral effects of tonic vibratory stimuli to dorsal neck and extraocular muscles in the cat. Experimental Neurology: 74; 67–85, 1981.

Barbas H, Ghashghaei H, Dombrowski SM, Rempel-Clower NL: Medial prefrontal cortices are unified by common connections with superior temporal cortices and distinguished by input from memory-related areas in the rhesus monkey. Journal of Comparative Neurology: 410; 343–367, 1999.

Barbas H, Henion TH, Dermon CR: Diverse thalamic projections to the prefrontal cortex in the rhesus monkey. Journal of Comparative Neurology: 313; 65–94, 1991.

Barbas H, Mesulam M-M.: Organization of afferent input to subdivisions of area 8 in the rhesus monkey. Journal of Comparative Neurology: 200; 407–431, 1981.

Barbas H, Mesulam M-M.: Cortical afferent input to the principalis region of the rhesus monkey. Neuroscience: 15; 619–637, 1985.

Barbas H, Pandya DN: Architecture and frontal cortical connections of the premotor cortex (area 6) in the rhesus monkey. Journal of Comparative Neurology: 256; 211–218, 1987.

Barbas H, Pandya DN: Architecture and intrinsic connections of the prefrontal cortex in the rhesus monkey. Journal of Comparative Neurology: 286; 353–375, 1989.

Barbas H, Rempel-Clower N: Cortical structure predicts the pattern of corticocortical connections. Cerebral Cortex: 7; 635–646, 1997.

Barris RW, Schuman HR: Bilateral anterior cingulate gyrus lesions. Syndrome of the anterior cingulate gyri. Neurology: 3; 44–52, 1953.

Bechara A, Tranel D, Damasio H, Damasio AR: Failure to respond autonomically to anticipated future outcomes following damage to prefrontal cortex. Cerebral Cortex: 6; 215–225, 1996.

Bisiach E, Capitani E, Luzzatti C, Perani D: Brain and conscious representation of outside reality. Neuropsychologia: 19; 543–551, 1981.

Botvinick M, Nystrom LE, Fissell K, Carter CS, Cohen JD: Conflict monitoring versus selection-for-action in anterior cingulate cortex. Nature: 402; 179–181, 1999.

Braak H, Braak E: Alzheimer's disease affects limbic nuclei of the thalamus. Acta Neuropathologica: 81; 261–268, 1991.

Breiter HC, Rauch SL, Kwong KK, Baker JR, Weisskoff RM, Kennedy DN, Kendrick AD, Davis TL, Jiang A, Cohen MS, Stern CE, Belliveau JW, Baer L, O'Sullivan RL, Savage CR, Jenike MA, Rosen BR: Functional magnetic resonance imaging of symptom provocation in obsessive–compulsive disorder. Archives of General Psychiatry: 53; 595–606, 1996.

Brodmann K: Vergleichende Lokalizationslehre der Grosshirnrinde in ihren Prinizipien dargestelt auf Grund der Zellenbaues. Leipzig: Barth, 1909.

Bruce CJ, Goldberg ME, Bushnell MC, Stanton GB: Primate frontal eye fields. II. Physiological and anatomical correlates of electrically evoked eye movements. Journal of Neurophysiology: 54; 714–734, 1985.

Buge A, Escourolle R, Rancurel G, Poisson M: Akinetic mutism and bicingular softening. 3 anatomo-clinical cases. Revue Neurologique: 131; 121–131, 1975.

Cahill L, McGaugh JL: Mechanisms of emotional arousal and lasting declarative memory. Trends in Neuroscience: 21; 294–299, 1998.

Carlsen J, Zaborszky L, Heimer L: Cholinergic projections from the basal forebrain to the basolateral amygdaloid complex: a combined retrograde fluorescent and immunohistochemical study. Journal of Comparative Neurology: 234; 155–167, 1985.

Carmichael ST, Clugnet M-C, Price JL: Central olfactory connections in the macaque monkey. Journal of Comparative Neurology: 346; 403–434, 1994.

Carmichael ST, Price JL: Limbic connections of the orbital and medial prefrontal cortex in macaque monkeys. Journal of Comparative Neurology: 363; 615–641, 1995a.

Carmichael ST, Price JL: Sensory and premotor connections of the orbital and medial prefrontal cortex of macaque monkeys. Journal of Comparative Neurology: 363; 642–664, 1995b.

Cassell MD, Wright DJ: Topography of projections from the medial prefrontal cortex to the amygdala in the rat. Brain Research Bulletin: 17; 321–333, 1986.

Cavada C, Company T, Tejedor J, Cruz-Rizzolo RJ, Reinoso-Suarez F: The anatomical connections of the macaque monkey orbitofrontal cortex. A review. Cerebral Cortex: 10; 220–242, 2000.

Cavada C, Goldman-Rakic PS: Posterior parietal cortex in rhesus monkey: II. Evidence for segregated corticocortical networks linking sensory and limbic areas with the frontal lobe. Journal of Comparative Neurology: 287; 422–445, 1989.

Chao LL, Knight RT: Prefrontal deficits in attention and inhibitory control with aging. Cerebral Cortex: 7; 63–69, 1997.

Chao LL, Knight RT: Contribution of human prefrontal cortex to delay performance. Journal of Cognitive Neuroscience: 10; 167–177, 1998.

Chavis DA, Pandya DN: Further observations on corticofrontal connections in the rhesus monkey. Brain Research: 117; 369–386, 1976.

Chu C-C., Tranel D, Damasio AR, Van Hoesen GW: The autonomic-related cortex: pathology in Alzheimer's disease. Cerebral Cortex: 7; 86–95, 1997.

Crowell RM, Morawetz RB: The anterior communicating artery has significant branches. Stroke: 8; 272–273, 1977.

D'Esposito M, Alexander MP, Fischer R, McGlinchey-Berroth R, O'Connor M: Recovery of memory and executive function following anterior communicating artery aneurysm rupture. Journal of the International Neuropsychological Society: 2; 565–570, 1996.

Damasio AR: Descarte's Error: Emotion, Reason, and the Human Brain. New York: G.P. Putnam's Sons, 1994.

Damasio AR, Tranel D, Damasio H: Individuals with sociopathic behavior caused by frontal damage fail to respond autonomically to social stimuli. Behavioral Brain Research: 41; 81–94, 1990.

Davis M: The role of the amygdala in fear and anxiety. Annual Review of Neuroscience: 15; 353–375, 1992.

De Lima AD, Voigt T, Morrison JH: Morphology of the cells within the inferior temporal gyrus that project to the prefrontal cortex in the macaque monkey. Journal of Comparative Neurology: 296; 159–172, 1990.

De Olmos JS: Amygdala. In Paxinos G (Ed), The Human Nervous System. San Diego: Academic Press, pp. 583–710, 1990.

De Olmos JS, Heimer L: The concepts of the ventral striatopallidal system and extended amygdala. Annals of the New York Academy of Sciences: 877; 1–32, 1999.

DeFelipe J, Hendry SH, Jones EG: Visualization of chandelier cell axons by parvalbumin immunoreactivity in monkey cerebral cortex. Proceedings of the National Academy of Sciences of the United States of America: 86; 2093–2097, 1989.

Delong MR, Georgopoulos AP: Motor functions of the basal ganglia. In Brooks VG, Brookhart JM, Mountcastle VB (Eds), The Handbook of Physiology. Bethesda, MD: American Physiological Society, pp. 1017–1061, 1981.

Dermon CR, Barbas H: Contralateral thalamic projections predominantly reach transitional cortices in the rhesus monkey. Journal of Comparative Neurology: 344; 508–531, 1994.

Desimone R: Neural mechanisms for visual memory and their role in attention. Proceedings of the National Academy of

Sciences of the United States of America: 93; 13494–13499, 1996.

Desimone R, Gross CG: Visual areas in the temporal cortex of the macaque. Brain Research: 178; 363–380, 1979.

Distler C, Boussaoud D, Desimone R, Ungerleider LG: Cortical connections of inferior temporal area TEO in macaque monkeys. Journal of Comparative Neurology: 334; 125–150, 1993.

Dolan RJ, Fletcher P, Frith CD, Friston KJ, Frackowiak RSJ, Grasby PM: Dopaminergic modulation of impaired cognitive activation in the anterior cingulate cortex in schizophrenia. Nature: 378; 180–182, 1995.

Dombrowski SM, Barbas H: Distinction of prefrontal architectonic areas using stereologic procedures. Neuroscience Abstracts: 24; 1163, 1998.

Dombrowski SM, Hilgetag CC, Barbas H: Quantitative architecture distinguishes prefrontal cortical systems in the rhesus monkey. Cerebral Cortex: 11; 975–988, 2001.

Douglas RJ, Martin KA, Whitteridge D: An intracellular analysis of the visual responses of neurones in cat visual cortex. Journal of Physiology (London): 440; 659–696, 1991.

Foote SL, Morrison JH: Extrathalamic modulation of cortical function. Annual Review of Neuroscience: 10; 67–95, 1987.

Frith C, Dolan R: The role of the prefrontal cortex in higher cognitive functions. Cognitive Brain Research: 5; 175–181, 1996.

Funahashi S, Bruce CJ, Goldman-Rakic PS: Mnemonic coding of visual space in the monkey's dorsolateral prefrontal cortex. Journal of Neurophysiology: 61; 331–349, 1989.

Funahashi S, Bruce CJ, Goldman-Rakic PS: Dorsolateral prefrontal lesions and occulomotor delayed-response performance: Evidence for mnemonic 'scotomas'. Journal of Neuroscience: 13; 1479–1497, 1993.

Fuster JM: The Prefrontal Cortex. New York: Raven Press, 1989.

Fuster JM: Frontal lobes. Current Opinion in Neurobiology: 3; 160–165, 1993.

Fuster JM, Alexander GE: Firing changes in cells of the nucleus medialis dorsalis associated with delayed response behavior. Brain Research: 61; 79–91, 1973.

Fuster JM, Bauer RH, Jervey JP: Functional interactions between inferotemporal and prefrontal cortex in a cognitive task. Brain Research: 330; 299–307, 1985.

Gabbott PLA, Bacon SJ: Local circuit neurons in the medial prefrontal cortex (areas 24a,b,c, 25 and 32) in the monkey: II. Quantitative areal and laminar distributions. Journal of Comparative Neurology: 364; 609–636, 1996.

Gaffan EA, Aggleton JP, Gaffan D, Shaw C: Concurrent and sequential pattern discrimination learning by patients with Korsakoff amnesia. Cortex: 26; 381–397, 1990.

Gallagher M, McMahan RW, Schoenbaum G: Orbitofrontal cortex and representation of incentive value in associative learning. Journal of Neuroscience: 19; 6610–6614, 1999.

Gaykema RP, Luiten PG, Nyakas C, Traber J: Cortical projection patterns of the medial septum-diagonal band complex. Journal of Comparative Neurology: 293; 103–124, 1990.

Gaykema RP, Vanweeghel R, Hersh LB, Luiten PGM: Prefrontal cortical projections to the cholinergic neurons in the basal forebrain. Journal of Comparative Neurology: 303; 563–583, 1991.

Ghashghaei H, Barbas H: Neural interaction between the basal forebrain and functionally distinct prefrontal cortices in the rhesus monkey. Neuroscience: in press, 2001.

Goldberg ME, Bushnell MC: Behavioral enhancement of visual responses in monkey cerebral cortex. II. Modulation in frontal eye fields specifically related to saccades. Journal of Neurophysiology: 46; 773–787, 1981.

Goldman-Rakic PS: Topography of cognition: Parallel distributed networks in primate association cortex. Annual Review of Neuroscience: 11; 137–156, 1988.

Goldman-Rakic PS, Lidow MS, Gallager DW: Overlap of dopaminergic, adrenergic, and serotoninergic receptors and complementarity of their subtypes in primate prefrontal cortex. Journal of Neuroscience: 10; 2125–2138, 1990.

Goldman-Rakic PS, Porrino LJ: The primate mediodorsal (MD) nucleus and its projection to the frontal lobe. Journal of Comparative Neurology: 242; 535–560, 1985.

Groenewegen HJ, Berendse HW, Wolters JG, Lohman AH: The anatomical relationship of the prefrontal cortex with the striatopallidal system, the thalamus and the amygdala: evidence for a parallel organization. Progress in Brain Research: 85; 95–116, 1990.

Groenewegen HJ, Wright CI, Uylings HB: The anatomical relationships of the prefrontal cortex with limbic structures and the basal ganglia. Journal of Psychopharmacology: 11; 99–106, 1997.

Gross CG, Bender DB, Rocha-Miranda CE: Visual receptive fields of neurons in inferotemporal cortex of the monkey. Science: 166; 1303–1306, 1969.

Grove EA: Efferent connections of the substantia innominata in the rat. Journal of Comparative Neurology: 277; 347–364, 1988a.

Grove EA: Neural associations of the substantia innominata in the rat: afferent connections. Journal of Comparative Neurology: 277; 315–346, 1988b.

Heffner HE, Heffner RS: Effect of unilateral and bilateral auditory cortex lesions on the discrimination of vocalizations by Japanese macaques. Journal of Neurophysiology: 56; 683–701, 1986.

Heimer L, Harlan RE, Alheid GF, Garcia MM, De Olmos J: Substantia innominata: a notion which impedes clinical-anatomical correlations in neuropsychiatric disorders. Neuroscience: 76; 957–1006, 1997.

Heizmann CW: Calcium-binding proteins: Basic concepts and clinical implications. General Physiology and Biophysics: 11; 411–425, 1992.

Hof PR, Nimchinsky EA, Morrison JH: Neurochemical phenotype of corticocortical connections in the macaque monkey: Quantitative analysis of a subset of neurofilament protein-immunoreactive projection neurons in frontal, parietal, temporal, and cingulate cortices. Journal of Comparative Neurology: 362; 109–133, 1995.

Holstege G: Descending motor pathways and the spinal motor system: limbic and non-limbic components. Progress in Brain Research: 87; 307–421, 1991.

Hreib KK, Rosene DL, Moss MB: Basal forebrain efferents to the medial dorsal thalamic nucleus in the rhesus monkey. Journal of Comparative Neurology: 277; 365–390, 1988.

Irle E, Markowitsch HJ: Afferent connections of the substantia innominata/basal nucleus of Meynert in carnivores and primates. Journal für Hirnforschung: 27; 343–367, 1986.

Jacobsen CF: Studies of cerebral function in primates: I. The functions of the frontal association area in monkeys. Computer Psychological Monographing: 13; 3–60, 1936.

Jacobson S, Butters N, Tovsky NJ: Afferent and efferent subcortical projections of behaviorally defined sectors of prefrontal granular cortex. Brain Research: 159; 279–296, 1978.

Joel D, Weiner I: The organization of the basal ganglia–thalamocortical circuits: open interconnected rather than closed segregated. Neuroscience: 63(2); 363–379, 1994.

Jones EG: Laminar distribution of cortical efferent cells. In Jones EG, Peters A (Eds), Cerebral Cortex: Vol. 1, Cellular Components of the Cerebral Cortex. New York: Plenum Press, pp. 521–553, 1984.

Jones EG: The Thalamus. New York: Plenum Press, 1985.

Jürgens U, Müller-Preuss P: Convergent projections of different limbic vocalization areas in the squirrel monkey. Experimental Brain Research: 29; 75–83, 1977.

Kaitz SS, Robertson RT: Thalamic connections with limbic cortex. II. Corticothalamic projections. Journal of Comparative Neurology: 195; 527–545, 1981.

Kapp BS, Schwaber JS, Driscoll PA: Frontal cortex projections to the amygdaloid central nucleus in the rabbit. Neuroscience: 15; 327–346, 1985.

Kawaguchi Y, Kubota Y: GABAergic cell subtypes and their synaptic connections in rat frontal cortex. Cerebral Cortex: 7; 476–486, 1997.

Kennard MA: Focal autonomic representation in the cortex and its relation to sham rage. Journal of Neuropathology and Experimental Neurology: 4; 295–304, 1945.

Kievit J, Kuypers HGJM: Basal forebrain and hypothalamic connections to frontal and parietal cortex in the rhesus monkey. Science: 187; 660–662, 1975.

Kievit J, Kuypers HGJM: Organization of the thalamo-cortical connexions to the frontal lobe in the rhesus monkey. Experimental Brain Research: 29; 299–322, 1977.

Kling A, Steklis HD: A neural substrate for affiliative behavior in nonhuman primates. Brain Behavior and Evolution: 13; 216–238, 1976.

Koechlin E, Basso G, Pietrini P, Panzer S, Grafman J: The role of the anterior prefrontal cortex in human cognition. Nature: 399; 148–151, 1999.

Koechlin E, Corrado G, Pietrini P, Grafman J: Dissociating the role of the medial and lateral anterior prefrontal cortex in human planning. Proceedings of the National Academy of Sciences of the United States of America: 97; 7651–7656, 2000.

Kosaki H, Hashikawa T, He J, Jones EG: Tonotopic organization of auditory cortical fields delineated by parvalbumin immunoreactivity in macaque monkeys. Journal of Comparative Neurology: 386; 304–316, 1997.

Kubota K, Niki H, Goto A: Thalamic unit activity and delayed alternation performance in the monkey. Acta Neurobiologiae Experimentalis: 32; 177–192, 1972.

Kubota K, Tonoike M, Mikami A: Neuronal activity in the monkey dorsolateral prefrontal cortex during a discrimination task with delay. Brain Research: 183; 29–42, 1980.

Laemle LK, Sharma SC: Bilateral projections of neurons in the lateral geniculate nucleus and nucleus lateralis posterior to the visual cortex in the neonatal rat. Neuroscience Letters: 63; 207–214, 1986.

LeDoux J: The Emotional Brain. New York: Simon and Schuster, 1996.

Leichnetz GR, Astruc J: The efferent projections of the medial prefrontal cortex in the squirrel monkey (Saimiri sciureus). Brain Research: 109; 455–472, 1976.

Leonard BW, Amaral DG, Squire LR, Zola-Morgan S: Transient memory impairment in monkeys with bilateral lesions of the entorhinal cortex. Journal of Neuroscience: 15; 5637–5659, 1995.

Lipton PA, Alvarez P, Eichenbaum H: Crossmodal associative memory representations in rodent orbitofrontal cortex. Neuron: 2; 349–359, 1999.

Luiten PG, Spencer DG Jr, Traber J, Gaykema RP: The pattern of cortical projections from the intermediate parts of the magnocellular nucleus basalis in the rat demonstrated by tracing with Phaseolus vulgaris-leucoagglutinin. Neuroscience Letters: 57; 137–142, 1985.

MacDonald AW, Cohen JD, Stenger VA, Carter CS: Dissociating the role of the dorsolateral prefrontal and anterior cingulate cortex in cognitive control. Science: 288; 1835–1888, 2000.

Markowitsch HJ: Thalamic mediodorsal nucleus and memory: A critical evaluation of studies in animals and man. Neuroscience and Biobehavioral Reviews: 6; 351–380, 1982.

Matelli M, Camarda R, Glickstein M, Rizzolatti G: Afferent and efferent projections of the inferior area 6 in the Macaque monkey. Journal of Comparative Neurology: 251; 281–298, 1986.

McGaugh JL, Cahill L, Roozendaal B: Involvement of the amygdala in memory storage: Interaction with other brain systems. Proceedings of the National Academy of Sciences of the United States of America: 93; 13508–13514, 1996.

McGuire PK, Bates JF, Goldman-Rakic PS: Interhemispheric integration: I. symmetry and convergence of the corticocortical connections of the left and the right principal sulcus (PS) and the left and the right supplementary motor area (SMA) in the rhesus monkey. Cerebral Cortex: 1; 390–407, 1991.

McGuire PK, Silbersweig DA, Wright I, Murray RM, David AS, Frackowiak RSJ, Frith CD: Abnormal monitoring of inner speech: a physiological basis for auditory hallucinations. Lancet: 346; 596–600, 1995.

McGuire PK, Silbersweig DA, Wright I, Murray RM, Frackowiak RSJ, Frith CD: The neural correlates of inner speech and auditory verbal imagery in schizophrenia: relationship to auditory verbal hallucinations. British Journal of Psychiatry: 169; 148–159, 1996.

McKinney M, Coyle JT, Hedreen JC: Topographic analysis of the innervation of the rat neocortex and hippocampus by the basal forebrain cholinergic system. Journal of Comparative Neurology: 217; 103–121, 1983.

Mesulam MM: A cortical network for directed attention and unilateral neglect. Annals of Neurology: 10; 309–325, 1981.

Mesulam MM: The cholinergic contribution to neuromodulation in the cerebral cortex. Seminars in the Neurosciences: 7; 297–307, 1995.

Mesulam MM: Spatial attention and neglect: parietal, frontal and cingulate contributions to the mental representation and attentional targeting of salient extrapersonal events. Philosophical Transactions of the Royal Society of London Series B: Biological Sciences: 354; 1325–1346, 1999.

Mesulam MM, Mufson EJ: Neural inputs into the nucleus basalis of the substantia innominata (Ch4) in the rhesus monkey. Brain: 107; 253–274, 1984.

Mesulam MM, Mufson EJ, Levey AI, Wainer BH: Cholinergic innervation of cortex by the basal forebrain: cytochemistry and cortical connections of the septal area, diagonal band nuclei, nuclei, nucleus basalis (substantia innominata), and hypothalamus in the rhesus monkey. Journal of Comparative Neurology: 214; 170–197, 1983a.

Mesulam MM, Mufson EJ, Wainer BH, Levey IA: Central cholinergic pathways in the rat: an overview based on an alternative nomenclature (Ch1–Ch6). Neuroscience: 10; 1185–1201, 1983b.

Mesulam MM, Mufson EJ, Wainer BH: Three-dimensional representation and cortical projection topography of the nucleus basalis (Ch4) in the macaque: concurrent demonstration of choline acetyltransferase and retrograde transport with a stabilized tetramethylbenzidine method for horseradish peroxidase. Brain Research: 367; 301–308, 1986a.

Mesulam MM, Van Hoesen GW: Acetylcholinesterase-rich projections from the basal forebrain of the rhesus monkey to neocortex. Brain Research: 109; 152–157, 1976.

Mesulam MM, Volicer L, Marquis JK, Mufson EJ, Green RC: Systematic regional differences in cholinergic innervation of the primate cerebral cortex: Distribution of enzyme activities and some behavioral implications. Annals of Neurology: 281; 611–633, 1986b.

Middleton FA, Strick PL: Basal ganglia and cerebellar loops: motor and cognitive circuits. Brain Research Reviews: 31; 236–250, 2000.

Miller EK: The prefrontal cortex: complex neural properties for complex behavior. Neuron: 22; 15–17, 1999.

Miller EK, Erickson CA, Desimone R: Neural mechanisms of visual working memory in prefrontal cortex of the macaque. Journal of Neuroscience: 16; 5154–5167, 1996.

Minciacchi D, Granato A: Development of the thalamocortical system: Transient-crossed projections to the frontal cortex in neonatal rats. Journal of Comparative Neurology: 281; 1–12, 1989.

Mizuno N, Uemura-Sumi M, Yasui Y, Konishi A, Matsushima K: Direct projections from the extrathalamic forebrain structures to the neocortex in the macaque monkey. Neuroscience Letters: 29; 13–17, 1982.

Morecraft RJ, Geula C, Mesulam M-M: Cytoarchitecture and neural afferents of orbitofrontal cortex in the brain of the monkey. Journal of Comparative Neurology: 323; 341–358, 1992.

Müller-Preuss JD, Newman JD, Jürgens U: Anatomical and physiological evidence for a relationship between the cingular vocalization area and the auditory cortex in the squirrel monkey. Brain Research: 202; 307–315, 1980.

Müller-Preuss P, Ploog D: Inhibition of auditory cortical neurons during phonation. Brain Research: 215; 61–76, 1981.

Nakamura K, Kubota K: Mnemonic firing of neurons in the monkey temporal pole during a visual recognition memory task. Journal of Neurophysiology: 74; 162–178, 1995.

Neafsey EJ: Prefrontal cortical control of the autonomic nervous system: Anatomical and physiological observations. Progress in Brain Research: 85; 147–166, 1990.

Nielsen JM, Jacobs LL: Bilateral lesions of the anterior cingulate gyri. Bulletin of the Los Angeles Neurological Societies: 16; 231–234, 1951.

Nishijo H, Ono T, Nishino H: Single neuron responses in amygdala of alert monkey during complex sensory stimulation with affective significance. Journal of Neuroscience: 8; 3570–3583, 1988.

Ongur D, An X, Price JL: Prefrontal cortical projections to the hypothalamus in macaque monkeys. Journal of Comparative Neurology: 401; 480–505, 1998.

Oyoshi T, Nishijo H, Asakura T, Takamura Y, Ono T: Emotional and behavioral correlates of mediodorsal thalamic neurons during associative learning in rats. Journal of Neuroscience: 16; 5812–5829, 1996.

Pandya DN, Seltzer B, Barbas H: Input–output organization of the primate cerebral cortex. In Steklis HD, Erwin J (Eds), Comparative Primate Biology, Vol. 4: Neurosciences. New York: Alan R. Liss, pp. 39–80, 1988.

Papez JW: A proposed mechanism of emotion. American Medical Association Archives of Neurology and Psychiatry: 38; 725–743, 1937.

Parker A, Gaffan D: The effect of anterior thalamic and cingulate cortex lesions on object-in-place memory in monkeys. Neuropsychologia: 35; 1093–1102, 1997.

Pearson RC, Gatter KC, Brodal P, Powell TP: The projection of the basal nucleus of Meynert upon the neocortex in the monkey. Brain Research: 259; 132–136, 1983.

Penfield W, Jasper H: Epilepsy and the Functional Anatomy of the Human Brain. Boston: Little, and Brown, 1954.

Peters A, Sethares C: The organization of double bouquet cells in monkey striate cortex. Journal of Neurocytology: 26; 779–797, 1997.

Petrides M: Lateral frontal cortical contribution to memory. Seminars in the Neurosciences: 8; 57–63, 1996a.

Petrides M: Specialized systems for the processing of mnemonic information within the primate frontal cortex. Philosophical Transactions of the Royal Society of London. Series B: Biological Sciences: 351; 1455–1462, 1996b.

Petrides M, Alivisatos B, Evans AC: Functional activation of the human ventrolateral frontal cortex during mnemonic retrieval of verbal information. Proceedings of the National Academy of Sciences of the United States of America: 92; 5803–5807, 1995.

Petrides M, Pandya DN: Projections to the frontal cortex from the posterior parietal region in the rhesus monkey. Journal of Comparative Neurology: 228; 105–116, 1984.

Petrides M, Pandya DN: Association fiber pathways to the frontal cortex from the superior temporal region in the rhesus monkey. Journal of Comparative Neurology: 273; 52–66, 1988.

Picard N, Strick PL: Motor areas of the medial wall: a review of their location and functional activation. Cerebral Cortex: 6; 342–353, 1996.

Porrino LJ, Crane AM, Goldman-Rakic PS: Direct and indirect pathways from the amygdala to the frontal lobe in rhesus monkeys. Journal of Comparative Neurology: 198; 121–136, 1981.

Potter H, Nauta WJH: A note on the problem of olfactory associations of the orbitofrontal cortex in the monkey. Neuroscience: 4; 361–367, 1979.

Preuss TM, Goldman-Rakic PS: Crossed corticothalamic and thalamocortical connections of macaque prefrontal cortex. Journal of Comparative Neurology: 257; 269–281, 1987.

Price JL, Russchen FT, Amaral DG: The limbic region. II. The amygdaloid complex. In Björklund A, Hökfelt T, Swanson LW (Eds), Handbook of Chemical Neuroanatomy. Vol.5, Integrated Systems of the CNS, Part I. Amsterdam: Elsevier, pp. 279–381, 1987.

Rainer G, Asaad WF, Miller EK: Selective representation of relevant information by neurons in the primate prefrontal cortex. Nature: 393; 577–579, 1998.

Rainer G, Rao SC, Miller EK: Prospective coding for objects in primate prefrontal cortex. Journal of Neuroscience: 19; 5493–5505, 1999.

Rapoport JL, Fiske A: The new biology of obsessive–compulsive disorder: implications for evolutionary psychology. Perspectives in Biology and Medicine: 41; 159–175, 1998.

Rauschecker JP: Parallel processing in the auditory cortex of primates. Audiology and Neuro-otology: 3; 86–103, 1998.

Rauschecker JP, Tian B, Hauser M: Processing of complex sounds in the macaque nonprimary auditory cortex. Science: 268; 111–114, 1995.

Ray JP, Price JL: The organization of projections from the mediodorsal nucleus of the thalamus to orbital and medial prefrontal cortex in macaque monkeys. Journal of Comparative Neurology: 337; 1–31, 1993.

Rempel-Clower N, Barbas H: Topographic organization of connections between the hypothalamus and prefrontal cortex in the rhesus monkey. Journal of Comparative Neurology: 398; 393–419, 1998.

Rempel-Clower NL, Barbas H: The laminar pattern of connections between prefrontal and anterior temporal cortices in the rhesus monkey is related to cortical structure and function. Cerebral Cortex: 10; 851–865, 2000.

Rempel-Clower NL, Ghashghaei H, Barbas H: Serial pathways from orbital and medial prefrontal cortices reach hypothalamic, brainstem, and spinal autonomic centers in the rhesus monkey. Neuroscience Abstracts: 25; 362, 1999.

Robbins TW, Granon S, Muir JL, Durantou F, Harrison A, Everitt BJ: Neural systems underlying arousal and attention.

Implications for drug abuse. Annals of the New York Academy of Sciences: 846; 222–237, 1998.

Robinson DA, Fuchs AF: Eye movements evoked by stimulation of the frontal eye fields. Journal of Neurophysiology: 32; 637–648, 1969.

Rolls ET: The orbitofrontal cortex. Philosophical Transactions of the Royal Society of London. Series B: Biological Sciences: 351; 1433–143, 1996.

Rolls ET, Critchley HD, Browning AS, Hernadi I, Lenard L: Responses to the sensory properties of fat of neurons in the primate orbitofrontal cortex. Neuroscience Letters: 2; 117–120, 1999.

Romanski LM, Bates JF, Goldman-Rakic PS: Auditory belt and parabelt projections to the prefrontal cortex in the rhesus monkey. Journal of Comparative Neurology: 403; 141–157, 1999.

Romanski LM, LeDoux JE: Equipotentiality of thalamo-amygdala and thalamo-cortico-amygdala circuits in auditory fear conditioning. Journal of Neuroscience: 12; 4501–4509, 1992.

Room P, Russchen FT, Groenewegen HJ, Lohman AH: Efferent connections of the prelimbic (area 32) and the infralimbic (area 25) cortices: an anterograde tracing study in the cat. Journal of Comparative Neurology: 242; 40–55, 1985.

Rosene DL, Heimer L, Van Hoesen GW: Centrifugal efferents to the olfactory bulb in the rhesus monkey. Neuroscience Abstracts: 4; 91, 1978.

Rosene DL, Van Hoesen GW: The hippocampal formation of the primate brain. A review of some comparative aspects of cytoarchitecture and connections. In Jones EG, Peters A (Eds), Cerebral Cortex, Vol. 6. New York: Plenum, pp. 345–455, 1987.

Roth RM, Baribeau J: Performance of subclinical compulsive checkers on putative tests of frontal and temporal lobe memory functions. Journal of Nervous and Mental Disorders: 184; 411–416, 1996.

Russchen FT, Amaral DG, Price JL: The afferent connections of the substantia innominata in the monkey, *Macaca fascicularis*. Journal of Comparative Neurology: 242; 1–27, 1985.

Russchen FT, Amaral DG, Price JL: The afferent input to the magnocellular division of the mediodorsal thalamic nucleus in the monkey, *Macaca fascicularis*. Journal of Comparative Neurology: 256; 175–210, 1987.

Rye DB, Wainer BH, Mesulam MM, Mufson EJ, Saper CB: Cortical projections arising from the basal forebrain: a study of cholinergic and noncholinergic components employing combined retrograde tracing and immunohistochemical localization of choline acetyltransferase. Neuroscience: 13; 627–643, 1984.

Sandell JH, Schiller PH: Effect of cooling area 18 on striate cortex cells in the squirrel monkey. Journal of Neurophysiology: 48; 38–48, 1982.

Saper CB, Loewy AD, Swanson LW, Cowan WM: Direct hypothalamo-autonomic connections. Brain Research: 117; 305–312, 1976.

Sarter M, Bruno JP: Cortical cholinergic inputs mediating arousal, attentional processing and dreaming: differential afferent regulation of the basal forebrain by telencephalic and brainstem afferents. Neuroscience: 95; 933–952, 2000.

Schall JD, Morel A, King DJ, Bullier J: Topography of visual cortex connections with frontal eye field in macaque: Convergence and segregation of processing streams. Journal of Neuroscience: 15; 4464–4487, 1995.

Schiller PH: The neural control of visually guided eye movements. In Richards JE (Ed), Cognitive Neuroscience of Attention. Hillsdale, NJ: Lawrence Erlbaum, pp. 3–50, 1998.

Schiller PH, Chou I: The effects of frontal eye field and dorsomedial frontal cortex lesions on visually guided eye movements. Nature Neuroscience: 1; 248–253, 1998.

Schnider A, Ptak R: Spontaneous confabulators fail to suppress currently irrelevant memory traces. Nature Neuroscience: 2; 677–681, 1999.

Schnider A, Treyer V, Buck A: Selection of currently relevant memories by the human posterior medial orbitofrontal cortex. Journal of Neuroscience: 20; 5880–5884, 2000.

Schoenbaum G, Chiba AA, Gallagher M: Orbitofrontal cortex and basolateral amygdala encode expected outcomes during learning. Nature Neuroscience: 1; 155–159, 1998.

Schoenbaum G, Chiba AA, Gallagher M: Neural encoding in orbitofrontal cortex and basolateral amygdala during olfactory discrimination learning. Journal of Neuroscience: 19; 1876–1884, 1999.

Schoenbaum G, Chiba AA, Gallagher M: Changes in functional connectivity in orbitofrontal cortex and basolateral amygdala during learning and reversal training. Journal of Neuroscience: 20; 5179–5189, 2000.

Selden NR, Gitelman DR, Salamon-Murayama N, Parrish TB, Mesulam MM: Trajectories of cholinergic pathways within the cerebral hemispheres of the human brain. Brain: 121; 2249–2257, 1998.

Shao Z, Burkhalter A: Role of $GABA_B$ receptor-mediated inhibition in reciprocal interareal pathways of rat visual cortex. Journal of Neurophysiology: 81; 1014–1024, 1999.

Sillito AM, Jones HE, Gerstein GL, West DC: Feature-linked synchronization of thalamic relay cell firing induced by feedback from the visual cortex. Nature: 369; 479–482, 1994.

Squire LR: Memory and the hippocampus: A synthesis from findings with rats, monkeys, and humans. Psychological Reviews: 99; 195–231, 1992.

Squire LR, Zola SM: Structure and function of declarative and nondeclarative memory systems. Proceedings of the National Academy of Sciences of the United States of America: 93; 13515–13522, 1996.

Squire LR, Zola-Morgan S: Memory: Brain systems and behavior. Trends in Neuroscience: 11; 170–175, 1988.

Steriade M: Neuromodulatory systems of thalamus and neocortex. Seminars in the Neurosciences: 7; 361–370, 1995.

Steriade M: Arousal: revisiting the reticular activating system. Science: 272; 225–226, 1996.

Steriade M, Jones EG, McCormick DA: Thalamus — Organisation and Function. Oxford: Elsevier Science, 1997.

Stewart DJ, MacFabe DF, Leung LWS: Topographical projection of cholinergic neurons in the basal forebrain to the cingulate cortex in the rat. Brain Research: 358; 404–407, 1985.

Suzuki WA, Miller EK, Desimone R: Object and place memory in the macaque entorhinal cortex. Journal of Neurophysiology: 78; 1062–1081, 1997.

Tagaki SF: Studies on the olfactory nervous system of the old world monkey. Progress in Neurobiology: 27; 195–250, 1986.

Takada M, Fishell G, Li ZK, Van Der Kooy D, Hattori T: The development of laterality in the forebrain projections of midline thalamic cell groups in the rat. Developmental Brain Research: 35; 275–282, 1987.

Talland GA, Sweet WH, Ballantine T: Amnesic syndrome with anterior communicating artery aneurysm. Journal of Nervous and Mental Disorders: 145; 179–192, 1967.

Thomas GJ, Gash DM: Mammillothalamic tracts and representational memory. Behavioral Neuroscience: 99; 621–630, 1985.

Tigges J, Tigges M, Cross NA, McBride RL, Letbetter WD, Anschel S: Subcortical structures projecting to visual cortical areas in squirrel monkey. Journal of Comparative Neurology: 209; 29–40, 1982.

Tigges J, Walker LC, Tigges M: Subcortical projections to the occipital and parietal lobes of the chimpanzee brain. Journal of Comparative Neurology: 220; 106–115, 1983.

Tomita H, Ohbayashi M, Nakahara K, Hasegawa I, Miyashita Y: Top-down signal from prefrontal cortex in executive control of memory retrieval. Nature: 401; 699–703, 1999.

Tremblay L, Schultz W: Relative reward preference in primate orbitofrontal cortex. Nature: 398; 704–708, 1999.

Turner BH, Mishkin M, Knapp M: Organization of the amygdalopetal projections from modality- specific cortical association areas in the monkey. Journal of Comparative Neurology: 191; 515–543, 1980.

Usher M, Cohen JD, Servan-Schreiber D, Rajkowski J, Aston-Jones G: The role of locus coeruleus in the regulation of cognitive performance. Science: 283; 549–554, 1999.

Van Hoesen GW, Pandya DN, Butters N: Some connections of the entorhinal (area 28) and perirhinal (area 35) cortices of the rhesus monkey. II. Frontal lobe afferents. Brain Research: 95; 25–38, 1975.

Vogt BA, Barbas H: Structure and connections of the cingulate vocalization region in the rhesus monkey. In Newman JD (Ed), The Physiological Control of Mammalian Vocalization. New York: Plenum, pp. 203–225, 1988.

Vogt BA, Pandya DN: Cingulate cortex of the rhesus monkey: II. Cortical afferents. Journal of Comparative Neurology: 262; 271–289, 1987.

Voytko ML: Cooling orbital frontal cortex disrupts matching-to-sample and visual discrimination learning in monkeys. Physiological Psychology: 13; 219–229, 1985.

Voytko ML, Olton DS, Richardson RT, Gorman LK, Tobin JR, Price DL: Basal forebrain lesions in monkeys disrupt attention but not learning and memory. Journal of Neuroscience: 14; 167–186, 1994.

Watanabe M: Cognitive and motivational operations in primate prefrontal neurons. Reviews in the Neurosciences: 9; 225–241, 1998.

Webster MJ, Bachevalier J, Ungerleider LG: Connections of inferior temporal areas TEO and TE with parietal and frontal cortex in macaque monkeys. Cerebral Cortex: 4; 470–483, 1994.

Weeks RA, Turjanski N, Brooks DJ: Tourette's syndrome: a disorder of cingulate and orbitofrontal function? Quarterly Journal of Medicine: 89; 401–408, 1996.

Weinberger DR: Schizophrenia and the frontal lobe. Trends in Neuroscience: 11; 367–370, 1988.

Whalen PJ, Rauch SL, Etcoff NL, McInerney SC, Lee MB, Jenike MA: Masked presentations of emotional facial expressions modulate amygdala activity without explicit knowledge. Journal of Neuroscience: 18; 411–418, 1998.

Wilson FA, Rolls ET: Neuronal responses related to reinforcement in the primate basal forebrain. Brain Research: 509; 213–231, 1990a.

Wilson FA, Rolls ET: Neuronal responses related to the novelty and familiarity of visual stimuli in the substantia innominata, diagonal band of broca and periventricular region of the primate basal forebrain. Experimental Brain Research: 80; 104–120, 1990b.

Wilson FA, Scalaidhe SP, Goldman-Rakic PS: Dissociation of object and spatial processing domains in primate prefrontal cortex. Science: 260; 1955–1958, 1993.

Witter MP, Van Hoesen GW, Amaral DG: Topographical organization of the entorhinal projection to the dentate gyrus of the monkey. Journal of Neuroscience: 9; 216–228, 1989.

Wouterlood FG, Steinbusch HWM, Luiten PGM, Bol JGJM: Projection from the prefrontal cortex to histaminergic cell groups in the posterior hypothalamic region of the rat. Anterograde tracing with *Phaseolus vulgaris* leucoagglutinin combined with immunocytochemistry of histidine decarboxylase. Brain Research: 406; 330–336, 1987.

Wurtz RH, Mohler CW: Enhancement of visual responses in monkey striate cortex and frontal eye fields. Journal of Neurophysiology: 39; 766–772, 1976.

Xiao D, Barbas H: Prefrontal cortical extension of the Papez circuit for emotion through the anterior thalamic nuclei in the rhesus monkey. Neuroscience Abstracts, 728.2, 2001.

Yasui Y, Breder CD, Saper CB, Cechetto DF: Autonomic responses and efferent pathways from the insular cortex in the rat. Journal of Comparative Neurology: 303; 355–374, 1991.

Zaborszky L, Pang K, Somogyi J, Nadasdy Z, Kallo I: The basal forebrain corticopetal system revisited. Annals of the New York Academy of Sciences: 877; 339–367, 1999.

Zola-Morgan S, Squire LR: Neuroanatomy of memory. Annual Review of Neuroscience: 16; 547–563, 1993.

Handbook of Neuropsychology, 2nd Edition, Vol. 7
J. Grafman (Ed)

CHAPTER 2

The prefrontal cortex: conjunction and cognition

Earl K. Miller * and Wael F. Asaad

*Center for Learning and Memory, RIKEN-MIT Neuroscience Research Center, and Department of Brain and Cognitive Sciences,
Massachusetts Institute of Technology, Cambridge, MA 02139, USA*

Introduction

Primates have a remarkably diverse and flexible repertoire of sophisticated behaviors. Our actions are not limited to reflexive or fixed reactions to external events. Rather, we can formulate and implement complex plans to achieve often far-removed goals. How we can orchestrate our thoughts and actions in concert with our intentions (and how we form these intentions) is one of the great mysteries of brain function. There is no doubt that it involves neural circuitry that extends over much of the brain, but it is commonly held that one region of cerebral cortex is particularly central to goal-directed, intended, behavior, the prefrontal cortex.

The prefrontal (PF) cortex is a neocortical region that is most elaborate in primates, animals known for their diverse and flexible behavioral repertoire (Fig. 1). Well positioned to coordinate a wide range of neural processes, the PF cortex is a collection of interconnected neocortical areas at the anterior end of the brain that sends and receives projections from virtually all cortical sensory systems, motor systems, and many subcortical structures (Fig. 2). Humans and monkeys with PF damage are impaired in situations that seem to require 'executive' control (Baddeley, 1986; Duncan, Emslie, Williams et al., 1996): when they need to ignore distractions, inhibit reflexive, prepotent responses, or in novel or

difficult situationswhen habitual behaviors cannot be used (Goldman-Rakic, 1987; Grafman, 1994; Fuster, 1989, 1995a; Wise, Murray and Gerfen, 1996). By contrast, PF damage leaves basic sensory, motor, and memory functions unimpaired.

Consider a classic test of PF impairment in humans, the Wisconsin Card Sorting (WCS) Task (Milner, 1963). Subjects must sort cards according to the shape, color, or number of symbols appearing on them. The sorting rule varies surreptitiously every few minutes. Thus, any given card can be associated with several possible actions; no single stimulus–response mapping will work; the correct one changes and is dictated by whichever principle or rule is currently in effect. Humans with PF damage cannot perform this task; they are unable to flexibly adapt their behavior to the changing rule (Milner, 1963). Monkeys with PF lesions are impaired on analogous tasks (Dias, Robbins and Roberts, 1996, 1997). This and other observations have suggested PF involvement in a number of high level functions: attention, working memory, response selection, recall, and inhibitory control. In fact, neural correlates of many of these functions have been found. But an understanding of the basic unifying principles of PF function has remained elusive. Our purpose here is to suggest some.

Task models, feedback and cognitive control

The PF cortex is not important for simple, automatic behaviors. These behaviors can be innate, such as our tendency to automatically orient to an unexpected sound or movement, or they can develop gradually

* Corresponding author. Bldg. E25, Room 236, MIT, Cambridge, MA 02139, USA. Tel.: +1 (617) 252-1584; Fax: +1 (617) 258-7978; E-mail: ekm@ai.mit.edu

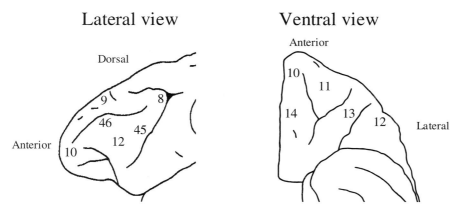

Fig. 1. Lateral and ventral views of the macaque monkey prefrontal cortex. Cytoarchitectonic areas are numbered and shown in their approximate location.

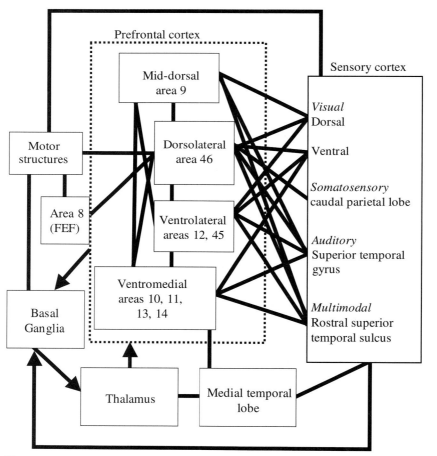

Fig. 2. Schematic diagram of some of the extrinsic and intrinsic connections of the prefrontal cortex. The partial convergence of inputs from many brain systems and internal connections of the PF cortex may allow it to play a central role in the synthesis of diverse information needed for complex behavior. Most connections are reciprocal; the exceptions are indicated by arrows.

with experience as learning mechanisms potentiate existing pathways or form new ones. Unambiguous sensory inputs in a highly familiar situation can automatically elicit a strongly associated behavior. Take, for example, a turn of our bicycle's handlebars (a learned behavior) that is reflexively evoked when a car suddenly veers into our path. These 'hard-wired' pathways are advantageous because they allow highly familiar behaviors to be executed quickly and automatically (i.e., without taxing our attention which, as we all know, is limited in capacity). But, these behaviors tend to be inflexible, stereotyped reactions elicited by just the right stimulus. They do not generalize well to novel situations and they take time to develop; we often need a great deal of experience before we learn something well enough to perform it automatically.

By contrast, the PF cortex is important when flexibility is needed. It is critical for the rapid acquisition of new goals and behaviors and in situations when the mapping between sensory inputs, thoughts and actions are complicated and changing (such as in the WCS task). It is also critical when the needed mappings are weakly established relative to others. Here, neural pathways mediating simple input–output relationships will not suffice. We seem to rely instead on an internal representation of the relevant features of the situation, the 'rules of the game' acquired from experience that specify the conditions and actions needed for achieving our goals (Abbott, Black and Smith, 1985; Barsalou and Sewell, 1985; Grafman, 1994; Norman and Shallice, 1986). Several investigators have argued acquisition of this information is a cardinal PF function (Cohen and Servan-Schreiber, 1992; Grafman, 1994; Miller, 1999, 2000b; Passingham, 1993; Wise et al., 1996).

In our view, the ability to form such representations stems from the position of the PF cortex at the top of the cortical hierarchy. The PF cortex is a network of neural circuits that is interconnected with cortical regions that analyze virtually all types of sensory inputs and with regions involved in generating motor outputs. It is also in direct contact with a wide array of subcortical structures that process, among other things, 'internal' information such as motivational state. The PF cortex thus provides a venue in which information from far-flung brain systems can interact through relatively local circuitry.

During learning, reward-related signals could act on the PF cortex to strengthen pathways — the associative links — between the neurons that processed the information that led to reward. As a result, the PF cortex rapidly constructs a neural representation of the constellation of data needed to achieve the goal at hand, i.e., a 'task model'. This provides executive control by sending excitatory signals back to the brain structures that provide the PF cortex with input. These signals enhance the activity of neurons that process task relevant information (that match the model), enabling the neural pathways needed for the task. With enough experience, task relevant neural pathways can be established independent of the PF cortex, in lower-level sensory and motor cortical and subcortical structures. When this happens, the behavior can become 'habitual' or 'automatic' and the PF cortex becomes less involved.

To help understand how this might work, consider the cartoon shown in Fig. 3. Processing units are shown that correspond to cues (C1, C2, C3). They can be thought of as neural representations of sensory events, internal states, stored memories, etc. in corresponding neural systems. Also shown are processing units that correspond to the motor circuits

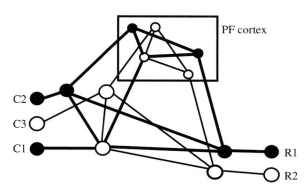

Fig. 3. Schematic diagram of a posited role for the PF cortex in cognitive control. Information such as sensory inputs, current motivational state, memories, etc. (e.g., 'cues' such as C1, C2, and C3) as well as information about behavior (e.g., 'responses' such as R1 and R2). Reward signals foster the formation of a task model, a neural representation that reflects the learned associations between task-relevant information. A subset of the information (e.g., C1 and C2) can then evoke the entire model, including information about the appropriate response (e.g., R1). Excitatory signals from the PF cortex feeds back to other brain systems to enable task-relevant neural pathways. Thick lines indicate activated pathways, thin lines indicate inactive pathways.

mediating two responses (R1 and R2). We have set up the sort of flexible situation for which the PF cortex is thought to be important. Namely, one cue (C1) can lead one of two responses (R1 or R2) depending on the situation (C2 or C3). Imagine that you suddenly decide that you want a beer (and let us consider that cue C1). If you are at home (C2), then you get up and get one (R1). But if you are in a bar (C3), you ask for one instead (R2). These conditional associations form the 'if–then' rules that are fundamental building blocks of voluntary behavior (Passingham, 1993). How does the PF cortex construct these representations?

In an unfamiliar situation, information flows into the PF cortex relatively unchecked. But then reward-related signals from successful (rewarded) experiences foster the formation of associations between the PF neurons that had processed the information immediately preceding reward. As this neural ensemble becomes established, it becomes self-reinforcing. It sends signals to back to other brain systems, biasing their processing toward matching information and thus refining the inputs to the PF cortex. As learning proceeds, reward-related signals, perhaps from midbrain dopamine (DA) neurons, appear progressively earlier as they become evoked by the events that first predict reward (Schultz, Dayan and Montague, 1997). Through repeated iterations of this process, more and more task-relevant information is linked into the PF representation; it 'bootstraps' from direct associations with reward to a multivariate network of associations that describes a task.

Once the PF representation is established, a subset of the information (such as the cues) can activate the remaining elements (such as the correct response). So, if we want a beer (C1) and we are at home (C2), the corresponding PF representation containing the correct response (R1) is activated. Excitatory bias signals from the PF cortex enhance the activity of task-relevant neurons in other brain regions, enabling the neural pathways needed for the task (e.g., C1–R1). A different pattern of cues (e.g., cues 1 and 3 in Fig. 3) evokes a different PF model and a different pattern of bias signals selects other neural pathways (C1–R2). With repeated selection of these pathways, they can become established independent of the PF cortex. As this happens, the PF cortex becomes less involved and the behavior becomes habitual or automatic.

Obviously, many details need to be added before we understand the complexity of mental life. But we believe that this general notion can explain many of the posited functions of the PF cortex. PF feedback signals to sensory systems may mediate its role in directing attention and in memory recall. Signals to the motor system may be responsible for response selection and inhibitory control. Without the PF cortex, the most frequently used (and thus best established) neural pathways would predominate. You would be unable to override these pathways and would, for example, try and get your own beer regardless of where you were (C1–R1). Such inappropriate and impulsive behaviors are hallmarks of PF dysfunction in humans.

In our view, the following attributes of the PF cortex are especially pertinent to its function:

Convergence of diverse information

The PF cortex receives input from a large number of cortical and subcortical structures. Highly processed sensory information arrives from posterior cortical areas, information about the motivational state of the animal is delivered directly or indirectly via the limbic system, and information about potential and current actions is available through connections with motor system structures. Thus the PF cortex has access to a wide variety of refined information concerning the external world and the internal drives of the organism; it is a microcosm of the data processed in the brain systems needed for voluntary goal-directed behavior.

Local interactions

The close apposition of diverse information within the PF cortex provides an excellent substrate for their association. The amount of 'wiring' needed is less than would be required to directly connect the brain systems responsible for initial processing. This may allow the PF cortex to quickly knit together behaviorally relevant associations into a representation of the relevant features of a task.

Reward-related signals and plasticity

A pattern of PF activity associated with achievement of a goal must be selectively reinforced. This may happen under the influence of reward-related dopamine signals that selectively potentiate connections between neurons that were active during the conditions that led to the reward. Reinforcing the activity pattern increases the probability that it will recur given a similar set of sensory inputs. Another result of this process may be that the PF neurons that are part of the reinforced circuit will express sustained activity. In many contexts, this would function as an active short-term, or working, memory trace. It is important for maintaining task-relevant information across temporal gaps between incoming sensory inputs and motor outputs.

Bias signals and top-down control
The PF cortex has reciprocal connections with the brain systems that provide it with input. This allows it to bias processing in those systems toward goal-relevant information. By selecting particular sensory inputs (attention), memories (recall) or motor outputs (response selection), this bias signal enables the neural pathways needed for the task at hand. Over time, these circuits can become established independent of the PF cortex, in the 'brain at large'. Then, the PF cortex becomes less involved and the behaviors become 'automatic'.

The view presented here is not unusual or unprecedented. The notion that the PF cortex acquires information about the formal demands of behavior is also captured by the hypothesis that the PF cortex conveys behavioral context (Cohen, Dunbar and McClelland, 1996; Cohen and Servan-Schreiber, 1992). Cohen and colleagues have also suggested that PF task information is used to select task-relevant neural pathways in other brain regions. Similarly, several investigators have interpreted the pattern of deficits following PF damage as a loss of the ability to acquire behavior-guiding rules or templates (Grafman, 1994; Passingham, 1993; Wise et al., 1996). Dehaene and Changeux have shown how rule-coding units can be used in a computational model of cognition (Dehaene, Kerszeberg and Changeux, 1998). Fuster has long advocated the role of the PF cortex in integration diverse information, which is central to our view (Fuster, 1995a). Goldman-Rakic has explored the ability of PF neurons to maintain task-relevant information (Goldman-Rakic, 1987). Finally, the view of PF function presented here draws on our other work on this topic (Miller, 1999, 2000a,b; Miller and Cohen, 2001).

Below, we elaborate.

The prefrontal cortex: anatomy and organization

The cytoarchitectonic areas that comprise the monkey PF cortex are often grouped into regional subdivisions, orbitofrontal, lateral, and mid-dorsal (see Figs. 1 and 2). Collectively, these areas have interconnections with virtually all sensory systems, with cortical and subcortical motor system structures, and with limbic and midbrain structures involved in affect, memory, and reward. The subdivisions have partly unique, but overlapping, patterns of connections with the rest of the brain, which suggests some regional specialization (Fig. 2). However, while different PF regions are likely to emphasize certain types of information or processes, we contend that any functional specialization is a matter of emphasis, not an absolute segregation of different types of information or processes into separate 'modules'. As in much of the neocortex, many PF connections are local; there are ample connections between different PF areas that are bound to result in an intermixing of disparate information. Complex behavior is not possible without synthesizing results from a wide variety of brain processes and thus a large degree of integration is likely to be evident in a brain area important for its orchestration.

Sensory inputs

The lateral and mid-dorsal PF cortex is more closely associated with sensory neocortex than is the orbitofrontal PF cortex (see Fig. 2). It receives visual, somatosensory, and auditory information from the occipital, temporal, and parietal cortices (Barbas and Pandya, 1989, 1991; Goldman-Rakic and Schwartz, 1982; Pandya and Barnes, 1987; Pandya and Yeterian, 1990; Petrides and Pandya, 1984, 1999; Seltzer and Pandya, 1989). Many PF areas receive converging inputs from at least two sensory modalities (Chavis and Pandya, 1976; Jones and Powell, 1970). For example, the dorsolateral (areas 8, 9, and 46), and ventrolateral (12 and 45) both receive projections from visual, auditory, and somatosensory cortex. Furthermore, the PF cortex is interconnected with other cortical regions that are themselves sites of multimodal convergence. Many PF areas (9, 12,

46,and 45) receive inputs from the rostral superior temporal sulcus, which has neurons with bimodal or trimodal (visual, auditory, and somatosensory) responses (Bruce, Desimone and Gross, 1981; Pandya and Barnes, 1987). The arcuate sulcus region (areas 8 and 45) and area 12 seem to be particularly multimodal. They contain zones that receive overlapping inputs from all three sensory modalities (Pandya and Barnes, 1987). In all of these cases, the PF cortex is not directly connected with primary sensory areas, but instead interconnected with secondary or 'association' sensory cortex.

Lesion studies also illustrate the polymodal nature of the PF cortex. This is particularly true of ventrolateral area 12, which receives multimodal sensory inputs and lies between the orbitofrontal and dorsal PF regions. Its damage produces deficits on tasks that use a wide variety of cues. Monkeys are impaired at visual matching tasks that use color or objects as cues (Mishkin and Manning, 1978; Passingham, 1975) and at tasks using visuospatial cues (Butter, 1969; Goldman, Rosvold, Vest and Galkin, 1971; Mishkin, Vest, Waxler and Rosvold, 1969). They are also impaired at auditory tasks that require learning to respond to one tone (go) and refrain from responding to another (no go) (Iversen and Mishkin, 1970). Passingham (1993) concluded that the "key to area 12 does not lie in the nature of the stimulus or response".

Neurophysiological studies using visual and auditory stimulation have found neurons throughout the lateral PF cortex that are sensitive to either or both modalities (Azuma and Suzuki, 1984; Watanabe, 1992; Vaadia, Benson, Hienz and Goldstein MH, 1986). For example, Vaadia et al. (1986) trained monkeys to make arm movements toward visual and auditory targets and found not only an intermixing of visual and auditory cells throughout the dorsolateral PF cortex, but also many bimodal cells. Azuma and Suzuki (1984) also found a similar intermixing of visual and auditory cells, but no bimodal cells. The difference between these studies may be due to the tasks employed. In the Azuma and Suzuki (1984) study, the visual and auditory stimuli were passively experienced by the monkeys, whereas in the Vaadia et al. (1986) study they were behaviorally relevant and instructed the same behaviors. In fact, Vaadia et al. found that when the cues were passively ex-

perienced, many of the bimodal neurons were no longer responsive. Similarly, Watanabe (1992) found that when visual and auditory cues signal the same behavioral event, up to half of lateral PF neurons tested were bimodal. This also illustrates that task demands can exert a strong modulatory influence on PF activity, a point that will be discussed below.

Motor system connections

The dorsal PF cortex, particularly dorsolateral area 46, has preferential connections with many motor system structures and may be primary regions by which the PF exerts control over behavior. Dorsolateral area 46 is interconnected with motor areas in the medial frontal lobe such as the supplementary motor area (SMA), pre-SMA, and the rostral cingulate, with the premotor cortex on the lateral frontal lobe, and with cerebellum and superior colliculus (Bates and Goldman-Rakic, 1993; Goldman and Nauta, 1976; Lu, Preston and Strick, 1994; Schmahmann and Pandya, 1997). DL area 46 also sends projections to area 8, which contains the frontal eye fields, a region important for voluntary shifts of gaze. As with sensory cortex, there are no direct connections between the PF cortex and primary motor cortex; it is instead connected with premotor areas that, in turn, send projections to primary motor cortex and the spinal cord. Also important are the dense interconnections between the PF cortex and basal ganglia, a structure that is likely to be crucial for automating behavior. The basal ganglia receives inputs from much of the cerebral cortex, but its major output (via the thalamus) is frontal cortex (see Fig. 2).

Limbic system connections

The orbitofrontal PF cortex is more closely associated with medial temporal limbic structures critical for long-term memory and 'internal' information, such as affect and motivation, than other PF regions. This includes direct and indirect (via the medial dorsal thalamus) connections with the hippocampus and associated neocortex, the amygdala and hypothalamus (Amaral and Price, 1984; Barbas and De Olmos, 1990; Barbas and Pandya, 1989; Goldman-Rakic, Selemon and Schwartz, 1984; Porrino, Crane and Goldman-Rakic, 1981; Van Hoesen, Pandya and

Butters, 1972). Other PF regions, however, have access to these systems both through interconnections with the orbitofrontal PF cortex and through intervening structures. There are also direct inputs from sensory cortex, but they tend to be weaker than their projections to lateral PF cortex.

Intrinsic connections

Like most of the neocortex, many PF connections are local. There are not only interconnections between all three major subdivisions (orbitofrontal, lateral, and mid-dorsal), but also interconnections between their constituent areas (Pandya and Barnes, 1987; Barbas and Pandya, 1991). The lateral PF cortex is particularly well connected. Ventrolateral areas 12 and 45 are interconnected with dorsolateral areas 46 and 8, with dorsal area 9 as well as with orbitofrontal areas 11 and 13. Intrinsic PF connections presumably allow information from a given PF afferent or from a given process within a PF subregion to be distributed to other parts of the PF cortex.

Functional organization

A major question of PF function concerns whether different processes can be mapped onto the distinct cytoarchitectonic areas and subdivisions of the PF cortex. The clearest example of a relative difference in function is between the orbitofrontal region and the remainder of the PF cortex. As described above, the orbitofrontal PF cortex is more closely associated with medial temporal limbic structures and 'internal' information such as memory and affect while lateral and mid-dorsal portions of the PF cortex are more associated with brain systems concerned with 'external' information such as sensory inputs and motor outputs. The orbitofrontal PF cortex is also phylogenically older than other PF regions, just as limbic structures are 'older' than neocortical sensory and motor cortex. Damage to the orbitofrontal PF cortex, like medial temporal damage, produces impairment in passive (automatic) forms of memory such as recognition while relatively sparing active short-term, or working, memory functions (Bachevalier and Mishkin, 1986). Lateral PF lesions tend to produce the opposite patterns of results. Also consistent with its relative emphasis on internal data are

observations that monkeys with orbitofrontal damage are impaired at learning and reversing stimulus–reward associations (Baylis and Gaffan, 1991) and that orbitofrontal PF neurons are concerned with information about rewards (Thorpe, Rolls and Maddison, 1983; Tremblay and Schultz, 1999). But this is likely to be a relative specialization; reward-related signals are also present in the lateral PF cortex (Leon and Shadlen, 1999; Watanabe, 1992, 1996). The orbitofrontal cortex thus may be a major interface between the limbic system and the rest of the PF cortex (Petrides, 1994b).

Different schemes have been suggested to explain the organization of the lateral and dorsal PF cortices. Some investigators suggest an organization based on information content. In their view, each PF region is thought to mediate a similar function but is specialized for processing a different stimulus modality. Others suggest an organization by function, rather than content. In this view, the PF cortex is polymodal and different subdivisions are specialized for different cognitive processes. Of course, these views are not mutually exclusive; intermediate organizational schemes are also possible.

Theories to explain the organization of the lateral PF cortex have often been guided by a particular view of the functional specialization of the visual cortex. In the visual cortex, there are at least two relatively segregated anatomical pathways, a dorsal pathway leading from V1 to the parietal cortex and a ventral pathway leading from V1 to the temporal cortex (Felleman and Van Essen, 1991; Maunsell and Newsome, 1987; Ungerleider and Mishkin, 1982). One hypothesis proposes that the ventral pathway is specialized for the sensory analysis of the form and color information needed to visually identify objects (*what*) while the dorsal pathway is specialized for the analysis of the visuospatial information needed to discern their location (*where*).

Because the ventrolateral (VL) and dorsolateral (DL) PF cortices receive preferential inputs from the ventral and dorsal visual pathways, respectively, it has been likewise suggested that these areas are specialized for short-term memory for *what* and *where* (Goldman-Rakic, 1996). Tests of this 'domain-specific' theory of PF function have focused on contrasting the roles of the VL and DL PF cortex. Wilson, O Scalaidhe and Goldman-Rakic (1993) examined the activity of VL PF neurons during a conditional association (object–saccade) task. In their 'object' task, each of two cue objects was uniquely associated with one of two saccades. The monkey made a saccade to the right following one object and a saccade to the left following the other object. They found that the majority of VL PF neurons were activated following one or the other cue. By contrast, they were not engaged during a 'pure' spatial task in which a spot was flashed to the right or left and the monkey made an eye movement to its remembered location. Wilson et al. (1993) concluded that this reflected a specialization for processing object information and thus, domain-specificity. They also found that when monkeys passively viewed visual stimuli, more neurons sensitive to object identity than expected by chance were found in the VL PF cortex (O Scalaidhe, Wilson and Goldman-Rakic, 1999).

While it seems clear that the dorsal and ventral visual pathways have relatively unique functions, it is not clear how they are specialized. First, it is becoming obvious that there is a substantial intermixing of *what* and *where* in both the visual and prefrontal cortices. Sereno and Maunsell (1998) have found that many neurons in dorsal visual area LIP are indeed sensitive to object form, a finding recently corroborated by other laboratories (Anderson, Asaad, Wallis and Miller, 1999; Toth and Assad, 1999). Conversely, neurons in temporal cortical areas TEO and TE often have relatively discrete receptive fields and thus can convey spatial information (Tanaka, Saito, Fukada and Moriya, 1991). Second, the conclusions that can be drawn from the studies purporting to find neurophysiological evidence for a regional specialization for *what* and *where* in the PF cortex are limited by their methodology. The 'object' task in the Wilson et al. (1993) study confounded the cue object with the saccade direction; object 'A' always meant 'saccade right', for example. Therefore, neurons selectively activated by this task could actually have been processing the object–saccade (*what–where*) association rather than *what* per se, as the authors conclude. In fact, several studies have shown that many lateral PF neurons do process such associations (Asaad, Rainer and Miller, 1998; Asaad, Rainer and Miller, 2000; Bichot, Schall and Thompson, 1996; Ferrera, Cohen and Lee, 1999). The VL PF cortex may even show some specialization for associative learning (to be

discussed below), which would explain why Wilson et al. (1993) found that more VL PF neurons were activated by their conditional association task than by their spatial task. In their other studies supporting PF regional specialization, monkeys viewed stimuli passively; they were not performing a task that engaged PF functions. Experience-dependent plasticity is evident in many neocortical regions (Merzenich and Sameshima, 1993; Recanzone, Merzenich, Jenkins et al., 1992) and, as will be argued below, may be central to PF function. Furthermore, studies which have examined this issue directly by training monkeys to perform PF-dependent tasks that require integration of *what* and *where* have found substantial intermixing of these signals in the lateral PF cortex (Asaad et al., 1998, 2000; Ferrera et al., 1999; Fuster, Bauer and Jervey, 1982; Rainer, Asaad and Miller, 1998a,b; Rao, Rainer and Miller, 1997; White and Wise, 1999).

For example, the prefrontal neuron shown in Fig. 4 was recorded while the monkey performed a task that required it to remember both what the object looked like and where it was (Rainer et al., 1998a). Monkeys were trained on a go/no go delayed match-to-object–place task that required them to remember, over a brief delay, which of 2–5 sample objects had appeared in which of 25 visual field locations. They released a lever when a test object matched a sample in both identity and location. During the delay, about half of the neurons in the lateral PF cortex simultaneously conveyed information about the identity of the sample object and its precise location (Fig. 4). In fact, the average diameter of the receptive field derived from activity during the memory delay of the task (i.e., 'memory fields', or MFs) of these neurons was only about 9°. Furthermore, unlike inferior temporal neurons, object-selective PF neurons did not emphasize central vision. Rather, they seemed well suited to the task demand to remember an object throughout a wide portion of the visual field. Many object-and-spatial selective neurons had MFs that were entirely extrafoveal and many were maximally activated by peripheral locations. Thus, across the population, these neurons could simultaneously identify and localize objects throughout a wide area of the visual field, both near the fovea and in the periphery. It has also been found that many neurons develop this

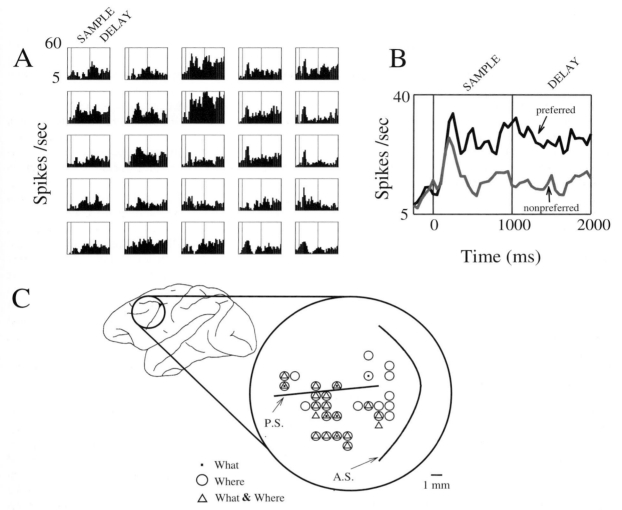

Fig. 4. (A) Histograms of a single PF neuron's activity to an object appearing at each of the 25 tested locations. The line to the left of each histogram shows time of sample onset and the line in the middle denotes sample offset. Bin width, 40 ms. The *y*-axis indicates firing rate in spikes per second and the *x*-axis, time. The time scale for each histogram is identical to the histogram shown in B. Note that this neuron is highly spatial selective. It only shows sustained activation when the object appears at two extrafoveal locations. The remaining locations may elicit brief bursts of activity at sample onset, but they do not elicit robust sustained activity. (B) Average activity of the same neuron to a preferred and nonpreferred object appearing at the two locations that elicited delay activity. Note that this neuron is also highly object selective. It shows strong activity to a preferred object but little or no activity to a non-preferred object. (C) Recording sites. Each symbol represents a recording site where neurons with object-selective delay activity (*what*), location selective delay activity (*where*), or both object and location selective delay activity (*what & where*) were found. Typically, several neurons were found at the same site. About half of the 149 neurons with task-related properties showed activity selective for both *what* and *where*. Adapted from (Rainer et al., 1998a).

sensitivity through experience (Asaad et al., 1998; Bichot et al., 1996).

Imaging studies have also shown that similar regions of the human PF cortex are activated during object and spatial tasks (Courtney, Petit, Maisog et al., 1998; Owen, Milner, Petrides and Evans, 1996b; Owen, Stern, Look et al., 1998; Postle and D'Esposito, 1998; Prabhakaran, Narayanan, Zhao and Gabrieli, 1999). In fact, one study found, not only that overlapping regions of the PF cortex were activated by *what* and *where*, but also that the PF cortex was more activated when subjects needed to integrate these attributes in working memory compared to when they could remember each attribute separately. This suggests that the PF cortex is particularly engaged in tasks that require integration.

Similarly, lesion studies of the monkey PF cortex, also, do not support a complete separation of *what* and *where*. DL PF lesions (especially those restricted to area 46), impair spatial memory, but not object memory, tasks and VL PF lesions can impair some object memory tasks (Goldman et al., 1971; Gross and Weiskrantz, 1962; Mishkin, 1957; Mishkin and Manning, 1978; Passingham, 1975). However, the dissociation is not complete. Spatial reversal tasks can show little or no impairment after DL PF lesions (Gaffan and Harrison, 1989; Goldman et al., 1971; Passingham, 1975). Dorsal PF lesions can impair certain types of object tasks, such as those requiring memory for the order of a sequence of objects (Petrides, 1995), and VL lesions can impair some spatial tasks (Mishkin et al., 1969; Passingham, 1975). Monkeys with VL PF lesions can show normal short-term memory for object information (Rushworth, Nixon, Eacott and Passingham, 1997). DL PF lesions can impair tasks that have no spatial component and, for that matter, no dependence on short-term memory (Passingham, 1978).

Furthermore, the observation that DL area 46 lesions do affect spatial tasks more than object-based tasks could be due to the pattern of inputs to the PF cortex, rather than a functional specialization per se. As Petrides (1994b) has pointed out, the DL area 46 has preferential connections with the posterior parietal cortex (and the motor system). Damage to it might preferentially disrupt *where* tasks because it destroys a major pathway through which information from these structures reaches the entire PF cortex, not because it is the one PF region responsible for processing *where* (Petrides, 1994b).

An alternative for the *what* versus *where* hypothesis has been offered by Goodale and Milner (1992). They have argued that dorsal and ventral visual cortex pathways are not specialized for processing different types of visual information but instead process similar information for different purposes. In their view, the dorsal visual pathway is specialized for guiding actions while the ventral pathway is specialized for perception. *What* information, for example, is used in the dorsal pathway to guide grasping of an object, while in the ventral pathway it is used to identify it. Support for this view comes from observations that humans with dorsal visual cortex damage have difficulty orienting their hands to accommodate an object's shape but have no difficulty in recognizing the object (Goodale and Milner, 1992). The dichotomy of perception vs. action is also consistent with some of the non-human primate work. For example, neurons in parietal cortex selective for the grips required to manipulate particular objects have been reported (Murata, Fadiga, Fogassi et al., 1996; Sakata, Taira, Kusunoki et al., 1998) and muscimol injections into parietal cortex impair monkeys' abilities to properly shape their hands for grasping objects (Gallese, Murata, Kaseda et al., 1994). Neurophysiological studies showing an intermixing of *what* and *where* signals in the visual and prefrontal cortices (described above) are also consistent with this model.

Regardless of which functional scheme best describes the ventral and dorsal systems, the differences are bound to be relative. Whether they are specialized for *what* versus *where* or perception versus action, the results of their processes need to be brought together. Indeed, there are interconnections between the ventral and dorsal visual pathways and between the VL and DL PF cortices (Barbas and Pandya, 1989; Felleman and Van Essen, 1991). The temporal (ventral) and parietal (dorsal) projections to the lateral PF cortex are not completely separate, either. While the VL PF receives a substantial input from the inferior temporal cortex, its more posterior portion also receives projections from the posterior parietal cortex (Bullier, Schall and Morel, 1996; Webster, Bachevalier and Ungerleider, 1994). Likewise, the inferior temporal cortex sends projections not only to the VL PF cortex, but also sends projections to DL PF area 46 (Barbas, 1988).

The domain-specificity model posits that a cardinal PF function is the active maintenance of recent sensory inputs, a process termed working memory. While other theories of cognitive function do recognize that maintenance is essential, they emphasize the higher-order, executive functions that are central to cognitive control (Baddeley, 1986; Johnson and Hirst, 1991; Norman and Shallice, 1986). For example, Baddeley's model of working memory included not only maintenance mechanisms but also control mechanisms, a central executive for which the information is maintained. Other views of PF function and organization have focused on these higher-order mechanisms.

Because virtually all complex behavior (and thus the executive mechanisms that control them) is acquired, many models of PF function emphasize its role in learning. While this probably involves most, if not all, of the PF cortex, there is some recent data suggesting that the VL PF cortex may be particularly important for a particular type of learning, namely, conditional associative learning. Conditional association tasks require learning a set of rules about which actions are to be made following presentation of each of a set of cues. VL PF damage (of disruption of its inputs) in monkeys impairs performance of a wide range of conditional tasks (Passingham, 1993). Neurophysiological studies in monkeys indicate that VL PF neurons are modulated during conditional learning (Asaad et al., 1998; Asaad et al., 2000; Boussaoud and Wise, 1993; White and Wise, 1999). But again, the localization seems relative; such cells are found throughout the lateral PF cortex (Asaad et al., 1998). The VL PF cortex may be particularly important for conditional learning because of its position at the 'crossroads' of the PF cortex. It receives sensory inputs from posterior cortex and lies between the orbitofrontal PF cortex (and its preferential limbic connections) and the more dorsal portions of the PF cortex (which by virtue of its connections with the motor system, seems to be a major 'output' pathway).

Another view posits a PF organization based on a regional separation of lower-order functions such as simple maintenance and higher-order mechanisms central for organizing behavior. It suggests that the lateral and orbitofrontal PF cortices mediate initial stages of PF processing and that the mid-dorsal PF cortex is a further stage of elaboration (Owen, Evans and Petrides, 1996a; Petrides, 1991, 1992, 1994a, 1996; Petrides, Alivisatos, Evans and Meyer, 1993). The lateral PF cortex is thought to subserve relatively basic functions such as encoding, retrieval and short-term maintenance. The mid-dorsal PF cortex

(areas 9 and dorsal area 46), by contrast, is thought to mediate second-order functions. It recodes the signals from lateral PF cortex for more intentional processes such as manipulation and monitoring of the maintained information. Note that, in this view, the PF cortex is organized much like sensory cortex, with a primary area (the lateral PF cortex), where information is initially processed, and higher-order areas, where it is further elaborated.

Support for this model comes from studies showing that mid-dorsal (area 9) PF cortex disrupts ordering tasks while simpler maintenance tasks such as delayed matching to sample are relatively spared (Petrides, 1995). In the self-ordered task, for example, the subject is presented with a non-varying set of stimuli. They have to select a new stimulus on each trial until every stimulus in the set has been chosen. These tasks require "active decisions about occurrence or non-occurrence of stimuli from a given set" (Owen et al., 1996a). That is, the subjects must constantly compare selected items with those still to be carried out. Mid-dorsal lesions disrupt both spatial and nonspatial versions of this task. Further support comes from functional imaging studies in humans. Mid-dorsal PF cortex shows evidence of greater activation during self-ordered tasks than during 'simpler' tasks such as visual matching or associative memory tasks (Petrides et al., 1993). Lateral PF areas are activated during maintenance tasks whereas both lateral and mid-dorsal areas are activated during tasks requiring updating and manipulating the contents of working memory (D'Esposito, Postle, Ballard and Lease, 1999).

In sum, consistent with their connections with a wide variety of brain regions, PF neurons have multimodal properties that indicate access to the diverse information needed to coordinate complex behavior. Different PF regions may emphasize different types of information or processes; the orbitofrontal PF cortex is associated more with internal information while more lateral and dorsal PF regions are associated with external information, motor output, and perhaps, in the mid-dorsal region, combining multiple items into higher-order representations. There also appears to be some functional division between the ventrolateral and dorsolateral PF cortices (and between ventral and dorsal visual cortical pathways), but the degree of separation between them and the nature of their difference remain unclear. In any case, to reiterate a point made at the start of this section, any regional specialization in the PF cortex is bound to be relative. Overlapping connections with other brain structures as well as intrinsic PF connections are likely to result in a substantial degree of integration, as neurophysiological studies do suggest. This does not preclude the possibility of some regional biases. Sometimes, a certain type of information or process needs to be isolated from others. This is presumably easier if there is some preferential representation in certain PF networks. But complex behavior cannot be realized without pulling together results from a wide variety of neural processes. The infrastructure of the PF cortex seems well equipped for this role. The multimodal properties of PF neurons, or more to the point, their ability to acquire them through experience, may be key to their role in cognitive control. This is discussed in the next section.

Associative learning and acquisition of task information

Of course, normal thoughts and actions do not come about from a haphazard intermixing of data. We need to learn how it is organized — which stimuli, memories and actions 'belong together'. This, and indeed most, learning depends on the ability to acquire information about associative relationships. Neurophysiological studies suggest that PF neurons are concerned with learning a wide range of associations between stimuli, expected events, and actions. But, as noted earlier, the PF cortex seems particularly important when flexibility is needed, when the links between sensory inputs and actions are continually changing. This is exemplified by studies indicating PF involvement in learning so-called conditional associations.

Conditional associations are the 'if–then' rules that are fundamental to voluntary goal-directed behavior (Passingham, 1993). For example, if we are sitting at home, we answer when the phone rings. If we are a guest in someone else's house, we do not. The same stimulus (ringing) evokes a different action depending on the environmental context. Conditional associations, then, are more than just a simple one-to-one mapping between two items.

They depend on a multidimensional network of links that include information about the situation in which a given input–output relationship is relevant. This engenders flexibility.

PF damage in humans and monkeys disrupts a wide variety of conditional association tasks. For example, Petrides found that following PF damage, patients could no longer learn conditional associations between visual patterns and hand gestures (Petrides, 1985, 1990). In monkeys, damage to ventrolateral area 12 or to the arcuate sulcus region also impairs the ability to learn conditional associations (Halsband and Passingham, 1985; Murray, Bussey and Wise, 2000; Petrides, 1982, 1985). Learning of visual stimulus–response conditional associations is also impaired by damage that disconnects the PF cortex from its inputs from the temporal cortex (Eacott and Gaffan, 1992; Gaffan and Harrison, 1988; Parker and Gaffan, 1998). Functional imaging studies in humans show high levels of PF activation during conditional learning, particularly the VL PF cortex (Passingham, personal communication). In fact, Passingham (1993) argues that most, if not all, tasks that are disrupted following PF damage depend on acquiring conditional associations.

A recent neurophysiological example was provided by Asaad et al. (1998). Lateral PF neural activity was recorded while monkeys learned to associate particular movements with particular visual stimuli. They learned to saccade to the right after one object was presented and saccade to the left after another object was presented. Once this was acquired, the monkeys then learned the reverse, that is, to saccade leftward after presentation of the object that had previously been associated with a rightward saccade and vice-versa. Many lateral PF neurons had properties consistent with representation of the learned associations. For example, a given cell might only be strongly activated when object 'A' instructed 'saccade left' and not when object 'B' instructed the same saccade or when object 'A' instructed another saccade (Fig. 4A). Thus, neural activity was not simply the result of a straightforward, linear combination of activity related to the object and saccade. Instead, their selectivity was nonlinear, that is, tuned to specific object–saccade associations.

The properties of these neurons are reminiscent of those of 'hidden units' in simulated neural networks. Hidden units are not directly connected to the outside; they lie between the input and output layers of the network. Neural networks with hidden units can solve much more complicated problems than networks that map directly between input and output layers. Hidden units provide computational power because they generalize over input patterns and come to represent meaningful conjunctions of the inputs. For example, a neural network trained to categorize 'cats' versus 'dogs' might develop some hidden units that are activated only to the particular combination of input features that define a 'cat'. The nonlinear nature of this associative process allows them to represent more complicated situations than are possible by a simple linear combination of inputs. This property is key to the extraction of the general principles needed to orchestrate complex behavior.

Other studies also indicate that individual neurons can acquire selectivity for features to which they are initially insensitive. Bichot et al. (1996) recorded from neurons in the frontal eye fields (FEF), a region in the bow of the arcuate sulcus. It is involved in voluntary shifts of gaze and has neurons whose activity is tuned to saccade direction. Ordinarily, FEF neurons are not selective to the form and color of stimuli. However, Bichot et al. (1996) found that when monkeys were trained to saccade to a target of a particular color, FEF neurons also acquired selectivity for that color. They suggested that this reflected a form of experience-dependent plasticity that mediates the learning of arbitrary stimulus–response associations.

In a series of experiments, Watanabe has shown that PF neurons can acquire information about a wide variety of arbitrary associations (Watanabe, 1990, 1992). Monkeys were taught that certain visual and auditory stimuli signaled, on different trials, that a reward (a drop of juice) would or would not be delivered. He found that the majority of neurons studied in the lateral PF cortex (around the arcuate sulcus and posterior end of the principal sulcus) reflected cue–reward associations. Their response to the cue was highly modulated by the status of its current association with reward. For example, a given neuron could show strong activation to one of the two auditory (and none of the visual) cues, but only when it signaled reward. Other neurons were bimodal, activated by both visual and auditory cues

but also strongly modulated by their reward status. In fact, only a minority of the neurons showed activity that reflected the physical properties of the cues alone.

Perhaps the most compelling evidence for a role of the PF cortex in acquiring a multivariate representation of the features of a task are observations of PF activity that is tuned to abstract task information such as the rule currently being used to guide behavior. White and Wise (1999) trained a monkey to orient to a visual target according to two different rules. A visual cue appeared at one of the four potential target locations. For the spatial rule, the cue's location indicated where the target would appear and thus where the monkey should direct its attention. For the conditional rule, the form and color of the cue instructed the monkey where the target would appear (e.g., if the cue was stimulus A, the target would appear on the left, if it was stimulus B, the target would be on the right, etc.). They found that up to half of PF neurons studied showed significant differences in activity that were attributed to which rule was currently being used. Such cells were also found by Hoshi, Shima and Tanji (1998). Monkeys were trained to reach for a target that matched either the shape of a cue or its location. Over a third of neurons that showed selectivity for location of the target were also modulated by which rule (matching shape or location) was being used.

Similarly, Asaad et al. (2000) recorded neural activity from the lateral prefrontal cortex while monkeys alternated between three tasks, an 'Object task', an 'Associative task', and a 'Spatial task'. The first two tasks shared common cues, but differed in how these cues were used to guide behavior, whereas the latter two used different cues to instruct the same behavior. All three required the same motor responses. The Associative task required the animals to associate a foveally presented cue object with a saccade either to the right or left on alternate blocks of trials. The Object task used the same cue objects as the Associative task; however, in the former case, they needed only to remember the identity of the cue and then saccade to the test object that matched it. The Spatial task used small spots of light to explicitly cue a saccade to the right or left. The activity of nearly half of PF neurons was task-dependent. For example, a neuron might be activated by a cue object during the Associative task, but unresponsive when the same cue appeared under identical sensory conditions in the Object task. This difference in neural activity between these tasks presumably reflects the fact that the monkeys needed to do different things with the cue, in one case (Object task) remember it while in the other case (Associative task) recall which action is currently associated with it. Also, many PF neurons studied showed changes in their baseline firing rates that depended on the task currently being performed. These results indicate that PF neurons do not simply reflect single stimuli or forthcoming actions. Rather, they also reflect their behavioral context, the pattern of associated information that is unique to a particular task.

Another study from our laboratory explored the role of PF neurons in representing abstract rules that were not tied to any specific stimuli (Wallis, Anderson and Miller, 2000). We recorded activity in a monkey trained to switch between two abstract rules, 'match' and 'nonmatch'. A sample object appeared followed by a brief delay and then by presentation of a test object. For the match rule, the monkey released a lever if the test object matched the sample. For the nonmatch rule, the monkey released the lever if the test was different. The monkeys could not predict the identity of the nonmatch. Many neurons in the dorsal, lateral, and orbitofrontal PF cortices seemed to represent these rules. A given neuron might only be active on 'match' trials while another neuron might only be active 'nonmatch' trials, irrespective of which stimulus was used as the sample. This abstracted representation of information within the PF cortex may provide the necessary foundation for the complex forms of behavior observed in primates, in whom this structure is most elaborate.

Reward-related signals, dopamine, and plasticity

Central to the formation of task representations are the signals that control the plastic changes that underlie learning. Needed are signals that specify which neural connections should be established, strengthened, or weakened and thus which information should be included in the task model representation. Ultimately, this must be derived from our successes and the rewards that result. Information about reward does seem to have a pervasive influence on PF activity. Neurons in both the lateral and orbitofrontal PF cortices convey information about expected rewards and show enhanced activity as the size and desirability of an expected reward increases (Leon and Shadlen, 1999; Tremblay and Schultz, 1999; Watanabe, 1990, 1992, 1996). A major source of reward-related signals may be the dopaminergic innervation of the PF cortex from a group of cells situated in the ventral tegmental area (VTA) of the midbrain. These cells have properties that are ideal for a role in modulating neural plasticity.

Midbrain DA neurons exhibit relatively low levels of spontaneous firing but give bursts of activity to behaviorally salient events, especially the delivery of unpredicted, appetitive stimuli, such as food or juice rewards (Mirenowicz and Schultz, 1994, 1996). As learning progresses, however, they become activated instead by events that predict reward. For example, as monkeys learn that a conditioned stimulus (CS), such as a light, will predict delivery of a reward, the DA neuronal response to the reward itself diminishes and is replaced by a response to the CS (Schultz, Apicella and Ljungberg, 1993). Interestingly, the transferred neural responses to the CS event also wane with further training. This may result from the further transfer of their responses to environmental cues that are sufficient to predict that reward is imminent (Schultz, 1998). If the predicted reward fails to appear, activity is inhibited at the expected time of its delivery (Hollerman and Schultz, 1998) and if the predicted reward appears earlier than expected, it will again elicit DA neural responses. Thus, midbrain DA neurons seem to be coding the error in the temporal prediction of reward (Schultz, 1998).

The backward transfer of the DA response from the reward to the stimuli that predict it mirrors the acquisition of task-relevant knowledge. Training animals on a complex task invariably starts with establishing 'first order' associations between reward and the events and actions that lead directly to them. As training progresses, further refinements and contingencies are added. Thus, the task is built from the reward backwards. Midbrain DA activity that likewise moves backwards through a learned chain of associations is likely to be temporally correlated with the plastic changes that accompany learning. One model (Schultz, 1998) proposes that early in learning, the arrival of DA from the midbrain selectively potentiates synapses that were activated by the events immediately preceding the reward (including their associations with reward). At the same time the events that elicited this activity become predictors of the reward and acquire the ability to activate, directly or indirectly, midbrain DA neurons. The iteration of this process will cause the DA burst to creep backward in time, 'latching' onto previous activity states in an incremental fashion (Schultz et al., 1997).

A possible correlate of this mechanism was observed in the learning paradigm described above, in which monkeys learned to associate visual cues each with one of two saccadic responses (Asaad et al., 1998). Initially, the monkeys chose their responses at random, but learned the correct cue–response pairing over a few (5–15) trials. As they learned the association, neural activity representing the forthcoming saccadic response appeared progressively earlier on successive trials (Fig. 5B). In other words, the initiation of response-related delay activity gradually shifted with learning — from a point in time just before the execution of the response and reward delivery to an earlier point in time, nearly coincident with the presentation of the cue (the CS).

We are just beginning to understand the effect the DA influx from the VTA might have on neocortical circuits, but several lines of evidence do suggest that they play a role in fostering plasticity. The wide divergence of the midbrain DA projection, along with the observation that as many as 75–80% of the neurons participate in reward-related activity bursts, suggests that DA functions more as a neuromodulator than as a carrier of specific information. This is further supported by observations that the resulting DA pulse can linger for several hundred milliseconds (Chergui, Suaud-Chagny and Gonon, 1994), allowing DA to spread beyond the particular synapse in which it was released (Gonon, 1997). Indeed, DA receptors are widespread and often found in close apposition to glutaminergic inputs onto PF layer V pyramidal neurons (Carr and Sesack, 1996; Goldman-Rakic, Leranth, Williams et al., 1989), suggesting the capacity to directly modulate the action of glutamate at these sites. In particular, DA has recently been shown to potentiate the NMDA component of the post-synaptic response to glutamate (Seamans, Gorelova, Durstewitz and Yang, 2001). This is

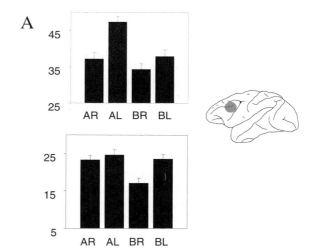

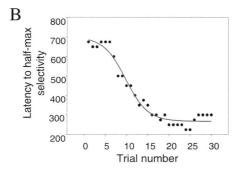

Fig. 5. (A) PF neurons tuned to associations between cue objects and saccade directions. The bar graphs show the average delay activity for two different lateral PF neurons. The error bars show the standard error of the mean. Note, for example, that the neuron in the top bar graph shows a high level of delay activity when sample object A is associated with a saccade to the left. By contrast, activity is lower when the same object is associated with a saccade to the left or when another object ('B') is associated with a saccade to the left or to the right. Thus, this neuron appears to be tuned for 'A — go left'. (B) Change in latency of saccade-direction-selective activity with learning. The time at which half of the maximal selectivity was reached within each trial is plotted along with the fitting sigmoid function. Activity related to the saccade direction appears progressively earlier with learning, merging with activity elicited by the cue object seen at the start of the trial. Note that the largest change occurs between trials 5 and 15, when the monkeys were acquiring the associations. Adapted from Asaad et al. (1998).

relevant because the activation of NMDA receptors, causing the subsequent influx of calcium, is regarded by many to be the crucial first-step toward the induction of synaptic plasticity (Malenka and Nicoll,

1999). DA could also initiate the mechanisms of synaptic plasticity through the activation of voltage-gated calcium conductances at relatively depolarized membrane potentials (Hernandez-Lopez, Bargas, Surmeier et al., 1997). Also, DA receptors are found on local inhibitory interneurons in the PF cortex (Mrzljak, Bergson, Pappy et al., 1996), and thus DA could modulate activity via an inhibitory GABAergic intermediary (Pirot, Godbout, Mantz et al., 1992). The net effect of such actions may be a strengthening of the most active inputs to a neuron accompanied by a winnowing of inactive or weakly activated inputs (Durstewitz, Seamans and Sejnowski, 1999). Support comes from observations that DA application as well as VTA stimulation improves 'signal to noise' in the PF cortex; spontaneous activity decreases while evoked responses are enhanced (Doyere, Burette, Negro and Laroche, 1993; Ferron, Thierry, Ledouarin and Glowinski, 1984; Pirot et al., 1992).

These data, along with theoretical considerations, provide a working model of how dopamine in the PF cortex might contribute to the development of task-specific neuronal activity during learning: in this scenario, dopamine is responsible for gating prefrontal afferents, that is, selecting the most active ones and inhibiting 'noise'. NMDA receptors are then recruited to establish the relevant neural connectivity under the temporal control of this dopamine pulse. As an animal learns, the dopamine pulse arrives progressively earlier, causing the successively earlier appearance of these reinforced patterns of activity during a particular behavioral sequence. Soon, the early expression of these learned patterns of activity could anticipate salient behavioral events relatively distant in the future, and this process, therefore, might endow PF activity with its prospective characteristics (Asaad et al., 1998; Rainer, Rao, Miller, 1999).

A study that used simultaneous single-unit recording and iontophoretic application of DA antagonists has found evidence for differential effects of DA during different phases of a spatial working memory task (Williams and Goldman-Rakic, 1995). Monkeys were rewarded for remembering the spatial location of a cue for 3 s and then saccading to its remembered location. Williams and Goldman-Rakic (1995) found that D1 antagonists selectively potentiated the spatial information conveyed by the neurons during the memory delay with greater effects for the

neuron's preferred orientation. They suggested that DA played a role in the direct gating of selective excitatory synaptic inputs to prefrontal neurons.

Congruently, a recent example of the ability of DA to modulate PF–hippocampal interactions was provided by a study of spatial memory retrieval in rats (Seamans, Floresco and Phillips, 1998). Rats were allowed to forage in an eight-arm maze for food located at the ends of the arms. In their delayed foraging task, only half of the arms were initially accessible. After a 30-min delay the rats had the opportunity to visit all the arms, but only the previously inaccessible arms contained reward. In a previous study (Floresco, Seamans and Phillips, 1997), these authors had demonstrated that the unilateral injection of lidocaine into the hippocampus and the contralateral PF cortex, but neither injection alone, impaired the ability of the rats to remember which arms they had visited before the delay. In this study, injection of a DA receptor antagonist into the PF cortex paired with a lidocaine injection into the contralateral hippocampus disrupted performance. None of these injections, however, produced impairment on a non-delayed version of the task. This suggests that the action of DA at the D1 receptor is required for task-relevant interactions between the hippocampus and PF cortex and provides a hint of how DA might modulate the interaction of different brain structures during specific behaviors.

In addition to these neurophysiological and pharmacological studies, circumstantial evidence for the plasticity of the prefrontal cortex and its role in behavioral learning is provided by recent observations of two structurally distinctive properties of this region. First, NMDA receptors have been found to be highly expressed in the prefrontal cortex (Huntley, Vickers and Morrison, 1997; Scherzer, Landwehrmeyer, Kerner et al., 1998). Their presence might facilitate the synaptic plasticity presumed to mediate learning. Second, continued neurogenesis has been observed in the prefrontal cortex (as well as in posterior associational cortices) even in the mature non-human primate (Gould, Reeves, Graziano and Gross, 1999). One might speculate, therefore, that the rate of neurogenesis can be modified by behavioral demands, or at least that the pattern of incorporation of these new neurons into the neocortex is biased towards those circuits undergoing modification.

Maintenance

The PF cortex also makes an important contribution to cognition by maintaining task-relevant information 'on-line' for a short while in the absence of further stimulation. This property has long been associated with the PF cortex. Maintenance mechanisms are critical because behavior often extends over time and thus fleeting events and other task information must be kept available while we wait for further inputs or to make a response. Maintenance is also important for building tasks models; it allows the forging of associations between temporally separate information (Fuster, 1990, 1995b).

Almost all neurophysiological studies of the maintenance functions have focused on the lateral PF areas (areas 9, 46, 12, 45). They have employed delay tasks in which a single stimulus is presented as a cue and then, after a delay, monkeys make a response based on it. During this delay, many lateral PF neurons show high levels of often cue-specific activity, as if they are actively sustaining cue-related information (Asaad, Rainer and Miller, 1997; Di Pellegrino and Wise, 1991; Funahashi, Bruce and Goldman-Rakic, 1989; Fuster, 1973; Fuster and Alexander, 1971; Kubota and Niki, 1971; Rainer et al., 1998a,b, 1999; Rao et al., 1997). Functional imaging studies in humans have yielded compatible results (Cohen, Perlstein, Braver et al., 1997; Courtney, Ungerleider, Keil and Haxby, 1997). This 'delay activity' can maintain all sorts of important task-related information, such as recent sensory events (Funahashi, Chafee and Goldman-Rakic, 1993; Miller, Erickson and Desimone, 1996; Rainer et al., 1998a; Rao et al., 1997), forthcoming actions (Asaad et al., 1998, 2000; Ferrera et al., 1999; Quintana and Fuster, 1992), recalled memories (Rainer et al., 1999), and expected rewards (Leon and Shadlen, 1999; Tremblay, Hollerman and Schultz, 1998; Watanabe, 1996).

The mechanisms involved in sustaining task-relevant activity in the PF cortex have been explored through the use of two classes of models, which we refer to as *cellular* and *circuit-based*. Cellular models propose that this ability is dependent on the biophysical properties of individual cells. Their activity is suggested to be bi- (or multi-) stable. The transitions between states are triggered by inputs to the PF cortex, but maintained via the activa-

tion of specific voltage-dependent conductances, for example, through NMDA receptors (Wang, 1999). Circuit-based models, on the other hand, propose that closed loops of interconnected neurons might sustain activity (Beiser and Houk, 1998; Zipser, Kehoe, Littlewort and Fuster, 1993). These loops could be intrinsic to the PF cortex or they might involve other structures (such as the cortex–striatal–globus pallidus–thalamus–cortex loop). Each type of model has its advantages and disadvantages with respect to accounting for neuronal delay activity. The models positing an intrinsic bistability of neuronal activity states must be expanded to include the possibility of multistability in order to accommodate recent neurophysiological findings showing that PF neurons can show graded levels of delay activity to reflect a parametrically varying stimulus (Romo, Brody, Hernandez and Lemus, 1999). These models must also consider the complex temporal dynamics of PF neuronal activity; many neurons do not simply 'turn on' or 'turn off' during a delay, but show variations in activity across time, and many often show anticipatory 'ramping activity' as expected events draw near. Those models suggesting recurrent loops of connected neurons as the mechanism generating sustained activity may be more capable of accounting for the complex temporal dynamics, but whether the macroscopically identified pathways (particularly the cortex–basal ganglia–thalamus–cortex loop) have the requisite microscopic specificity to generate stimulus- or response-selective delay activity is still unknown.

Virtually all cortical areas seem to have some sort of short-term storage or buffering ability. In many cortical visual areas, a brief visual stimulus will evoke activity that outlasts it for several hundred milliseconds to several seconds. What sets working memory apart as being more 'cognitive' is its ability to maintain task-related information in the face of distractions, that is, as new information enters the cortex and threatens to displace the results of previous processing. Our retention intervals, after all, are seldom empty but rather are filled with new sensory inputs, actions, etc. and one of the classic signs of PF damage is increased distractibility; subjects seem unable to keep 'on task' in the face of distractions. Miller et al. (1996) tested whether PF neurons had this ability. They used a delayed matching to sample task with intervening stimuli. Following presentation of a sample object, the monkey viewed a sequence of one to five test objects, all separated by a 1-s delay. The monkey could not predict when the match would appear in the sequence and it was rewarded for releasing a lever when one of the test objects matched the sample. This revealed that PF activity was indeed robust. Delay activity reflecting the sample object was maintained throughout the trial across intervening objects. Di Pellegrino and Wise (1993) also found a similar maintenance of PF delay activity across intervening visual inputs. This ability is not unique to the PF cortex, however. Suzuki, Miller and Desimone (1997) found that neurons in the entorhinal cortex, another region critical for visual memory, can also maintain sample-specific delay activity across intervening stimuli, but these neurons seem less prevalent in entorhinal than in the PF cortex.

By contrast, the extrastriate visual areas responsible for analysis of sensory information do not appear to have this property. Stimulus-specific delay activity has been reported in the inferior temporal (IT) cortex and in the posterior parietal (PP) (Constantinidis and Steinmetz, 1996; Fuster and Jervey, 1981; Gnadt and Andersen, 1988; Miller, Li and Desimone, 1993; Miyashita and Chang, 1988; Sereno and Maunsell, 1998). However, delay activity in these areas appears labile and easily disrupted by intervening inputs. Stimulus-specific activity seen in the delay immediately following a to-be-remembered visual cue was abolished by intervening stimuli (Constantinidis and Steinmetz, 1996; Miller et al., 1993, 1996).

How does the PF cortex 'latch' onto task-relevant information and maintain it in the face of distractions? Here, dopamine may again play a role. Braver and Cohen (2000) suggest that it acts as a gating signal; its neuromodulatory effects could strengthen task-relevant representations, protecting them against interference from disruption by irrelevant, distracting information until another dopamine influx reinforces another task-relevant representation. For a detailed discussion, we refer the reader to work by Cohen and colleagues (Braver and Cohen, 2000; Cohen and Servan-Schreiber, 1992; Servan-Schreiber, Bruno, Carter and Cohen, 1998a; Servan-Schreiber, Carter, Bruno and Cohen, 1998b).

Maintenance mechanisms are fundamental to the associative learning processes important for acquiring task information. Often, associations must be made between temporally separated events. For example, consider the Asaad et al. (1998) study discussed above. In that case, the monkey needed to associate a cue object with a saccade direction across a 1-s delay. PF activity bridged this gap; activity related to the cue object seen before the delay was maintained until the saccade was made at the end of the delay. Furthermore, as learning progressed, activity related to the forthcoming saccade direction was triggered progressively earlier. Thus, even though the cue and action were separated in time, information about each was simultaneously present in the PF cortex, which is advantageous for forming an association between them. Thus, delay activity in the PF cortex is fundamental to associative learning. It allows information about fleeting events and actions that would otherwise be separated in time to co-mingle (Fuster, 1990).

Bias signals and top-down control

Task information maintained by the PF cortex is of little use unless it can somehow be used to control processing in the brain systems that analyze sensory inputs, store long-term memories, and produce behavior. We posit that this control stems from feedback signals the PF cortex provides to those systems. This notion of feedback signals transmitting acquired knowledge is in line with another familiar concept in cognitive science, namely top-down signals. Top-down (or conceptually driven) signals are distinct from the bottom-up (or data-driven) signals that convey the information streaming into our sensory receptors. Top-down signals are thought to come into play whenever we use acquired knowledge to guide what we should pay attention to, determine what things are, how they go together, and how we should act.

To understand how this might work, consider our best understood example of top-down signals, selective visual attention. In vision, neurons processing different aspects of the visual scene compete with each other via mutually inhibitory interactions. This inhibition is thought to be central to visual processing by enhancing contrast and segmenting figure from ground. The neurons that 'win' the competition and remain active are those that incur a higher level of activity than those with which they share inhibitory interactions. The biased competition model posits that visual attention exploits these mechanisms (Desimone and Duncan, 1995). In voluntary shifts of attention a competitive advantage is incurred from excitatory top-down signals that represent the to-be-attended stimulus. They increase the activity of visual cortical neurons that process the stimulus and, by virtue of the mutual inhibition, suppress activity of neurons processing other stimuli. We suggest that this notion of bias signals that resolve local competition can be extended to include a wide variety of brain systems and processes. The PF cortex has the neural machinery to provide these feedback signals; virtually all projections to the PF cortex are matched by reciprocal connections that could convey feedback. And as we have seen above, PF neurons do indeed acquire and convey knowledge about behavioral tasks.

Several studies have yielded results consistent with this notion. Fuster and colleagues used cooling probes to deactivate the lateral PF cortex while recording neural activity from the inferior temporal (IT) cortex (Fuster, Bauer and Jervey, 1985). They found that cooling the PF cortex attenuated the ability of IT neurons to communicate information about a to-be-remembered visual stimulus. Perhaps the most direct demonstration of functional interactions between the PF cortex and posterior sensory cortex was recently provided by Tomita, Ohbayashi, Nakahara et al. (1999). They explored the role of PF top-down signals in the recall of visual images stored in the IT cortex. Appearance of a cue object instructed monkeys to recall and then choose another object that was associated with the cue during training. In the intact brain, information is shared between IT cortices in the two cerebral hemispheres. By severing the connecting fibers, each IT cortex could only 'see' (receive bottom-up inputs from) visual stimuli in the contralateral visual field. The fibers connecting the two PF cortices were left intact. When Tomita et al. examined activity of single neurons in an IT cortex that could not 'see' the cue, it nonetheless reflected the recalled object, albeit with a long latency. It appeared that visual information took a circuitous route, traveling from the opposite IT cortex (which could 'see' the cue), to the still-connected PF cortices and then down to the 'blind' IT cortex.

This was confirmed by severing the two PF cortices and eliminating the feedback, which abolished the IT activity and disrupted task performance.

Other evidence suggestive of PF–IT interactions also comes from Miller and Desimone's (Miller and Desimone, 1994; Miller et al., 1996) investigations of the respective roles of the PF cortex and IT cortex in working memory. On each trial, monkeys were shown first a sample stimulus. Then, one to four test stimuli appeared in sequence. If a test stimulus matched the sample, the monkey indicated so by releasing a lever. Sometimes, one of the intervening nonmatch stimuli could be repeated. For example, the sample stimulus 'A' might be followed by 'B … B … C … A'. The monkey was only rewarded for responding to the final match ('A') and thus had to use working memory to specifically retain the sample rather than just detect repetition of any stimulus. As noted above (in 'Maintenance'), only sustained activity in the PF cortex seemed to maintain the memory of the sample; intervening stimuli abolished sample-related sustained activity in IT cortex. However, many IT neurons (and PF neurons) did show an enhancement of their neural responses to the match (e.g., 'A') that was not evident when the intervening stimuli were repeated. Thus, PF neurons sustained activity that maintained the sample and IT neurons showed enhanced responses when a test stimulus matched that memory trace. This suggested that sustained activity from the PF cortex might have been feeding back to the IT cortex to produce the enhanced responses (Miller et al., 1996).

A few words about experience and automaticity

As task-relevant neural pathways are repeatedly selected by PF bias signals, activity-dependent plasticity mechanisms can strengthen them. With enough experience, these pathways can become established independent of the PF cortex. When this happens, the PF cortex becomes less involved and the task becomes less taxing on our limited attentional resources, that is, we can perform the task automatically. This is consistent with neuroimaging and neurophysiological studies that show greater PF activation during initial learning and then weaker activity to familiar stimuli or well-practiced tasks (Asaad et al., 1998; Shadmehr and Holcomb, 1997). Also,

PF damage can impair new learning but spare well-practiced tasks (Rushworth et al., 1997). The basal ganglia (BG), like the prefrontal cortex, is a site of extensive, multimodal convergence, receiving inputs from practically all cortical regions. Individual neurons within its dorsal subdivision, the striatum, have been estimated to receive up to 10,000 afferents. The prefrontal cortex, meanwhile, has a privileged anatomical relationship with the basal ganglia, for while many cortical regions provide input to the striatum, its output via the subthalamic nucleus and thalamus is directed primarily back to the prefrontal cortex (Alexander and Crutcher, 1990). The basal ganglia receive a dopaminergic projection from the substantia nigra in the midbrain. Just as DA in the PF cortex is thought to selectively gate cortico-cortical and subcortical connections in favor of those conveying reward-predicting activity, DA in the striatum is thought to gate cortico-striatal inputs. This may also allow reinforcement and consolidation of task-relevant neural pathways. Indeed, in monkeys, striatal neurons have been shown to undergo a variety of changes during learning (Tremblay et al., 1998). Furthermore, a structure involved in the learning of new patterns of behavior would be expected to have a significant response to re-enforcing events, such as the delivery of reward. Indeed, the reward-related response of striatal neurons is robust (Hollerman and Schultz, 1998).

A relatively common view of the role of the basal ganglia holds that this structure is involved in the establishment of habitual, well-learned patterns of behavior. Consistent with this, neurons in striatum of rats show responses to various behavioral events during early learning, but eventually, after learning, respond only to the beginning of a behavioral sequence (Jog, Kubota, Connolly et al., 1999). These responses then remain stable for at least many weeks thereafter. The PF cortex, on the other hand, seems to remain critical (at least for a while) for situations that require flexible switching between well-practiced behaviors. A recent example was provided by Rossi, Rotter, Desimone and Ungerleider (1999). Monkeys had to detect the presence of one of several visual targets in a display cluttered with distractors. When the monkey could predict which stimulus would be the target (because the same stimulus was the target for several trials in a row), PF lesions had little ef-

fect on performance. However, when the target could not be predicted because it was randomly switched from trial to trial, PF damage impaired performance. Thus, bias signals from the PF cortex may remain important for deciding between alternate stimulus–response mappings that are equally salient.

These functional properties have suggested to some that, although they are both involved in the acquisition of new behaviors, the prefrontal cortex and BG act in a somewhat antagonistic manner to either reject or establish patterns of action in response to particular environmental contexts, respectively (Wise et al., 1996). Along these lines, one might imagine that the crucial difference between the prefrontal cortex and BG lies in the time course of plastic changes; the neocortex may be rapidly modifiable by experience, whereas changes in the BG might occur over longer spans of time and only after repeated exposure to a particular situation. The result might be the rapid acquisition of a rough approximation of a re-enforced behavior by the cortex, which is then fine-tuned by the BG. The extent to which these structures act in parallel or in series to accomplish this, or even if such a conceptual distinction is reasonable, remains to be elucidated.

Conclusions

One of the great mysteries of the brain is cognitive control. How can interactions between millions of neurons result in behavior that appears willful and voluntary? There is consensus that it depends on the PF cortex, but there has been little understanding of the neural mechanisms that endow it with the properties needed for executive control.

Here, we have suggested that this stems from the extensive connections of the PF cortex and the ability of its circuitry to be modified by experience. Reward signals acting on the PF cortex cause it to knit together a neural representation of the goal-relevant features of a given situation. The progressively earlier influx of DA during learning may allow the PF cortex to 'bootstrap' from acquisition of simple associations with reward to a multivariate network of associative links that describe a complex task. Once established, this PF task model can be evoked by the appropriate cues to provide bias signals that select task-relevant neural pathways throughout the neocortex. As a result, our attention is directed, memories are recalled, and responses are selected. Eventually, the needed pathways can become established independent of the PF cortex and the behavior becomes automatic. We have reviewed a wide variety of studies that have provided data consistent with this model. They have shown that the PF cortex has the necessary infrastructure and that its neurons have properties that are highly dependent on, and shaped by, task demands.

This is not meant to be a comprehensive and detailed account of the role of the PF cortex in mental life; needless to say, many questions need to be addressed and many of the details of our model will need elaboration and modification. For example, while we posit a central role for DA in potentiating existing pathways, there are other, less explored, mechanisms that could play a major role, such as the contribution of other neuromodulatory systems such as serotonin and acetylcholine to PF function, and incorporation of newly grown neurons into pathways (Gould et al., 1999). Instead, this model is intended to convey a general view of the type of mechanisms that might underlie cognitive control. Something of this flavor seems reasonable to us. Virtually all complex behavior is acquired and involves piecing together relationships between diverse, arbitrary bits of information that have no intrinsic association. Any brain region centrally involved in cognition is likely to be centrally involved in this process.

Acknowledgements

Our work has been supported by an N.I.N.D.S. grant, the RIKEN-MIT Neuroscience Research Center, The Pew Charitable Trusts, The Mcknight Foundation, The Whitehall Foundation, The John Merck Fund, and The Alfred P. Sloan Foundation. We thank Mark Histed, Jonathan Wallis, Richard Wehby, and Marlene Wicherski for valuable comments on this manuscript.

References

Abbott V, Black JB, Smith EE: The representation of scripts in memory. Journal of Memory and Language: 24; 179–199, 1985.

Alexander GE, Crutcher MD: Preparation for movement: neural representations of intended direction in three motor areas of the monkey. Journal of Neurophysiology: 64; 133–150, 1990.

Amaral DG, Price JL: Amygdalo-cortical projections in the monkey (*Macaca fascicularis*). Journal of Comparative Neurology: 230; 465–496, 1984.

Anderson KC, Asaad WF, Wallis JD, Miller EK: Simultaneous recordings from monkey prefrontal (PF) and posterior parietal (PP) cortices during visual search. Society for Neuroscience Abstracts: 25; 885, 1999.

Asaad WF, Rainer G, Miller EK: Activity of neurons in monkey prefrontal (PF) cortex during learning of object–spatial associations. Society for Neuroscience Abstracts: 23; 1614, 1997.

Asaad WF, Rainer G, Miller EK: Neural activity in the primate prefrontal cortex during associative learning. Neuron: 21; 1399–1407, 1998.

Asaad WF, Rainer G, Miller EK: Task-specific activity in the primate prefrontal cortex. Journal of Neurophysiology: 84; 451–459, 2000.

Azuma M, Suzuki H: Properties and distribution of auditory neurons in the dorsolateral prefrontal cortex of the alert monkey. Brain Research: 298; 343–346, 1984.

Bachevalier J, Mishkin M: Visual recognition impairment follows ventromedial but not dorsolateral prefrontal lesions in monkeys. Behavioural Brain Research: 20; 249–261, 1986.

Baddeley A: Working Memory. Oxford: Clarendon Press, 1986.

Barbas H: Anatomic organization of basoventral and mediodorsal visual recipient prefrontal regions in the rhesus monkey. Journal of Comparative Neurology: 276; 313–342, 1988.

Barbas H, De Olmos J: Projections from the amygdala to basoventral and mediodorsal prefrontal regions in the rhesus monkey. Journal of Comparative Neurology: 300; 549–571, 1990.

Barbas H, Pandya D: Patterns of connections of the prefrontal cortex in the rhesus monkey associated with cortical architecture. In Levin HS, Eisenberg HM, Benton AL (Eds), Frontal Lobe Function and Dysfunction. New York: Oxford University Press, pp. 35–58, 1991.

Barbas H, Pandya DN: Architecture and intrinsic connections of the prefrontal cortex in the rhesus monkey. Journal of Comparative Neurology: 286; 353–375, 1989.

Barsalou LW, Sewell DR: Contrasting the representation of scripts and categories. Journal of Memory and Language: 24; 646–665, 1985.

Bates JF, Goldman-Rakic PS: Prefrontal connections of medial motor areas in the rhesus monkey. Journal of Comparative Neurology: 336; 211–228, 1993.

Baylis LL, Gaffan D: Amygdalectomy and ventromedial prefrontal ablation produce similar deficits in food choice and in simple object discrimination learning for an unseen reward. Experimental Brain Research: 86; 617–622, 1991.

Beiser DG, Houk JC: Model of cortical-basal ganglionic processing: encoding the serial order of sensory events. Journal of Neurophysiology: 79; 3168–3188, 1998.

Bichot NP, Schall JD, Thompson KG: Visual feature selectivity in frontal eye fields induced by experience in mature macaques. Nature: 381; 697–699, 1996.

Boussaoud D, Wise SP: Primate frontal cortex: effects of stimulus and movement. Experimental Brain Research: 95; 28–40, 1993.

Braver TS, Cohen JD: On the control of control: the role of dopamine in regulating prefrontal function and working memory. In Monsell S, Driver J (Eds), Attention and Performance 18. Cambridge: MIT Press, 2000.

Bruce C, Desimone R, Gross CG: Visual properties of neurons in a polysensory area in superior temporal sulcus of the macaque. Journal of Neurophysiology: 46; 369–384, 1981.

Bullier J, Schall JD, Morel A: Functional streams in occipito-frontal connections in the monkey. Behavioural Brain Research: 76; 89–97, 1996.

Butter CM: Perseveration in extinction and in discrimination reversal tasks following selective frontal ablations in *Macaca mulatta*. Physiology and Behavior: 4; 163–171, 1969.

Carr DB, Sesack SR: Hippocampal afferents to the rat prefrontal cortex: synaptic targets and relation to dopamine terminals. Journal of Comparative Neurology: 369; 1–15, 1996.

Chavis DA, Pandya DN: Further observations on cortico-frontal connections in the rhesus monkey. Brain Research: 117; 369–386, 1976.

Chergui K, Suaud-Chagny MF, Gonon F: Nonlinear relationship between impulse flow, dopamine release and dopamine elimination in the rat brain in vivo. Neuroscience: 62; 641–645, 1994.

Cohen JD, Dunbar K, McClelland JL: On the control of automatic processes: A parallel distributed processing model of the Stroop effect. Psychological Review: 97; 332–361, 1996.

Cohen JD, Perlstein WM, Braver TS, Nystrom LE, Noll DC, Jonides J, Smith EE: Temporal dynamics of brain activation during a working memory task. Nature: 386; 604–608, 1997.

Cohen JD, Servan-Schreiber D: Context, cortex, and dopamine: A connectionist approach to behavior and biology in schizophrenia. Psychological Review: 99; 45–77, 1992.

Constantinidis C, Steinmetz MA: Neuronal activity in posterior parietal area 7a during the delay periods of a spatial memory task. Journal of Neurophysiology: 76; 1352–1355, 1996.

Courtney SM, Petit L, Maisog JM, Ungerleider LG, Haxby JV: An area specialized for spatial working memory in human frontal cortex. Science: 279; 1347–1351, 1998.

Courtney SM, Ungerleider BG, Keil K, Haxby JV: Transient and sustained activity in a distributed neural system for human working memory. Nature: 386; 608–611, 1997.

Dehaene S, Kerszeberg M, Changeux JP: A neuronal model of a global workspace in effortful cognitive tasks. Proceedings of the National Academy of Science of the United States of America: 95; 14529–14534, 1998.

Desimone R, Duncan J: Neural mechanisms of selective visual attention. Annual Review of Neuroscience: 18; 193–222, 1995.

D'Esposito M, Postle BR, Ballard D, Lease J: Maintenance versus manipulation of information held in working memory: an event-related fMRI study [In Process Citation]. Brain and Cognition: 41; 66–86, 1999.

Di Pellegrino G, Wise SP: A neurophysiological comparison of three distinct regions of the primate frontal lobe. Brain: 114; 951–978, 1991.

Di Pellegrino G, Wise SP: Visuospatial versus visuomotor activity in the premotor and prefrontal cortex of a primate. Journal of Neuroscience: 13; 1227–1243, 1993.

Dias R, Robbins TW, Roberts AC: Primate analogue of the Wisconsin Card Sorting Test: effects of excitotoxic lesions of the prefrontal cortex in the marmoset. Behavioral Neuroscience: 110; 872–886, 1996.

Dias R, Robbins TW, Roberts AC: Dissociable forms of inhibitory control within prefrontal cortex with an analog of the Wisconsin Card Sort Test: Restriction to novel situations and independence from 'on-line' processing. Journal of Neuroscience: 17; 9285–9297, 1997.

Doyere V, Burette F, Negro CR, Laroche S: Long-term potentiation of hippocampal afferents and efferents to prefrontal cortex: implications for associative learning. Neuropsychologia: 31; 1031–1053, 1993.

Duncan J, Emslie H, Williams P, Johnson R, Freer C: Intelligence and the frontal lobe: The organization of goal-directed behavior. Cognitive Psychology: 30; 257–303, 1996.

Durstewitz D, Seamans JK, Sejnowski TJ: Dopaminergic modulation of activity states in the prefrontal cortex. Society for Neuroscience Abstracts: 25; 1216, 1999.

Eacott MJ, Gaffan D: Inferotemporal–frontal disconnection — the uncinate fascicle and visual associative learning in monkeys. European Journal of Neuroscience: 4; 1320–1332, 1992.

Felleman DJ, Van Essen DC: Distributed hierarchical processing in the primate cerebral cortex. Cerebral Cortex: 1; 1–47, 1991.

Ferrera VP, Cohen J, Lee BB: Activity of prefrontal neurons during location and color delayed matching tasks. NeuroReport: 10; 1315–1322, 1999.

Ferron A, Thierry AM, Ledouarin C, Glowinski J: Inhibitory influence of the mesocortical dopaminergic system on spontaneous activity or excitatory response induced from the thalamic mediodorsal nucleus in the rat medial prefrontal cortex. Brain Research: 302; 257–265, 1984.

Floresco SB, Seamans JK, Phillips AG: Selective roles for hippocampal, prefrontal cortical, and ventral striatal circuits in radial-arm maze tasks with or without a delay. Journal of Neuroscience 17; 1880–1890, 1997.

Funahashi S, Bruce CJ, Goldman-Rakic PS: Mnemonic coding of visual space in the monkey's dorsolateral prefrontal cortex. Journal of Neurophysiology: 61; 331–349, 1989.

Funahashi S, Chafee MV, Goldman-Rakic PS: Prefrontal neuronal activity in rhesus monkeys performing a delayed antisaccade task. Nature: 365; 753–756, 1993.

Fuster JM: Unit activity in prefrontal cortex during delayed-response performance: neuronal correlates of transient memory. Journal of Neurophysiology: 36; 61–78, 1973.

Fuster JM: The Prefrontal Cortex. New York: Raven Press, 1989.

Fuster JM: Prefrontal cortex and the bridging of temporal gaps in the perception–action cycle. 608: 318–329; discussion 330, 1990.

Fuster JM: Memory in the Cerebral Cortex. Cambridge, MA: MIT Press, 1995a.

Fuster JM: Temporal processing. Annals of the New York Academy of Sciences: 769; 173–181, 1995b.

Fuster JM, Alexander GE: Neuron activity related to short-term memory. Science: 173; 652–654, 1971.

Fuster JM, Bauer RH, Jervey JP: Cellular discharge in the dorsolateral prefrontal cortex of the monkey in cognitive tasks. Experimental Neurology: 77; 679–694, 1982.

Fuster JM, Bauer RH, Jervey JP: Functional interactions between inferotemporal and prefrontal cortex in a cognitive task. Brain Research: 330; 299–307, 1985.

Fuster JM, Jervey JP: Inferotemporal neurons distinguish and retain behaviorally relevant features of visual stimuli. Science: 212; 952–955, 1981.

Gaffan D, Harrison S: Inferotemporal–frontal disconnection and fornix transection in visuomotor conditional learning by monkeys. Behavioural Brain Research: 31; 149–163, 1988.

Gaffan D, Harrison S: A comparison of the effects of fornix transection and sulcus principalis ablation upon spatial learning by monkeys. Behavioural Brain Research: 31; 207–220, 1989.

Gallese V, Murata A, Kaseda M, Niki N, Sakata H: Deficit of hand preshaping after muscimol injection in monkey parietal cortex. Neuroreport: 5; 1525–1529, 1994.

Gnadt JW, Andersen RA: Memory related motor planning activity in posterior parietal cortex of macaque. Experimental Brain Research: 70; 216–220, 1988.

Goldman PS, Nauta WJ: Autoradiographic demonstration of a projection from prefrontal association cortex to the superior colliculus in the rhesus monkey. Brain Research: 116; 145–149, 1976.

Goldman PS, Rosvold HE, Vest B, Galkin TW: Analysis of the delayed-alternation deficit produced by dorsolateral prefrontal lesions in the rhesus monkey. 77; 212–220, 1971; Journal of Comparative Physiological Psychology.

Goldman-Rakic PS: Circuitry of primate prefrontal cortex and regulation of behavior by representational memory. In Plum F (Ed), Handbook of Physiology: The Nervous System. Bethesda, MD: American Physiological Society, pp. 373–417, 1987.

Goldman-Rakic PS: Prefrontal cortex revisited: a multiple-memory domain model of human cognition. In Caminiti R, Hoffman KP, Lacquaniti F, Altman J (Eds), Vision and Movement Mechanisms in the Cerebral Cortex. Strasbourg: HFSP, pp. 162–172, 1996.

Goldman-Rakic PS, Leranth C, Williams SM, Mons N, Geffard M: Dopamine synaptic complex with pyramidal neurons in primate cerebral cortex. Proceedings of the National Academy of Sciences of the United States of America: 86; 9015–9019, 1989.

Goldman-Rakic PS, Schwartz ML: Interdigitation of contralateral and ipsilateral columnar projections to frontal association cortex in primates. Science: 216; 755–757, 1982.

Goldman-Rakic PS, Selemon LD, Schwartz ML: Dual pathways connecting the dorsolateral prefrontal cortex with the hippocampal formation and parahippocampal cortex in the rhesus monkey. Neuroscience: 12; 719–743, 1984.

Gonon F: Prolonged and extrasynaptic excitatory action of dopamine mediated by D1 receptors in the rat striatum in vivo. Journal of Neuroscience: 17; 5972–5978, 1997.

Goodale MA, Milner AD: Separate visual pathways for perception and action. Trends in Neurosciences: 15; 20–25, 1992.

Gould E, Reeves AJ, Graziano MS, Gross CG: Neurogenesis in the neocortex of adult primates. Science: 286; 548–552, 1999.

Grafman J: Alternative frameworks for the conceptualization of prefrontal functions. In Boller F, Grafman J (Eds), Handbook of Neuropsychology, Vol. 9. Amsterdam: Elsevier, p. 187, 1994.

Gross CG, Weiskrantz L: Evidence for dissociation of impairment on auditory discrimination and delayed response following lateral frontal lesions in monkeys. Experimental Neurology: 5; 453–476, 1962.

Halsband U, Passingham RE: Premotor cortex and the conditions for movement in monkeys. Behavioural Brain Research: 18; 269–276, 1985.

Hernandez-Lopez S, Bargas J, Surmeier DJ, Reyes A, Galarraga E: D1 receptor activation enhances evoked discharge in neostriatal medium spiny neurons by modulating an L-type Ca^{2+} conductance. Journal of Neuroscience: 17; 3334–3342, 1997.

Hollerman JR, Schultz W: Dopamine neurons report an error in the temporal prediction of reward during learning. Nature Neuroscience: 1; 304–309, 1998.

Hoshi E, Shima K, Tanji J: Task-dependent selectivity of movement-related neuronal activity in the primate prefrontal cortex. Journal of Neurophysiology: 80; 3392–3397, 1998.

Huntley GW, Vickers JC, Morrison JH: Quantitative localization of NMDAR1 receptor subunit immunoreactivity in inferotemporal and prefrontal association cortices of monkey and human. Brain Research: 749; 245–262, 1997.

Iversen SD, Mishkin M: Perseverative interference in monkeys following selective lesions of the inferior prefrontal convexity. Experimental Brain Research: 11; 376–386, 1970.

Jog MS, Kubota Y, Connolly CI, Hillegaart V, Graybiel AM: Building neural representations of habits. Science: 286; 1745–1749, 1999.

Johnson MK, Hirst W: Processing subsystems of memory. In Lister RG, Weingartner HJ (Eds), Perspectives in Cognitive Neuroscience. New York: Oxford University Press, pp. 3–16, 1991.

Jones EG, Powell TPS: An anatomical study of converging sensory pathways within the cerebral cortex of the monkey. Brain: 93; 793–820, 1970.

Kubota K, Niki H: Prefrontal cortical unit activity and delayed alternation performance in monkeys. Journal of Neurophysiology: 34; 337–347, 1971.

Leon MI, Shadlen MN: Effect of expected reward magnitude on the response of neurons in the dorsolateral prefrontal cortex of the macaque. Neuron: 24; 415–425, 1999.

Lu MT, Preston JB, Strick PL: Interconnections between the prefrontal cortex and the premotor areas in the frontal lobe. Journal of Comparative Neurology: 341; 375–392, 1994.

Malenka RC, Nicoll RA: Long-term potentiation–a decade of progress? Science: 285; 1870–1874, 1999.

Maunsell JH, Newsome WT: Visual processing in monkey extrastriate cortex. Annual Review of Neuroscience: 10; 363–401.

Merzenich MM, Sameshima K: Cortical plasticity and memory. Current Opinion in Neurobiology: 3; 187–196, 1993.

Miller EK: The prefrontal cortex: complex neural properties for complex behavior. Neuron: 22; 15–17, 1999.

Miller EK: The neural basis of top-down control of visual attention in the prefrontal cortex. In Monsell S, Driver J (Eds), Attention and Performance, Vol. 18. Cambridge, MA: MIT Press, 2000a.

Miller EK: The prefrontal cortex and cognitive control. Nature Reviews Neuroscience: 1; 59–65, 2000b.

Miller EK, Cohen JD: An integrative theory of prefrontal function. Annual Review of Neuroscience, in press, 2001.

Miller EK, Desimone R: Parallel neuronal mechanisms for short-term memory. Science: 263; 520–522, 1994.

Miller EK, Erickson CA, Desimone R: Neural mechanisms of visual working memory in prefrontal cortex of the macaque. Journal of Neuroscience: 16; 5154–5167, 1996.

Miller EK, Li L, Desimone R: Activity of neurons in anterior inferior temporal cortex during a short-term memory task. Journal of Neuroscience: 13; 1460–1478, 1993.

Milner B: Effects of different brain lesions on card sorting. Archives of Neurology: 9; 90, 1963.

Mirenowicz J, Schultz W: Importance of unpredictability for reward responses in primate dopamine neurons. Journal of Neurophysiology: 72; 1024–1027, 1994.

Mirenowicz J, Schultz W: Preferential activation of midbrain dopamine neurons by appetitive rather than aversive stimuli. Nature: 379; 449–451, 1996.

Mishkin M: Effects of small frontal lesions on delayed alternation in monkeys. Journal of Neurophysiology: 20; 615–622, 1957.

Mishkin M, Manning FJ: Non-spatial memory after selective prefrontal lesions in monkeys. 143; 313–323, 1978.

Mishkin M, Vest B, Waxler M, Rosvold HE: A re-examination of the effects of frontal lesions on object alternation. 7; 357–364, 1969.

Miyashita Y, Chang HS: Neuronal correlate of pictorial short-term memory in the primate temporal cortex. Nature: 331; 68–70, 1988.

Mrzljak L, Bergson C, Pappy M, Huff R, Levenson R, Goldman-Rakic PS: Localization of dopamine D4 receptors in GABAergic neurons of the primate brain. Nature: 381; 245–248, 1996.

Murata A, Fadiga L, Fogassi L, Gallese V, Raos V, Rizzolatti G: Object representation in the ventral premotor cortex (area F5) of the monkey. Journal of Neurophysiology: 78; 2226–2230, 1997.

Murray EA, Bussey TJ, Wise SP: Role of prefrontal cortex in a network for arbitrary visuomotor mapping. Experimental Brain Research: 133; 114–129, 2000.

Norman DA, Shallice T: Attention to action: willed and automatic control of behavior. In Davidson RJ, Schwartz GE, Shapiro D (Eds), Consciousness and Self-Regulation: Advances in Research and Theory. New York: Plenum, pp. 1–48, 1986.

O Scalaidhe SP, Wilson FAW, Goldman-Rakic PS: Face-selective neurons during passive viewing and working memory performance of Rhesus monkeys: Evidence for intrinsic specialization of neuronal coding. Cerebral Cortex: 9; 459–475, 1999.

Owen AM, Evans AC, Petrides M: Evidence for a two-stage model of spatial working memory processing within the lateral frontal cortex: A positron emission tomography study. Cerebral Cortex: 6; 31–38, 1996a.

Owen AM, Milner B, Petrides M, Evans AC: Memory for object features versus memory for object location: A positron-emission tomography study of encoding and retrieval processes. Proceedings of the National Academy of Sciences of the United States of America: 93; 9212–9217, 1996b.

Owen AM, Stern CE, Look RB, Tracey I, Rosen BR, Petrides M: Functional organization of spatial and nonspatial working memory processing within the human lateral frontal cortex. Proceedings of the National Academy of Sciences of the United States of America: 95; 7721–7726, 1998.

Pandya DN, Barnes CL: Architecture and connections of the frontal lobe. In Perecman E (Ed), The Frontal Lobes Revisited. New York: IRBN Press, pp. 41–72, 1987.

Pandya DN, Yeterian EH: prefrontal cortex in relation to other cortical areas in rhesus monkey. Architecture and Connections: 85; 63–94, 1990.

Parker A, Gaffan D: Memory after frontal/temporal disconnection in monkeys: conditional and non-conditional tasks, unilateral and bilateral frontal lesions. Neuropsychologia: 36; 259–271, 1998.

Passingham R: Delayed matching after selective prefrontal lesions in monkeys (*Macaca mulatta*). Brain Research: 92; 89–102, 1975.

Passingham R: Information about movements in monkeys (*Macaca mulatta*) with lesions of dorsal prefrontal cortex. Brain Research: 152; 313–328, 1978.

Passingham R: The Frontal Lobes and Voluntary Action. Oxford: Oxford University Press, 1993.

Petrides M: Motor conditional associative-learning after selective prefrontal lesions in the monkey. Behavioural Brain Research: 5; 407–413, 1982.

Petrides M: Deficits in non-spatial conditional associative learning after periarcuate lesions in the monkey. Behavioural Brain Research: 16; 95–101, 1985.

Petrides M: Nonspatial conditional learning impaired in patients with unilateral frontal but not unilateral temporal lobe excisions. Neuropsychologia: 28; 137–149, 1990.

Petrides M: Functional specialization within the dorsolateral frontal cortex for serial order memory. Proceedings of the Royal Society of London: 246; 299–306, 1991.

Petrides M: Functional specialization within the primate dorsolateral frontal cortex. Advances in Neurology: 57; 379–388, 1992.

Petrides M: Frontal lobes and behaviour. Current Opinion in Neurobiology: 4; 207–211, 1994a.

Petrides M: Frontal lobes and working memory: Evidence from investigations of the effects of cortical excisions in nonhuman primates. In Boller F, Grafman J (Eds), Handbook of Neuropsychology, Vol. 9. Elsevier, Amsterdam, pp. 59–82, 1994b.

Petrides M: Impairments on nonspatial self-ordered and externally ordered working memory tasks after lesions of the mid-dorsal part of the lateral frontal cortex of the monkey. Journal of Neuroscience: 15; 359–375, 1995.

Petrides M: Specialized systems for the processing of mnemonic information within the primate frontal cortex. Philosophical Transactions of the Royal Society of London Series B: Biological Sciences: 351; 1455–1461, 1996.

Petrides M, Alivisatos B, Evans AC, Meyer E: Dissociation of human mid-dorsolateral from posterior dorsolateral frontal cortex in memory processing. 90; 873–877, 1993.

Petrides M, Pandya DN: Projections to the frontal cortex from the posterior parietal region in the rhesus monkey. Journal of Comparative Neurology: 228; 105–116, 1984.

Petrides M, Pandya DN: Dorsolateral prefrontal cortex: comparative cytoarchitectonic analysis in the human and the macaque brain and corticocortical connection patterns. European Journal of Neuroscience: 11; 1011–1036, 1999.

Pirot S, Godbout R, Mantz J, Tassin JP, Glowinski J, Thierry AM: Inhibitory effects of ventral tegmental area stimulation on the activity of prefrontal cortical neurons: evidence for the involvement of both dopaminergic and GABAergic components. Neuroscience: 49; 857–865, 1992.

Porrino LJ, Crane AM, Goldman-Rakic PS: Direct and indirect pathways from the amygdala to the frontal lobe in rhesus monkeys. Journal of Comparative Neurology: 198; 121–136, 1981.

Postle BR, D'Esposito M: Analysis of spatial and object delayed response with event-related fMRI. Proceedings of the Cognitive Neuroscience Society: 5; 85, 1998.

Prabhakaran V, Narayanan K, Zhao Z, Gabrieli JDE: Frontal-lobe integration of diverse information within working memory. Nature Neuroscience: 3; 85–90, 1999.

Quintana J, Fuster JM: Mnemonic and predictive functions of cortical neurons in a memory task. NeuroReport: 3; 721–724, 1992.

Rainer G, Asaad WF, Miller EK: Memory fields of neurons in the primate prefrontal cortex. Proceedings of the National Academy of Science of the United States of America: 95; 15008–15013, 1998a.

Rainer G, Asaad WF, Miller EK: Selective representation of relevant information by neurons in the primate prefrontal cortex. Nature: 393; 577–579, 1998b.

Rainer G, Rao SC, Miller EK: Prospective coding for objects in the primate prefrontal cortex. Journal of Neuroscience: 19; 5493–5505, 1999.

Rao SC, Rainer G, Miller EK: Integration of what and where in the primate prefrontal cortex. Science: 276; 821–824, 1997.

Recanzone GH, Merzenich MM, Jenkins WM, Grajski KA, Dinse HR: Topographic reorganization of the hand representation in cortical area 3b owl monkeys trained in a frequency-discrimination task. Journal of Neurophysiology: 67; 1031–1056, 1992.

Romo R, Brody CD, Hernandez A, Lemus L: Neuronal corre-lates of parametric working memory in the prefrontal cortex. Nature: 399; 470–473, 1999.

Rossi AF, Rotter PS, Desimone R, Ungerleider LG: Prefrontal lesions produce impairments in feature-cued attention. Society for Neuroscience Abstracts: 25; 3, 1999.

Rushworth MF, Nixon PD, Eacott MJ, Passingham RE: Ventral prefrontal cortex is not essential for working memory. Journal of Neuroscience: 17; 4829–4838, 1997.

Sakata H, Taira M, Kusunoki M, Murata A, Tanaka Y, Tsutsui K: Neural coding of 3D features of objects for hand action in the parietal cortex of the monkey. Philosophical Transactions of the Royal Society of London B: Biological Sciences: 353; 1363–1373, 1998.

Scherzer CR, Landwehrmeyer GB, Kerner JA, Counihan TJ, Kosinski CM, Standaert DG, Daggett LP, Velicelebi G, Penney JB, Young AB: Expression of N-methyl-D-aspartate receptor subunit mRNAs in the human brain: hippocampus and cortex. Journal of Comparative Neurology: 390; 75–90, 1998.

Schmahmann JD, Pandya DN: Anatomic organization of the basilar pontine projections from prefrontal cortices in rhesus monkey. Journal of Neuroscience: 17; 438–458, 1997.

Schultz W: Predictive reward signal of dopamine neurons. Journal of Neurophysiology: 80; 1–27, 1998.

Schultz W, Apicella P, Ljungberg T: Responses of monkey dopamine neurons to reward and conditioned stimuli during successive steps of learning a delayed response task. Journal of Neuroscience: 13; 900–913, 1993.

Schultz WPD, Dayan P, Montague PR: A neural substrate of prediction and reward. Science: 275; 1593–1599, 1997.

Seamans JK, Floresco SB, Phillips AG: D1 receptor modulation of hippocampal-prefrontal cortical circuits integrating spatial memory with executive functions in the rat. Journal of Neuroscience: 18; 1613–1621, 1998.

Seamans JK, Gorelova N, Durstewitz D, Yang CR: Bidirectional dopamine modulation of GABAergic inhibition in prefrontal cortical pyramidal neurons. Journal of Neuroscience; 21; 3628–3638, 2001.

Seltzer B, Pandya DN: Frontal lobe connections of the superior temporal sulcus in the rhesus monkey. Journal of Comparative Neurology: 281; 97–113, 1989.

Sereno AB, Maunsell JHR: Shape selectivity in primate lateral intraparietal cortex. Nature: 395; 500–503, 1998.

Servan-Schreiber D, Bruno RM, Carter CS, Cohen JD: Dopamine and the mechanisms of cognition: Part I. A neural network model predicting dopamine effects on selective attention. Biological Psychiatry: 43; 713–722, 1998a.

Servan-Schreiber D, Carter CS, Bruno RM, Cohen JD: Dopamine and the mechanisms of cognition: Part II. D-amphetamine effects in human subjects performing a selective attention task. Biological Psychiatry: 43; 723–729, 1998b.

Shadmehr R, Holcomb H: Neural correlates of motor memory consolidation. Science: 277; 821–824, 1997.

Suzuki WA, Miller EK, Desimone R: Object and place memory in the macaque entorhinal cortex. Journal of Neurophysiology: 78; 1062–1081, 1997.

Tanaka K, Saito H, Fukada Y, Moriya M: Coding visual images of objects in the inferotemporal cortex of the macaque monkey. Journal of Neurophysiology: 66; 170–189, 1991.

Thorpe SJ, Rolls ET, Maddison S: The orbitofrontal cortex: neuronal activity in the behaving monkey. Experimental Brain Research: 49; 93–115, 1983.

Tomita H, Ohbayashi M, Nakahara K, Hasegawa I, Miyashita Y: Top-down signal from prefrontal cortex in executive control of memory retrieval [see comments]. Nature: 401; 699–703, 1999.

Toth LJ, Assad JA: Visual stimulus selectivity during an association task in LIP. Society for Neuroscience Abstracts: 25; 1546, 1999.

Tremblay L, Hollerman JR, Schultz W: Modifications of reward expectation-related neuronal activity during learning in primate striatum. Journal of Neurophysiology: 80; 964–977, 1998.

Tremblay L, Schultz W: Relative reward preference in primate orbitofrontal cortex [see comments]. Nature: 398; 704–708, 1999.

Ungerleider LG, Mishkin M: Two cortical visual systems. In Ingle J, Goodale MA, Mansfield RJW (Eds), Analysis of Visual Behavior. Cambridge, MA: MIT Press, pp. 549–586, 1982.

Vaadia E, Benson DA, Hienz RD, Goldstein MH Jr: Unit study of monkey frontal cortex: active localization of auditory and of visual stimuli. Journal of Neurophysiology: 56; 934–952, 1986.

Van Hoesen GW, Pandya DN, Butters N: Cortical afferents to the entorhinal cortex of the Rhesus monkey. Science: 175; 1471–1473, 1972.

Wallis JD, Anderson KC, Miller EK: Single neurons in the prefrontal cortex encode abstract rules. Nature: 411; 953–956.

Wang XJ: Synaptic basis of cortical persistent activity: the importance of NMDA receptors to working memory. Journal of Neuroscience: 19; 9587–9603, 1999.

Watanabe M: Prefrontal unit activity during associative learning in the monkey. Experimental Brain Research: 80; 296–309, 1990.

Watanabe M: Frontal units of the monkey coding the associative significance of visual and auditory stimuli. Experimental Brain Research: 89; 233–247, 1992.

Watanabe M: Reward expectancy in primate prefrontal neurons. Nature: 382; 629–632, 1996.

Webster MJ, Bachevalier J, Ungerleider LG: Connections of inferior temporal areas TEO and TE with parietal and frontal cortex in macaque monkeys. Cerebral Cortex: 4; 470–483, 1994.

White IM, Wise SP: Rule-dependent neuronal activity in the prefrontal cortex. Experimental Brain Research: 126; 315–335, 1999.

Williams GV, Goldman-Rakic PS: Modulation of memory fields by dopamine D1 receptors in prefrontal cortex. Nature: 376; 572–575, 1995.

Wilson FAW, O Scalaidhe SP, Goldman-Rakic PS: Dissociation of object and spatial processing domains in primate prefrontal cortex. Science: 260; 1955–1958, 1993.

Wise SP, Murray EA, Gerfen CR: The frontal-basal ganglia system in primates. Critical Reviews in Neurobiology: 10; 317–356, 1996.

Zipser D, Kehoe B, Littlewort G, Fuster J: A spiking network model of short-term active memory. Journal of Neuroscience: 13; 3406–3420, 1993.

Handbook of Neuropsychology, 2nd Edition, Vol. 7
J. Grafman (Ed)

CHAPTER 3

Working memory: findings from neuroimaging and patient studies

Edward E. Smith *, Christy Marshuetz and Anat Geva

Department of Psychology, University of Michigan, 525 East University Avenue, Ann Arbor, MI 48109-1109, USA

Introduction

What is working memory and why is it important?

The notion of 'working memory' first appeared in the cognitive literature in a classic book by Miller, Galanter and Pribram (1960), but the idea only took hold when the seminal studies of Baddeley and Hitch (1974) were published. The latter work argued that the primary purpose of keeping information active for a brief period of time was to allow the information to be processed in some way, as for example when we keep numbers active while we mentally multiply them. The concept of short-term memory gave way to that of working memory.

Working memory may be defined as a system used for the temporary storage and manipulation of information. With regard to the storage aspect, it has long been known that: the capacity of the system is on the order of 5–7 items (e.g., Miller, 1956); the decay or forgetting rate is measured in seconds (e.g., Peterson and Peterson, 1959); and the information in store can be retrieved or accessed in a matter of milliseconds (e.g., Sternberg, 1966). Presumably the information is as accessible as it is because it is in an active state.

This kind of system is needed for many if not most acts of higher-level cognition. We have already noted that you need working memory to do mental multiplication. Mental calculation involves the generation of partial products — outputs of a process that must be kept active until they are used by some subsequent process (when you multiply two 2-digit numbers, for example, you first have to multiply the right-most digits, and hold onto this partial product until you execute the next process). This generation and temporary storage of partial products seems to be a general feature of higher-level cognition, making working memory essential for all kinds of mental activities (see Jonides, 1995). Thus, when solving a complex problem, routinely we have to set up sub-goals and their associated plans, and then keep track of which sub-goals and plans have been satisfied; the sub-goals and plans are partial products that have to be maintained in working memory for some period of time. A similar story applies for understanding language. When we participate in a conversation, or read a book, we have to keep prior propositions in an active state so that we can readily connect them to what we are currently hearing or reading; if we do not do this, we will not be able to truly understand the message.

In light of the above, it is not surprising that individual differences in working-memory capacity correlate highly with individual differences in reasoning ability. As one example, performance on a standard measure of working memory (Daneman and Carpenter, 1980) is highly correlated with performance on Raven's Progressive Matrices test, a widely used non-verbal test of intelligence (Carpenter, Just and Shell, 1990). As another example, it is well known that performance on verbal reasoning tasks declines

* Corresponding author. Tel.: +1 (734) 764-0186; Fax: +1 (734) 763-7480; E-mail: eesmith@umich.edu

with normal aging, and that much of that decline can be shown to be due to a decrease in the capacity of working memory (e.g. Salthouse, Legg, Palmon and Mitchell, 1990).

Understanding working memory, then, is important not just for what it can tell us about another memory system, but for what it can tell us about thought itself. In this chapter, we are concerned with understanding working memory at both the cognitive and the neurological levels. Studying a human memory system at the neurological level forces us to deal with certain issues about the nature of the data, to which we turn next.

The need to consider both neuroimaging and neuropsychological data

Recent research on the neural bases of working memory has used neuroimaging techniques, both positron emission tomography (PET) and functional magnetic resonance imaging (fMRI). Typically, participants have their brains scanned while engaging in working-memory and control conditions, and then the conditions are compared to determine those neural areas that are selectively activated (or deactivated) during the storage and processing aspects of the working-memory condition. Such neuroimaging data provide a map of brain regions that mediate task performance. But note that the data are essentially correlational. All we really know is that area A is active when cognitive process P is required, not that A is necessary or sufficient for P to occur. To establish that A is necessary for P, we need to show that if A is damaged (lesioned), then performance on any task requiring P will be compromised. Hence, for the logic of scientific inference, it is critical to connect imaging data to lesion or patient data (e.g., Smith, 1997).

Making such connections is currently difficult because neuroimaging and neuropsychological approaches have tended to use different tasks to study the same cognitive domain. In part, this difference comes about because neuropsychologists favor tasks that have diagnostic value — they can distinguish prefrontal from anterior-temporal lesions, for example — but such tasks typically involve multiple processes. An example of this situation is the Wisconsin Card Sort Task, which is widely used to diagnose

damage in prefrontal cortex, but by anyone's account is a complex task that involves multiple cognitive processes (see e.g., Dehaene and Chanqeux, 1991, for a computational model of this task). In contrast, neuroimagers favor tasks that have the potential to isolate specific cognitive processes — paradigms that isolate the storage component of verbal working memory, for example, with little consideration for the diagnostic value of the task.

Thus the need to connect neuroimaging and neuropsychological results is paramount — without doing it, there can be no strong conclusions about the causal relations between areas activated in imaging studies and cognitive processes, but there are methodological obstacles in forging these connections. In this chapter, we endeavor to consider both kinds of data for each topic considered.

Agenda

The rest of this chapter is devoted to three major issues, each of which occupies a separate section. In the next, or second section, we take up the issue of whether the storage and processing aspects of working memory can be dissociated. We will provide neuroimaging and neuropsychological data for such a dissociation. In the course of treating this issue, it will become evident that some of the key processes involved in working memory are meta-processes that regulate the operation of lower-level operations. These meta-processes are standardly referred to as 'executive process', and two of them will be the foci of the remaining sections. In the third section we consider the executive process of 'attention and inhibition', and in the fourth section we discuss the executive process of 'temporal coding', which is responsible for coding working-memory representations with respect to their time of occurrence. While these two meta-processes do not exhaust the set of executive processes, they figure centrally in most discussions of the topic.

One caveat is in order before proceeding. Although we have talked of working memory as it were a single system, the evidence indicates that there are different working memories for different kinds of information. In particular, there appear to be different systems for verbal, spatial, and visual–object information (e.g., Smith, Jonides and Koeppe, 1996;

Smith and Jonides, 1999). In this chapter, we restrict our analysis to verbal working memory. This system has been the most explored, particularly in connection with executive processes, which are of major concern to us.

Dissociations between storage and executive processes

Nature of the issue

Earlier we defined working memory as a system used for the temporary storage *and* manipulation of information. Given this conjunctive definition, it is tempting to divide the system into two general components: short-term storage (roughly, the old concept of short-term memory) and a set of executive processes that modulate the more specific processes that operate on the contents of short-term storage (Smith and Jonides, 1999). There is a viable alternative to this qualitative division, namely that tasks that involve only short-term storage and those that also require executive processes may draw on the very same pool of computational and neural resources

(Carpenter, Just, Keller et al., 1999). This common resources hypothesis seems quite plausible once we note that tasks that presumably involve 'only storage' frequently require rehearsal processes to maintain the contents of storage, so some sort of processing is called for in both types of tasks. A variant of the common resources hypothesis motivated the seminal studies of Baddeley and Hitch (1974). Thus making a distinction between storage and processing in working memory, as we do, amounts to taking a stance on a fundamental issue about the structure of working memory. It behooves us to supply some empirical evidence for our stance.

A comparison of specific tasks

It is convenient to begin with a contrast between the results associated with two frequently used working-memory tasks in neuroimaging studies of working memory. One is the item-recognition task, introduced by Sternberg (1966; see Fig. 1A). In this task on each trial a target set of items (digits, letters, words, etc.) is presented, followed by a brief delay (usually a few seconds), followed in turn by a probe

A. Item Recognition Task

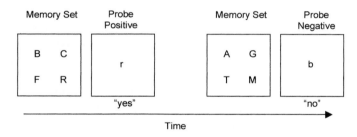

B. Two-Back Task

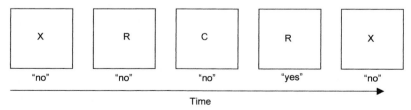

Fig. 1. Schematic drawing of two working-memory backs. (A) Item recognition task. Participants have to determine whether probe item appeared in memory set. (B) Two-back task. Participants have to determine whether each letter is the same as the one two-back in the sequence.

item; the participant's task is to decide as quickly as possible whether the probe is the same as one of the targets. In addition to perceptual and response processes, the item-recognition task is assumed to involve only those process needed to maintain information in working memory. That is, the task involves short-term storage, but no executive processes. The contrasting paradigm is a 'back' task, e.g., '2-back' (see Fig. 1B). A relatively long sequence of items is presented with a few seconds delay between successive items, and for each item the participant must decide whether it is identical to the one that occurred two back in the sequence. For example, given the sequence P T T X T, one should respond positively only to the last item, because only this letter matches the one two back. In addition to requiring the maintenance and updating of two items in working memory, this task also requires that participants: (a) code each item in short-term storage in terms of its relative temporal position; and (b) inhibit any tendency to respond positively to repeated items that are separated by less than two elements (e.g., inhibit responding positively to the immediately repeated T in the preceding example). Since (a) and (b) are instances of executive processes, the 2-back task is said to involve executive processes as well as short-term storage.

In neuroimaging experiments, the item-recognition and back tasks produce overlapping but distinctive patterns of results. Consider first item recognition. Activations in this task are generally lateralized to the left hemisphere, and the areas most frequently involved are in the posterior parietal cortex (in Brodmann Area, BA 40), supplementary motor cortex (BA 6), premotor area (BA 6), and Broca's area (BA 44). Given that the latter three areas are known to be involved in the preparation of motor action, including speech, and given that all participants in these studies inevitably report rehearsing the target items during the delay period, it seems plausible that the supplementary motor, premotor and Broca's areas mediate a rehearsal loop. (See Awh, Jonides, Smith et al., 1996 for independent evidence for this interpretation of the three frontal regions just mentioned.) The posterior parietal area, in contrast, has often been interpreted as mediating some sort of passive storage (e.g., D'Esposito and Postle, 2000; Smith et al., 1996).

The typical pattern of activations for 2- and 3-back tasks include the left-hemisphere areas just mentioned — supplementary motor, premotor, Broca's area, and posterior parietal cortex. But the back tasks produce bilateral patterns of activation, and, most importantly for our purposes, they include other areas of activation as well. These other areas include regions in the anterior prefrontal cortex, particularly the dorsolateral prefrontal cortex (DLPFC) (which corresponds to BA 9 and 46). This anterior region is of particular interest because, as we will later see, there is abundant evidence linking it to executive processes. In short, the activation patterns for 2- and 3-back tasks look like the pattern obtained in a pure maintenance (or pure storage) task with a pattern for executive processes tacked on. At this rough level of description, the evidence supports a distinction between storage and executive processes. (In what follows, we often use the term 'maintenance' instead of 'storage' and the distinction of interest becomes that between maintenance and maintenance-plus-processing.)

Reviews of relevant studies

The preceding argument has been strengthened by recent reviews of neuroimaging studies of verbal working memory. In one such paper, Smith and Jonides (1999) contrasted 4 neuroimaging studies of tasks that presumably involved mainly maintenance (see Fig. 2A) and 10 studies in which the tasks presumably involved executive processes as well as maintenance (most of these were 2- and 3-back tasks, so the executive processes involved were attention/inhibition and temporal coding; see Fig. 2B). Inspection of Fig. 2 indicates that both maintenance and maintenance-plus-processing tasks revealed notable clusters of activations in left-hemisphere, posterior frontal cortex — the rehearsal circuit — and in left-hemisphere posterior parietal cortex — the passive storage area. But the maintenance-plus-processing tasks reveal another major cluster of activations as well, this one in the DLPFC (Fig. 2B). (A Statistical test reveal that the frontal activations in 2B are significantly more anterior than those in 2A.) The neural circuitry recruited for executive processes, therefore, seem to be somewhat distinct from that needed for just storage and rehearsal. Essentially

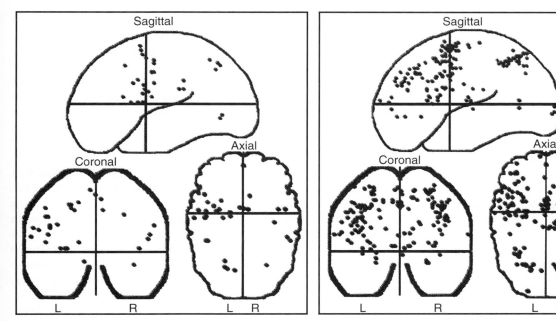

Fig. 2. Neuroimaging results for verbal working memory are summarized by sets of three projections, with each containing points and axes conforming to standard Talairach space. Each projection collapses one plane of view for each activation focus — that is, the sagittal view collapses across the x plane as though one were looking through the brain from the side; the coronal view collapses across the y plane as though one were looking through the brain from the front or back; and the axial view collapses across the z plane as though one were looking through the brain from the top. Included in the summary are published ^{15}O PET or fMRI studies of verbal working memory that reported coordinates of activation and had a memory load of six or fewer items.

the same conclusion was drawn by D'Esposito and colleagues (D'Esposito, Aguirre, Zarahan and Ballard, 1998) in their review of neuroimaging studies of working memory.

Given that we have some neuroimaging evidence for the distinction of interest, the next step is to seek converging evidence from neuropsychology. D'Esposito and Postle (2000) reviewed studies that tested patients with damage in prefrontal cortex on working-memory tasks. Each study also included an age-matched control group — either of intact participants or of patients who had damage in non-frontal regions — so that deficits could be assessed. Two kinds of tasks were considered: simple span tasks (like digit span), which the authors believe provide an estimate of sheer storage capacity; and delayed-response tasks, which presumably involve more than just passive storage. The delayed response tasks themselves divide into two kinds, those in which the delay period was filled with distracting items and those in which the delay period was unfilled; the

former kind almost certainly require the executive process of inhibition, to prevent the distracting information from entering working memory. In short, two of the tasks studied — simple-span and filled-delay — map onto our distinction between maintenance and maintenance-plus-processing.

D'Esposito and Postle (2000) found that across numerous studies, frontal patients showed no impairment at all on simple-span tasks but clear deficits on filled-delay tasks. (Unfilled-delay tasks produced equivocal results.) Thus frontal patients are impaired on maintenance-plus-processing tasks but not on maintenance tasks. This is a clear dissociation between storage and executive processing which converges nicely with that obtained in the neuroimaging studies previously reviewed. Furthermore, some studies reviewed by D'Esposito and Postle (2000) included patients who had only parietal damage, and some of these patients had deficits on simple-span but not delay tasks. When these parietal patients are contrasted with the frontal patients, there is a striking double dissociation between maintenance and maintenance-plus-processing.

A caveat is in order, though. As forewarned, the tasks used in the patient studies differ from those employed in the imaging experiments described earlier. In particular, in the patient studies the maintenance-only tasks is the familiar memory span paradigm, in which participants read a short sequence of items and then repeat them back *in order*. The requirement to maintain the input order requires some sort of temporal coding of the items, and such coding is considered an executive process. Still, maintaining a brief, sequentially presented sequence of items in their input order may require minimal executive processing (people will reproduce a brief ordered string in input order even when allowed to recall it in any order). So presumably there is still a difference in the *degree* of executive processing required between the simple-span and the filled-delay tasks surveyed by D'Esposito and Postle (2000).

Specific experimental tests of the distinction

The evidence reviewed thus far consists entirely of after-the-fact analyses of experiments performed for some reason other than testing the distinction between maintenance and maintenance-plus-processing. Recently, a couple of neuroimaging experiments have been performed to explicitly test the distinction of interest.[1]

Both studies involved a contrast between conditions that varied the amount of temporal coding required. In Collette, Salmon, Van der Linden et al. (1999), in one condition participants were presented a series of words and had to recall them in order (a word-span task); in the other condition participants were presented a series of words but had to recall them in alphabetic order (the 'alpha-span' task). The alpha-span task requires reordering the input order, and hence necessitates a greater degree of temporal coding and executive processing than the word-span task. Only the alpha-span task resulted in activation in DLPFC, a supposed 'home' of executive processing.

In the other experiment of interest (D'Esposito, Postle, Ballard and Lease, 1999), the relevant condition again required memory for input order. Now six target letters were presented simultaneously, followed by a brief delay, followed by a probe that consisted of a letter and a digit, e.g., K 3. The participant's task was to decide whether the probe letter had appeared in the target set in the serial position designated by the probe digit (e.g., K 3 means "Was K the third letter in the target set?"). In the other relevant condition the same target sets and probes were presented, but now participants had to reorder the target set into alphabetic order and decide whether the probe letter had the relative alphabetic position designated by the probe digit (e.g., K 3 now means "With respect to alphabetic order, was K the third letter in the target set?"). Again there is a variation in the amount of temporal coding needed, and again the more coding needed the greater was the activation in DLPFC.

In sum, there is a substantial amount of evidence, from neuropsychology as well as neuroimaging, for a neural distinction between working-memory tasks that require only maintenance and those that require some sort of processing as well. This in turn implies a neural difference between the storage and executive-processing components of working memory. Given this dissociation, one can justify a focus on either component. The remainder of this chapter is concerned only with executive processes, specifically the processes of (1) attention and inhibition and (2) temporal coding, which have figured so centrally in drawing the critical distinction.

Executive process of attention and inhibition

Early inhibition in working memory tasks

Recency and item recognition
If maintenance and executive processes are truly distinct, one should be able to start with a maintenance task and add an executive process to it. This kind of 'cognitive addition' was implemented in a PET study by Jonides, Smith, Marshuetz et al. (1998).

The maintenance task was the item-recognition task. In a 'standard condition', a set of 4 letters was presented briefly, followed by a 3-s delay, followed by a probe letter; participants decided as quickly as possible whether the probe was the same as

---

[1] The two relevant experiments will be described briefly here, and in more detail in the final section, which explicitly deals with temporal coding.

one of the targets. Importantly, the negative items — probes that required a No response — had not occurred on a recent trial. At the cognitive level, the participant must encode the letters, queue them into the rehearsal set, rehearse them sub-vocally during the delay, and make a decision about the probe. No executive processes are needed to complete the task. A simple change in some of the negative items, however, can bring the executive process of attention and inhibition into play. In a second version of the task, called the 'high-recency' condition, half of the negative probes had appeared in the previous target set of items. Because such 'recent negatives' had been rehearsed just a few seconds ago, they should have been in an active state, and highly familiar. This in turn should have set up a conflict between the familiarity of the probe, which points toward an incorrect Yes response, and the fact that the representation of the probe does not contain a marker indicating it is a member of the current target set, which points to a correct No response. Such a process-conflict is assumed to trigger the executive process of attention and inhibition, which involves inhibiting the outcome of the irrelevant familiarity process, and selectively attending to the process that inspects the relevant list marker.

The prediction, then, is that only the high-recency condition should invoke attention and inhibition. Behavioral support for this prediction is provided by the finding that participants took longer to correctly respond to recent negatives than to non-recent negatives (there was no difference in the time needed to respond to non-recent negatives in the standard and high-recency conditions). Of greater interest are the neuroimaging results. The critical PET comparison is that between the high-recency and the standard condition. This comparison revealed that a portion of left-hemisphere ventrolateral prefrontal cortex (BA 45) was the only significant focus of activation, implying that this area is critically involved in attentional and inhibitory processing.

This conclusion was strengthened by a follow-up experiment by D'Esposito and colleagues (D'Esposito et al., 1999), which employed fMRI (and permitted single-trial analysis). Only the high-recency condition was tested, and now the critical comparison was between No trials that are recent negatives and No trials that are non-recent negatives.

In the critical left-hemisphere ventrolateral prefrontal region (BA 45), there was more activation for recent than non-recent negatives. In addition, this difference between the two trial types occurred only during the interval when the probe was presented. These findings fit well with the hypothesis that the activation in question reflects inhibiting familiarity information and selectively attending to information about list membership when a response is being selected.

In keeping with the need to augment neuroimaging findings with converging evidence from neuropsychology, patient performance was examined in a version of these item-recognition tasks. Thompson-Schill and colleagues (Thompson-Schill, Jonides, Marshuetz et al., 1999) examined the contrast between standard and high-recency conditions in a patient (R.C.) who had damage to much of the prefrontal cortex (PFC) including left-hemisphere BA 45, the area identified by the original PET study. When compared with a group of neurological controls (i.e. they had PFC lesions sparing 45/46), there was nothing exceptional about R.C.'s performance on the standard task. However, R.C. had a marked impairment, reflected in both response time and accuracy, on trials with recent negatives. This impairment resulted in a significantly exaggerated difference between recent negatives and non-recent-negatives, which is line with R.C. having a deficit in attention and inhibition. It seems clear that left-hemisphere BA 45 is critically involved in the attention and inhibitory demands of this task.

Distracter effects

We have assumed that in the preceding tasks, inhibition occurs prior to the point at which a response is selected. (As we will later see, in paradigms in which inhibition occurs on already-selected responses there is a different neural signature.) We refer to inhibition prior to the response as 'early' inhibition. There are other relevant studies that seem to involve early inhibition, particularly patient studies due to Knight and his associates.[2]

---

[2] Until this point, we have referred to the executive process of interests as 'attention and inhibition'. In what follows, though, we refer to it simply as 'inhibition', partly for ease of exposition, and partly because many of the studies reviewed appear to emphasize inhibitory processes rather than attentional ones.

Chao and Knight (1995) examined the role of the PFC in inhibiting irrelevant information during an auditory item-recognition task. Participants had to indicate whether target and probe sounds were identical. They were tested on two separate conditions, a no-distracter and a distracter condition. In the no-distracter condition, there was a silent interval between the target and probe. In the distracter condition, distracter tones occurred during the interval between the target and probe sounds; such distracters would have to be inhibited in order to prevent their intrusion into working memory. So only the distracter condition should have involved inhibition, and it is an early inhibition that is needed. Patients with either left hemisphere damage to the DLPFC, posterior hippocampal region, or the temporal–parietal junction were tested. All patient groups performed somewhat worse than the normal controls in the no-distracter condition. However, the DLPFC patients were disproportionately impaired in the distracter condition, even at the shortest delays. Further evidence about this sort of early inhibition in a working-memory task has been obtained by Ptito, Crane, Leonard et al. (1995). They found that patients with lesions invading BA 46 (a major part of DLPFC) were disproportionately impaired in remembering the location of a light when distracting material had occurred between the initial presentation of the light and the memory probe.

How exactly does the DLPFC affect inhibition in the above distracter studies? According to Knight and his colleagues (Knight, Scabini and Woods, 1989; Yamaguchi and Knight, 1990), the DLPFC inhibits input to primary cortical regions. Thus, damage to this region results in "chronic leakage of irrelevant sensory inputs", which may account for frontal patient distractibility (Chao and Knight, 1995). Chao and Knight (1998) have found support for this view by demonstrating that in an auditory working-memory task, frontal patients generated enhanced primary-auditory-cortex evoked responses to tone pips serving as distracters. This view of the inhibition involved is consistent with the idea that the inhibition is produced early in the processing sequence.

All the studies reviewed on inhibition implicate the lateral surface of the PFC. But note that the studies that induced cognitive conflict by varying probe familiarity found the critical region was in ventrolateral PFC, whereas the studies that induced conflict by presenting distracters during the delay period focused on the dorsolateral PFC. This disparity in results could reflect numerous differences in how the two sets of studies were conducted. For example: the probe-familiarity studies are mainly neuroimaging experiments that involved visual stimuli and blank delay periods, whereas the distracter studies are neuropsychological experiments that used auditory inputs and filled delay periods. Also, in the distracter studies, the patients with DLPFC damage may have had some damage in ventrolateral PFC, and that may have been the source of their difficulties in inhibition. For now, there is no way to definitely resolve the disparity in results.

Response inhibition in tasks with minimal memory loads

Stroop task

The inhibition tasks described above can be contrasted with the Stroop task, a classic inhibition paradigm in which at least some of the inhibition presumably occurs late in processing, perhaps even after a response has been selected. In Stroop, participants are presented a set of color names printed in different colors and asked to name the print colors. Performance is poorer when the print color differs from the color name (an 'incongruent' condition) than when they are the same (a 'congruent condition'); e.g., it takes longer to say 'red' to the word 'blue' printed in red (incongruent) than to the word 'red' printed in red (congruent). This difference in latency between incongruent and congruent trials is one measure of the 'Stroop effect'. Another is the latency difference between incongruent and neutral trials (trials with non-color words).

It is generally agreed that the Stroop effect arises because there is a conflict between reading the word, which is an irrelevant process, and naming the print color, which is the relevant process. Since the irrelevant process is stronger than the relevant one, accurate performance can be achieved only by selectively attending to the print color while inhibiting word reading. Hence, the Stroop task, like some of the tasks mentioned earlier, engenders conflict between two processes, and requires that the stronger but irrelevant one be inhibited (Cohen, McClelland

and Dunbar, 1990; Dyer, 1973; for a review, see MacLeod, 1991).

But the Stroop task differs in important ways from the inhibition tasks described earlier. First, Stroop requires little in the way of short-term storage. Participants simply detect the color of the word and report it. (Perhaps there is some need to initially keep the instructions in working memory, but this should be a minimal memory load as trials progress.) Given the distinction between short-term storage and executive processes that we argued for earlier, there is no principled reason why *some* inhibitory process cannot be studied with or without a storage load. However, it may turn out that *some* inhibitory processes are operative only when there is a storage load (e.g., a process that inhibits *a* working-memory representation). Another potentially important difference between Stroop and our earlier paradigms is that at least some of the inhibition in Stroop likely occurs at the response end of processing. In behavioral studies with normal participants, the magnitude of the Stroop effect is significantly reduced when there is no overlap between the color names and the word meanings (e.g., the colors could be red, green, and yellow, while the words are 'blue', 'white', and 'orange'), even though there is still process interference between word reading and color naming (Klein, 1964). The implication is that part of the usual Stroop effect is due to response inhibition, perhaps even the inhibition of an already-selected response. For these reasons, the Stroop task may involve a different kind of inhibition than that present in the studies reviewed earlier.

Consistent with this analysis, neuroimaging studies of Stroop reveal a somewhat different picture than that described for the item-recognition task. Although PET studies of Stroop in normal individuals show substantial variation in regions of activation, two broad regions in frontal cortex appear to be involved when one compares incompatible to compatible (or control) conditions. These frontal regions are the anterior cingulate cortex and the PFC (Pardo, Pardo, Janer and Raichle, 1990; Taylor, Kornblum, Minoshima et al., 1994b). Activations in anterior cingulate cortex have also been obtained in other experiments that induce a conflict between response tendencies (Bush, Whalen, Rosen et al., 1998; Taylor, Kornblum, Lauber et al., 1994a). Taken together, these studies suggest that the anterior cingulate may be involved in response inhibition (Bench, Frith, Grasby et al., 1993; see Smith and Jonides, 1999, for a recent review).

As usual, we want to know how well the neuroimaging findings fit with relevant results from neuropsychology. There is evidence that patients with damage in PFC are selectively impaired in the Stroop task (Perret, 1974; Vendrell, Junque, Pujol et al., 1995). This is broadly compatible with the imaging results. Indeed, so widespread is Stroop impairment in frontal patients that an abnormality in Stroop performance has been used to diagnose decreases in frontal lobe integrity in a variety of patient populations, among them, schizophrenia (Cohen and Servan-Schreiber, 1992), Huntington's (Brandt, 1991) and Parkinson's diseases (Brown and Marsden, 1991; Henik, Singh, Beckley and Rafal, 1993).

What about more specific correspondences between the imaging and patient literatures? Patients with lesions localized to PFC seem to be selectively impaired in the incompatible condition of Stroop (Perret, 1974; Vendrell et al., 1995), confirming the role of this region in whatever inhibitory processes are operative. But the situation is a little less clear when it comes to the anterior cingulate (e.g., Vendrell et al., 1995).

Other response inhibition tasks
While we have suggested that the inhibition in Stroop is at least partly an inhibition of response, this interpretation is speculative because there is nothing about the Stroop task that forces a participant to prepare one response and then suppress it. There are, however, other paradigms that do have this character. One is the Go/No Go task (see Fig. 3A). In this paradigm, a sequence of letters is presented and participants have to respond to each letter, save one exception letter for which they must withhold their response (the single exception is designated in the instructions). On the face of it, this task involves a single prepotent response, which may be prepared in advance of each letter and which must be inhibited on exception trials. Importantly, frontal patients have difficulty in performing this task (Malloy, Bihrle, Duffy and Cimino, 1993). The obvious questions are: "What brain areas are activated during perfor-

A. Item Recognition and Inhibition Task

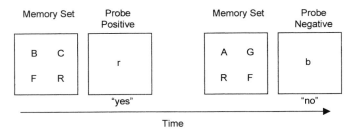

B. Go/No Go

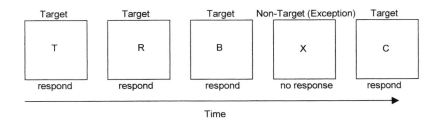

C. AX-CPT Task

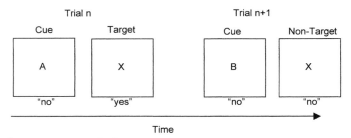

D. AX-CPT + Memory Task

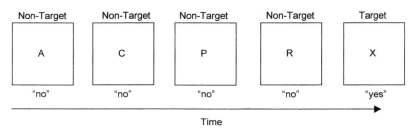

Fig. 3. Schematic drawings of working-memory tasks involving attention and inhibition. (A) Item-recognition inhibition task. Participants decide whether a probe matches one of the target items (Jonides et al., 1998). (B) Go/No Go task. Participants respond to each letter presented in a sequence, save one exception for which they must withhold their response (Casey et al., 1997). (C) AX-CPT task. Participants respond to a target letter (X) only when it is preceded by another target (A) (Barch et al., 1997). (D) AX-CPT memory-load task. Participants respond to the target letter, X, when it is preceded by the target letter, A, followed by three intervening letters (Seidman et al., 1998).

mance of the task?" and "How do they line up with Stroop activations?"

A recent fMRI study of the Go/No Go task provides some tentative answers (Casey, Trainor, Orendi et al., 1997). In comparison with the appropriate controls, the Go/No Go task activated bilateral DLPFC (BAs 9, 46/10) and the anterior cingulate (BAs 24 and 32), as well as more ventrolateral portions of the prefrontal cortex (BAs 45, 47) and some orbital-frontal areas as well (BA 11). Again, it seems that a task that requires the inhibition of response recruits a brain circuit that includes the DLPFC and anterior cingulate. But the results also show activation of ventrolateral areas that we have linked with early inhibition, so the picture is by no means entirely clear.

There are other fMRI studies of tasks emphasizing response inhibition that generally activate the DLPFC and anterior cingulate. Perhaps the most relevant results come from neuroimaging studies of the AX-CPT task (see Fig. 3B). Barch, Braver, Nystrom et al. (1997) and Carter, Braver, Barch et al. (1998) (both report results from the same data set) used a variant of this task, in which participants see a series of letter pairs and are instructed to make a target response to the second letter if it is an X, but only if the first letter is an A; in all other cases they are to withhold response. Since AX pairs occur the vast majority of the time, again the task includes a pre-potent response that must sometimes be withheld. Significant activation was found in DLPFC and the anterior cingulate. (See Garvan, Ross and Stein, 1999 for a related result.)

Taken together, the results from these response-suppression tasks support the claim that response inhibition is mediated by a circuit involving DLPFC and the anterior cingulate. The involvement of the anterior cingulate is what most clearly distinguishes this pattern of results from the one we described for item-recognition studies that presumably involved earlier inhibitory processes. But as already noted, the two sets of studies differ not only in when inhibition supposedly occurs, but also in that the item-recognition experiments involve substantial short-term storage loads whereas the Stroop-like tasks involve minimal storage loads. It is possible, then, that the different patterns of activation reflect the fact that different neural circuits mediate inhibition when there is a concurrent storage load than when there is not. One way to assess this possibility is to introduce a storage load into a response-inhibition task (another instance of 'cognitive addition').

Such a manipulation was implemented in an experiment by Seidman, Breiter, Goodman et al. (1998). Participants performed a task similar to the AX-CPT in addition to a condition that introduced a memory load (see Fig. 3C). In the memory-load condition, participants responded positively to any target letter, X that was preceded by A and three items (A3X condition), for example, A C P R X, and responded negatively to all other stimuli. Further complicating the task, letters could be interleaved (e.g., A A F A D X B), so participants could not simply drop the three items in memory once a response was made; they always had to hold 3–4 items in short-term store. In the comparison of the A3X versus AX conditions, activation was detected in all of the most commonly observed left-hemisphere storage and rehearsal sites: Broca's area (BA 44), precentral sites (BA 6, 8), and posterior parietal cortex (BAs 7 and 40). In addition, both conditions showed activations in response-inhibition sites: bilateral DLPFC (BAs 9 and 10) and anterior and posterior cingulate (BAs 32 and 24). The critical determinant of the activations in DLPFC and the cingulate, then, seems not to be a function of whether the task requires memory, but whether response inhibition is involved.

Summary

Neuroimaging studies indicate that in comparison to tasks requiring only short-term maintenance, tasks that require inhibition lead to activations in anterior prefrontal cortex. The regions involved are in the ventrolateral and dorsolateral prefrontal cortex, as well as in the anterior cingulate. This much seems clear, and is broadly consistent with patient studies. In addition, there is suggestive evidence that which specific regions are activated is determined by whether the inhibition occurs relatively early or late in processing. Thus far the only studies showing ventrolateral activations have involved inhibition relatively early in processing, whereas studies that report a combination of DLPFC and anterior-cingulate activations often involved response inhibition. It is important to know whether the supposed early–late differences in inhibition can be supported by patient studies, but that largely remains a task for future research. It is also important to know whether the anterior prefrontal regions of interest are unique to the executive process of attention and inhibition; some evidence about this issue will be forthcoming from our review of the executive process of temporal coding, to which we now turn.

Executive process of temporal coding

The second executive process of interest involves the coding of representations in working memory. Such representations can be coded for various contextual aspects, but the coding of greatest research interest has been for temporal aspects (e.g., when did this item occur, or when did it occur relative to another item?). On the face of it, coding seems to be a different kind of cognitive entity than inhibition. But both executive processes are meta-processes: Just as inhibition can modulate a lower-level process (like one that assesses familiarity), so coding modulates existing representations. The link between the two executive processes may be even stronger. Inhibition routinely involves selectively attending to some process or representation while inhibiting others, and perhaps the coding of representations requires selectively attending to the representation of interest.

Prefrontal patients and temporal coding

The early studies on this topic were all patient studies. For this reason we begin our review with patient studies and then discuss recent neuroimaging experiments. As we will see, the patient studies mainly show that the prefrontal cortex is involved in temporal coding, while the neuroimaging experiments are more specific about which frontal regions are implicated.

Recency judgments

The hypothesis that the frontal lobes are involved in recency judgments was first assessed by Corsi (cited in Milner, 1971, 1974). Participants were shown a series of pairs of items; occasionally, two items would appear with a question mark between them and the participant's task was to determine which of the two items had been viewed most recently. When only one of the items had been previously seen,

this was a simple task of item recognition. However, when both items were 'old', this was a test of recency memory. An innovative aspect of Corsi's experiment was that patient groups with unilateral removals from the frontal or temporal lobe were tested on three versions of the recency-discrimination task that differed in terms of item content — concrete words, representational drawings, and abstract paintings.

Corsi's results demonstrated the functional specialization of the frontal and temporal lobes: Selective damage to the former contributed to a deficit in memory for temporal order, whereas damage to the latter contributed to an impairment in item recognition. Moreover, there was evidence for hemispheric specialization for the material to-be-remembered. Patients with damage to the left hemisphere were impaired in processing verbal material, whereas right-hemisphere damage was associated with an inability to process nonverbal stimuli. Thus Corsi's experiment provides evidence that the frontal lobes are needed to temporally separate events in time and further shows that this process is material sensitive. These findings have been strengthened and extended by Milner, Corsi and Leonard (1991).

How might such recency judgements be performed? The cognitive literature speaks to this issue and offers at least three possibilities. First, the participant might simply check the sheer level of strength or familiarity of the items in working memory in order to determine their relative positions in time and sequence. Presumably, the representations that are less familiar have decayed somewhat, and hence likely occurred earlier (e.g. Cavanagh, 1976; Corballis, Kirby and Miller, 1972; Wickelgren and Norman, 1966). Essentially, the comparative familiarity of the representations is being used as a heuristic for estimating their temporal order, and when asked to choose the most recent of two items the participants simply selected the stronger or more familiar one. A second possible mechanism consists of forming directional associations between the items presented, e.g., Item 1 *precedes* Item 2, Item 2 *precedes* Item 3, and so on (e.g., Burgess and Hitch, 1999). Now when asked to judge the more recent of two test items, participants could answer directly if the items had occurred consecutively, else they could chain their pairwise associations to determine which one had preceded the other in the sequence. A third possible mechanism involves associating temporal tags to the items as they are presented. When asked to judge the more recent of two test items, participants could compare the items' temporal tags. The tags might be exact — 'first', 'second', 'third', etc. — in which case the determination of which of the two items occurred first is trivial. More plausibly, at least some of the temporal tags may be inexact — e.g., 'the early part of the sequence' — in which case the determination of which test item occurred first might sometimes be difficult (e.g., Banks, Clark and Lucy, 1975).

Since the first mechanism described — the heuristic use of relative familiarity — is sufficient to accomplish the recency-judgment task, we cannot be sure that the frontal patients' problem was in the handling of *temporal* information. To bolster the latter conclusion, we need to consider the performance of frontal patients in tasks that cannot readily be accomplished by a familiarity heuristic. We need to consider tasks that require the use of temporal associations or temporal tagging. Such tasks have not been used often in patient studies of short-term or working memory, but they have figured centrally in studies of temporal coding that involve long-term as well as short-term memory. It is worth taking a brief look at these studies, as the results have implications for temporal coding in working memory.

In experiments by Shimamura, Janowsky and Squire (1990), frontal patients and matched controls were presented sequentially a list of 15 words that they were to study. After a 15-s delay, the study words, one to a card, were presented in a random array, and the participants were asked to place the cards in the order corresponding to the initial order of presentation of the words. The task required only memory for order, and some of the stored order information may still have been in working memory. (The delay was unfilled, which permitted participants to keep active some of the information.) Frontal patients were impaired on this task, even though these same patients performed normally in other conditions that required memory only for item information. Thus the deficit seems to be selective to the encoding and maintenance of temporal information, and the nature of the task makes it unlikely that the temporal coding could have been accomplished by a familiarity heuristic (familiarity information

would have been too imprecise to permit the exact reproduction of 15 items).

Further research using tasks like the above indicates that the frontal deficit arises from the need to *explicitly* use temporal coding. Mangels (1997), for example, tested patients with DLPFC lesions and matched controls in a memory-for-order task like that described above, but had participants process the study list under either intentional instructions ("Remember the order of presentation of the items"), or incidental instructions ("Make a pleasantness judgement for each item"). Under intentional instructions, the frontal patients again manifested a deficit in memory for order information; but with unintentional instructions, the frontal patients performed normally on memory for order. Hence the prefrontal cortex, and the DLPFC in particular, seems to play a critical role in temporal coding and memory only when such coding is deliberate. Being 'deliberate' implies that the processes involved are controlled rather than automatic (Shiffrin and Schneider, 1977), and executive processes are routinely assumed to have this quality. Furthermore, as we are about to see, the neuroimaging studies of temporal coding and working memory involve tasks that seem to require deliberate or controlled processing.

Neuroimaging tasks of temporal coding

The patient studies just reviewed have numerous limitations: those tasks that involved only working memory may not have sufficiently taxed temporal-coding mechanisms; the studies using more demanding tasks dealt with long-term as well as short-term memory; and none of the studies implicated a particular region of prefrontal cortex (e.g., no studies used as controls patients with lesions in prefrontal cortex that spared DLPFC). These limitations have been overcome in some recent neuroimaging studies of temporal coding in normal participants.

In our earlier discussion of dissociations between storage and executive processes, we briefly discussed a PET experiment by Collette et al. (1999) that contrasted word span (recall a short string of words in order) and alpha span (recall a short string of words in alphabetic order; Fig. 4A). The critical result was that alpha span task activated DLPFC, whereas word span did not, implicating the DLPFC in temporal

A. Word Span and Alpha Span

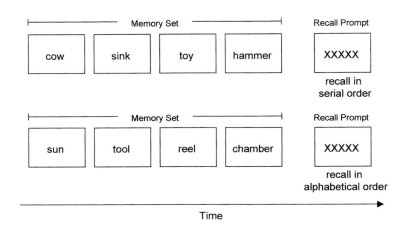

B. Input Order and Alphabetic Order

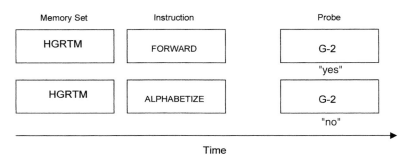

Fig. 4. Schematic drawings of temporal-coding tasks. (A) Word span and alpha span. In the first condition, participants recall words in serial order, and in the second condition, the string of words is recalled in alphabetic order (Collette et al., 1999). (B) Other memory tests for temporal sequence. In the first condition, participants are probed on their memory for the input order of the letters. In the second condition, memory for the serial position of a letter within the memory set is probed. In the final condition, the relative alphabetic position of a letter is probed (D'Esposito et al., 1999).

coding. From our current perspective, we further note that: since the task requires a reordering of the input order into an alphabetic order, it likely can be accomplished only by a mechanism that assigns temporal tags to items held in working memory, which seems like a controlled process. Hence the major result cited is in agreement with patient studies.

There are other results from this study that we did not mention earlier and that have implications for temporal coding. In addition to word span and alpha span, two other conditions were included in the experiment. In one of these, participants simply had to read the words in their order of presentation, whereas in the other condition participants had to read the words in alphabetical order. Neither of these additional conditions involved any storage load, but they still varied with respect to whether they required a reordering of the input order (read in alphabetical order) or not (read in input order). These two conditions, then, test for the neural bases of temporal coding when there is no storage load (just as some of the inhibition studies considered earlier tested for the neural bases of inhibition with minimal storage load). The important result was the alphabetic order task activated DLPFC, whereas the other task did not. This is further evidence that the DLPFC is involved in temporal coding, and that executive processes can be dissociated from storage processes.

Similar results were obtained by Wildgruber, Kischka, Ackermann et al. (1999). They required participants to covertly articulate the names of the months of the year in order (forward condition), or backward (backward condition). The backward condition requires a reordering of the order in which the months are presumably retrieved (the forward order), so it should effectively tax temporal-coding mechanisms. As expected, the backward condition (compared to the forward one) activated a number of regions in the prefrontal cortex, including DLPFC, anterior cingulate, and ventrolateral areas (BAs 45 and 47). In addition, the backward condition activated some of the standard storage regions, including posterior parietal areas (BAs 7, 39, and 40), presumably because one must hold the months (in forward order) in working memory while reordering them. Although this study led to more activations than the previous one (likely because the backward condition involves more components than the alpha task), the studies converge in showing that temporal coding activates DLPFC.

Further converging evidence for this conclusion is provided by another study that we mentioned in our earlier discussion of dissociations between storage and executive processing, the experiment by D'Esposito et al. (1999). To reiterate: a target set of six letters were presented simultaneously, followed by a brief delay, followed by a probe consisting of a letter and a digit (e.g., K3; see Fig. 4B). In one condition (input order), the participant's task was to decide whether the probe letter had appeared in the serial position designated by the probe digit ("Was K in the third position in the target set?"); in the other condition of interest (alphabetic order), the task was to determine whether the probe letter had the relative alphabetical position designated by the probe digit ("With respect to alphabetic order, was K the third letter?"). The alphabetic order condition requires a reordering of the input order, and consistent with the previous two studies, only the alphabetic order condition activated DLPFC.

Summary and comment

Our review lends support to four major generalizations about working memory and executive pro-

cesses. First, working memory has two distinct components, one corresponding to short-term storage, the other to executive processes. This distinction is supported by numerous dissociations, some involving neuroimaging results with normal participants, others involving data from neurological patients. In a number of cases, double dissociations were obtained; e.g., a neuroimaging task that requires storage but not executive processing activated the posterior parietal cortex but not the anterior prefrontal cortex, whereas a task that required the executive process of temporal coding but not storage showed the opposite pattern of activation (Collette et al., 1999); parietal patients performed below normal on storage tasks but were relatively normal on tasks like Stroop that required the executive process of attention and inhibition, whereas frontal patients manifested the exact opposite pattern (D'Esposito and Postle, 2000). Thus, while typically the short-term storage and executive components work together in the service of higher-level cognition, the two components are conceptually, behaviorally, and neurologically distinct.

Our second major generalization is that the executive process of attention and inhibition is mediated by various regions of anterior prefrontal cortex. Virtually every imaging study of inhibition — and by now there are many of them — reports activations in these regions. Furthermore, there is some evidence that the particular prefrontal regions activated differ for different kinds of inhibition. A few studies that have apparently tapped an inhibition that occurs relatively early in processing (at least prior to the preparation of a response) have found activations in ventrolateral cortex (e.g., Jonides et al., 1998), while studies that appear to involve the inhibition of prepared responses routinely activate the DLPFC and anterior cingulate (Carter et al., 1998).

The third conclusion is that the executive process of temporal coding is also mediated by regions of anterior prefrontal cortex, particularly the DLPFC. The strongest data on this point are imaging results. Tasks that require the reordering of a temporal sequence routinely activate the DLPFC (e.g., D'Esposito et al., 1999). While the patient data are less precise about the areas involved, they are in general agreement about the selective involvement of anterior prefrontal areas in temporal coding.

A fourth conclusion is implicit in the preceding three: There is reasonably good agreement between imaging results with normal participants and behavioral results with neurological patients, though of course the imaging results are generally more precise about the regions involved. We did not come across a single case in which patient data indicated a general area is necessary for accomplishment of some task — e.g., the prefrontal cortex is necessary for performance on Stroop — but imaging studies found no activation within this general area.

Finally, some comment should be made about the relation between the areas involved in the process of attention and inhibition on the one hand, and those involved in the processes of temporal coding on the other. The imaging studies of temporal coding all show activation in the DLPFC. This area is also activated in most studies of attention and inhibition, with the exception of those experiments that presumably tapped an early inhibition process (e.g., Jonides et al., 1998). Thus the DLPFC is strongly implicated in the two different executive processes. While some studies of temporal coding report areas of activation in addition to DLPFC, these additional areas are also found in some imaging studies of attention and inhibition. The upshot is that at this point in time we have little evidence for a neural dissociation between the two executive processes. This lack of a dissociation could reflect methodological problems, such as a lack of spatial resolution even in the imaging experiments, or that our conclusion of 'no difference' rests on a comparison of different studies. Alternatively, the null result may be telling us that we need to re-conceptualize our cognitive analysis of executive processes — e.g., perhaps the critical aspect of temporal coding is that it requires one to selectively attend to a representation or a temporal tag, and inhibit others, and it is this act of attention and inhibition that gives rise to the activations in DLPFC.

References

Awh E, Jonides J, Smith EE, Schumacher EH, Koeppe RA, Katz S: Dissociation of storage and rehearsal in verbal working memory: Evidence from PET. Psychological Science: 7; 25–31, 1996.

Baddeley AD, Hitch G: Working memory. In Bower GH (Ed), The Psychology of Learning and Motivation, Vol. 8. New York: Academic Press, 1974.

Banks WP, Clark HH, Lucy P: The locus of the semantic congruity effect. Journal of Experimental Psychology: Human Perception and Performance: 1; 35–47, 1975.

Barch DM, Braver TS, Nystrom LE, Forman SD, Noll DC, Cohen JD: Dissociating working memory from task difficulty in human prefrontal cortex. Neuropsychologia: 35; 1373–1380, 1997.

Bench CJ, Frith CD, Grasby PM, Friston KJ, Paulesu E, Frackowiak RSJ, Dolan J: Investigations of the functional anatomy of attention using the Stroop test. Neuropsychologia: 31; 907–922, 1993.

Brandt J: Cognitive impairment in Huntington's disease: Insight into the neuropsychology of the striatum. In Boller F, Grafman J (Eds), Handbook of Neuropsychology, Vol. 5. pp. 241–264, 1991.

Brown RG, Marsden CD: Dual task performance and processing resources in normal subjects and patients with Parkinson's disease. Brain: 114; 215–231, 1991.

Burgess N, Hitch GJ: Memory for serial order: a network model of the phonological loop and its timing. Psychological Review: 106; 551–581, 1999.

Bush G, Whalen PJ, Rosen BR, Jenike MA, McInerney SC, Rauch SL: The counting stroop: An interference task specialized for functional neuroimaging — Validation study with functional MRI. Human Brain Mapping: 6; 270–282, 1998.

Carter CS, Braver TS, Barch DM, Botvinick MM, Noll DC, Cohen JD: Anterior cingulate cortex, error detection, and the online monitoring of performance. Science: 280; 747–749, 1998.

Casey BJ, Trainor RJ, Orendi JL, Schubert AB, Nystrom LE, Giedd JN, Castellanos FX, Haxby JV, Noll DC, Cohen JD, Forman SD, Dahl RE, Rapoport JL: A developmental functional MRI study of prefrontal activation during performance of a go–no go task. Journal of Cognitive Neuroscience: 9; 835–847, 1997.

Chao LL, Knight RT: Human prefrontal lesions increase distractability to irrelevant sensory inputs. NeuroReport: 6; 1605–1610, 1995.

Chao LL, Knight RT: Contribution of human prefrontal cortex to delay performance. Journal of Cognitive Neuroscience: 10; 167–177, 1998.

Carpenter PA, Just MA, Keller TA, Eddy W, Thulborn K: Graded functional activation in the visuospatial system with the amount of task demand. Journal of Cognitive Neuroscience: 11; 9–24, 1999.

Carpenter PA, Just MA, Shell P: What one intelligence test measures: A theoretical account of the processing in the Raven Progressive Mattrices Test. Psychological Review: 97; 404–431, 1990.

Cavanagh JP: Holographic and trace-strength models of rehearsal effects in the item recognition task. Memory and Cognition: 4; 186–199, 1976.

Cohen J, McClelland JL, Dunbar K: On the control of automatic processes: a parallel distributed processing account of the Stroop effect. Psychological Review: 97; 332–361, 1990.

Cohen JD, Perlstein WM, Braver TS, Nystrom LE, Noll DC, Jonides J, Smith EE: Temporal dynamics of brain activation during a working memory task. Nature: 386; 604–608, 1997.

Cohen JD, Servan-Schreiber D: Context, cortex, and dopamine: a connectionist approach to behavior and biology in schizophrenia. Psychological Review: 99; 45–77, 1992.

Collette F, Salmon E, Van der Linden M, Chicherio C, Belleville S, Degueldre C, Delfiore G, Franck G: Regional brain activity during tasks devoted to the central executive of working memory. Cognitive Brain Research: 7; 411–417, 1999.

Corballis MC, Kirby J, Miller A: Access to elements of a memorized list. Journal of Experimental Psychology: 74; 185–190, 1972.

D'Esposito M, Aguirre GK, Zarahan E, Ballard D: Functional MRI studies of spatial and non-spatial working memory. Cognitive Brain Research: 7; 1–13, 1998.

D'Esposito M, Postle B: Neural correlates of component processes of working memory: Evidence from neuropsychological and pharmacological studies. In Attention and Performance XVIII. Cambridge, MA: MIT Press, 2000.

D'Esposito M, Postle BR, Ballard D, Lease J: Maintenance versus manipulation of information held in working memory: an fMRI study. Brain and Cognition: 41; 66–86, 1999.

Daneman M, Carpenter PA: Individual differences in working memory and reading. Journal of Verbal Learning and Verbal Behavior: 19; 450–466, 1980.

Dehaene S, Chanqeux JP: The Wisconsin Card Sorting Test: Theoretical analysis and modelling in a neuronal network. Cerebral Cortex: 1; 62–79, 1991.

Dyer FN: Stroop phenomenon and its use in the study of perceptual, cognitive, and response processes. Memory and Cognition: 1; 106–120, 1973.

Garvan H, Ross TJ, Stein EA: Right hemispheric dominance of inhibitory control: An event-related fMRI study. Proceedings of the National Academy of Sciences of the United States of America: 96; 8301–8306, 1999.

Henik A, Singh J, Beckley DJ, Rafal RD: Disinhibition of automatic word reading in Parkinson's disease. Cortex: 29; 589–599, 1993.

Jonides J: Working memory and thinking. In Smith EE, Osherson D (Eds), An Invitation to Cognitive Science, Vol. 3, 2nd ed. Cambridge, MA: MIT Press, 1995.

Jonides J, Schumacher EH, Smith EE, Lauber EJ, Awh E, Minoshima S, Koeppe RA: Verbal working memory load affects regional brain activation as measured by PET. Journal of Cognitive Neuroscience: 9; 462–475, 1996.

Jonides J, Smith EE, Marshuetz C, Koeppe RA, Reuter-Lorenz PA: Inhibition in verbal working memory revealed by brain activation. Proceedings of the National Academy of Sciences of the United States of America: 95; 8410–8413, 1998.

Klein GS: Semantic power measured through the interference of words with color naming. American Journal of Psychology: 77; 576–588, 1964.

Knight RT, Scabini D, Woods DL: Prefrontal cortex gating in auditory transmission in humans. Brain Research: 504; 338–342, 1989.

MacLeod CM: Half a century of research on the Stroop effect: an integrative review. Psychological Bulletin: 109; 163–203, 1991.

Malloy P, Bihrle A, Duffy J, Cimino C: The orbital frontal syndrome. Archives of Clinical Neuropsychology, 8, 185–201, 1993.

Mangels JA: Strategic processing and memory for temporal order in patients with frontal lobe lesions. Neuropsychology: 11; 207–221, 1997.

Miller GA: The magic number seven, plus or minus two: Some limits on our capacity for processing information. Psychological Review: 63; 81–97, 1956.

Miller GA, Galanter E, Pribram K: Plans and the Structure of Behavior. New York: Holt, 1960.

Milner B: Interhemispheric differences in the localization of psychological processes in man. British Medical Bulletin: 27; 272–277, 1971.

Milner B: Hemispheric specialization: scope and limits. In Schmitt FO, Worden FG (Eds), The Neurosciences: Third Study Program. Cambridge, MA: MIT Press, pp. 75–89, 1974.

Milner B, Corsi P, Leonard G: Frontal lobe contribution to recency judgements. Neuropsychologia: 29; 601–618, 1991.

Pardo JV, Pardo PJ, Janer KW, Raichle ME: The anterior cingulate cortex mediates processing selection in the Stroop attentional conflict paradigm. Proceedings of the National Academy of Sciences of the United States of America: 87; 256–259, 1990.

Perret E: The left frontal lobe of man and the suppression of habitual responses in verbal categorical behaviour. Neuropsychologia: 12; 323–330, 1974.

Peterson LR, Peterson M: Short-term retention of individual items. Journal of Experimental Psychology: 58; 193–198, 1959.

Ptito A, Crane J, Leonard G, Amsel R, Caramanos Z: Visual–spatial localization by patients with frontal-lobe lesions invading or sparing area 46. NeuroReport: 6; 1781–1784, 1995.

Salthouse TA, Legg S, Palmon R, Mitchell D: Memory factors in age-related differences in simple reasoning. Psychology and Aging: 5; 9–15, 1990.

Seidman LJ, Breiter HC, Goodman JM, Goldstein JM, Woodruff PWR, O'Craven K, Savoy R, Tsuang MT: A functional magnetic resonance imaging study of auditory vigilance with low and high information processing demands. Neuropsychology: 12; 505–518, 1998.

Shiffrin RM, Schneider W: Controlled and automatic human information processing: II. Perceptual learning, automatic attending, and a general theory. Psychological Review: 84; 127–190, 1977.

Shimamura AP, Janowsky JS, Squire LR: Memory for the temporal order of events in patients with frontal lobe lesions and amnesic patients. Neuropsychologia: 28; 803–813, 1990.

Smith EE: Research strategies for functional neuroimaging: a comment on the interview with R.G. Shulman. Journal of Cognitive Neuroscience: 9; 167–169, 1997.

Smith EE, Jonides J: Neuroimaging analyses of human working memory. Proceedings of the National Academy of Sciences of the United States of America: 95; 12061–12068, 1998.

Smith EE, Jonides J: Storage and executive processes in the frontal lobes. Science: 283; 1657–1661, 1999.

Smith EE, Jonides J, Koeppe RA: Dissociating verbal and spatial working memory using PET. Cerebral Cortex: 6; 11–20, 1996.

Smith EE, Jonides J, Marshuetz C, Koeppe RA: Components of verbal working memory: evidence from neuroimaging. Proceedings of the National Academy of Sciences of the United States of America: 95; 876–882, 1998.

Sternberg S: High-speed scanning in human memory. Science: 153(3736); 952–654, 1966.

Taylor SF, Kornblum S, Lauber EJ, Minoshima S, Koeppe R: Isolation of specific interference processing in the Stropp task: PET activation studies. Neuroimage: 6; 81–92, 1994a.

Taylor SF, Kornblum S, Minoshima S, Oliver LM, Koeppe R: Changes in medial cortical blood flow with a stimulus–response compatibility task. Neuropsychologia: 32; 249–255, 1994b.

Thompson-Schill SL, Jonides J, Marshuetz C, Smith EE,

D'Esposito M, Kan IP, Knight RT, Swick D: Impairments in the executive control of working memory following prefrontal damage: A case study. Poster presented at Society for Neuroscience, Miami FL, 1999.

Vendrell P, Junque C, Pujol J, Jurado MA, Molet J, Grafman J: The role of the prefrontal regions in the Stroop task. Neuropsychologia: 33; 341–352, 1995.

Wickelgren WA, Norman DA: Strength models and serial position in short-term recognition memory. Journal of Mathematical Psychology: 3; 316–347, 1966.

Wildgruber D, Kischka U, Ackermann H, Klose U, Grodd W: Dynamic pattern of brain activation during sequencing of word strings evaluated by fMRI. Cognitive Brain Research: 7; 285–294, 1999.

Yamaguchi S, Knight RT: Gating of somatosensory input by human prefrontal cortex. Brain Research: 521; 281–288, 1990.

Handbook of Neuropsychology, 2nd Edition, Vol. 7
J. Grafman (Ed)

CHAPTER 4

Age, cognition and emotion: the role of anatomical segregation in the frontal lobes

Louise H. Phillips *, Sarah E.S. MacPherson and Sergio Della Sala

Psychology Department, University of Aberdeen, Old Aberdeen AB24 2UB, UK

Introduction

Phineas Gage (as described by Harlow, 1868) is the prototypical frontal lobe patient. The overall effects of Gage's frontal lobe lesion have been described as relatively intact cognitive functioning, but impaired social and emotional functioning (e.g. Damasio, Grabowski, Frank et al., 1994). Recently, theories of cognitive aging have converged on the idea that normal adult aging affects the frontal lobes of the brain more than other areas (e.g. Moscovitch and Winocur, 1995; West, 1996). Yet the typical picture of the effects of normal aging on psychological function do not seem to match those described for Gage — older adults tend to have many impairments of cognitive function but maintain social and emotional function relatively well. In this chapter we aim to reconcile this anomaly by considering the effects of adult aging on different regions of the frontal lobes. In particular, we propose that adult aging impacts the dorsolateral prefrontal areas of the frontal lobes earlier and more severely than the ventromedial prefrontal areas, and review the neuroimaging, cognitive and behavioral data relevant to this hypothesis.

The frontal lobe hypothesis of aging

Recent neuropsychological theories of normal adult aging have converged upon the idea that the frontal lobes of the brain are affected by the process of normal adult aging both earlier than other brain areas, and at a more rapid pace (Daigneault and Braun, 1993; Mittenberg, Seidenburg, O'Leary and DiGiulio, 1989; Moscovitch and Winocur, 1995; Parkin, 1997; Raz, 1996; Shimamura, 1994; West, 1996; Whelihan and Lesher, 1985). This 'frontal theory of aging' has been very influential on recent models of cognitive and neuropsychological change with age. The theory proposes that many age-related changes in cognition are attributable to deterioration of the frontal lobes, although obviously age also affects a number of other brain regions that impact cognition, such as the hippocampus. Two main strands of evidence support the frontal theory of aging: neurophysiological findings that brain changes with age are most prominent in the frontal lobes, and claims that age changes in cognition parallel those found in patients with focal frontal lobe damage. There is now substantial agreement that the frontal lobes are involved in executive processes of cognition (e.g. Baddeley and Della Sala, 1996; Grafman, 1994; Shallice, 1988), therefore age-related changes should be most evident in cognitive tasks demanding supervisory control.

In this chapter, we provide a brief review of the cognitive and neuroanatomical evidence for age changes in the frontal lobes, then utilize recent advances in understanding about different subregions within the frontal lobes to propose a more specific dorsolateral prefrontal theory of aging.

* Corresponding author. Tel.: +44 (1224) 227-229;
E-mail: louise.phillips@abdn.ac.uk

Frontal lobe changes with age

The aging brain shows greater structural changes in the frontal lobes compared to other brain regions. Age decrements are evident in the size and number of neurons and cortical thickness (Albert, 1984; Haug and Eggers, 1991; Terry, De Teresa and Hansen, 1987). Earlier studies erroneously equated neuron density with neuron number therefore biasing the degree of age-related changes, and later interpretations suggest that the major age changes are in terms of the size of neurons in the frontal lobes (Coleman and Flood, 1987; Morrison and Hof, 1997). Age declines are also more prominent in the frontal lobes than other brain areas in terms of the density of presynaptic terminals (Masliah, Mallory, Hansen et al., 1993) as well as the quantity of normal tau protein (Mukaetova-Ladinska, Hurt and Wischik, 1995). Similarly, senile plaques, which reflect neuronal degeneration, are more numerous in the frontal lobes than elsewhere in the brain (Struble, Price, Cork and Price, 1985).

Moreover, neuroimaging studies show that the volume of the frontal lobes decreases more than that of other cerebral areas with age. For example, Coffey, Wilkinson, Parashos et al. (1992) report a frontal volume decrement equal to 0.55% per year, twice the rate of change in other areas of the brain. In a review of studies of magnetic resonance imaging (MRI), Raz (1996) found that the correlation between age and volume of cortical gray matter in the brain was more substantial for the frontal lobes than any other brain region. Studies which have examined age differences in the metabolic uptake of various regions of the brain using positron emission tomography (PET), indicate that blood flow to the frontal lobes is overwhelmingly reduced with age (Gur, Gur, Obrist et al., 1987; Shaw, Mortel, Stirling Meyer et al., 1984).

There is evidence that older adults perform poorly on cognitive tasks that are sensitive to frontal lobe damage. Mittenberg et al. (1989) examined age differences on a range of neuropsychological tests purported specifically to tap the functioning of right and left hemispheres of the brain in relation to temporal, parietal and frontal lobes. There was no age differentiation between right and left hemispheric functions. Frontal lobe measures showed stronger age-related declines than measures of temporal or parietal lobe function. Mittenberg et al. concluded that "degenerative changes in frontal lobe efficiency are the most pronounced and relatively specific sequelae of the aging process" (p. 926).

Some neuropsychological tests, such as fluency, Stroop, the Wisconsin card sort test (WCST), and the Tower of London, are commonly used to assess frontal lobe function (Lezak, 1995). Age differences are usually found on these tests, e.g. letter fluency (Phillips, 1999; Whelihan and Lesher, 1985), Tower of London (Allamanno, Della Sala, Laiacona et al., 1987; Gilhooly, Phillips, Wynn et al., 1999), the Stroop test (Boone, Miller, Lesser et al., 1990; Daigneault, Braun and Whitaker, 1992), and the WCST (Axelrod and Henry, 1992; Daigneault et al., 1992; Libon, Glosser, Malamut et al., 1994).

Age, memory and the frontal lobes

A major focus of interest in the study of the frontal lobe theory of aging has been the effect of age and frontal lobe damage on memory functions. Age differences have been found on a range of memory measures identified as sensitive to frontal lobe dysfunction. There is a large literature on age and memory in relation to the frontal lobe theory of aging. For more thorough reviews of this area see e.g. Chao and Knight, 1997; Mayes and Daum, 1997; Moscovitch and Winocur, 1995; West, 1996. Here we describe only a few of the best-known findings.

Moscovitch and Winocur (1995) review a large number of memory tasks that are impaired both by focal frontal lesions and adult aging. These tasks include remembering the temporal order of presented material, remembering the source of information, working memory, and free recall of organized material: in essence any memory task which involves strategic or inhibitory components. For example, Levine, Stuss and Milberg (1997) looked at the effects of normal adult aging and focal frontal damage on a test of conditional associative learning. In this task, there are fixed associations between members of a set of stimuli, and the participant must attempt to learn the associations through trial and error. Patients with damage to the frontal lobes, and in particular the dorsolateral regions of the prefrontal cortex, were impaired in associative learning, as were older adults.

Levine et al. (1997) argue that age-related changes in memory performance are attributable to frontal-lobe changes in the effectiveness of inhibition.

Age differences in memory may be influenced by both changes in the hippocampus and the frontal lobes (Moscovitch and Winocur, 1992). Glisky, Polster and Routhieaux (1995) report a double dissociation between age-related deficits in item memory and source memory (remembering where a particular item was learnt). Individuals who performed poorly on 'frontal lobe tests' (e.g. WCST, verbal fluency, backwards digit span) were poor at remembering the source from which information was learnt but could accurately remember item information. In contrast, those who performed poorly on 'temporal lobe tests' such as subtests from the Weschler Memory Scales and delayed recall performed well on source memory but poorly on item memory.

Age, problem-solving and the frontal lobes

Much of the research into the relationship between age and frontal lobe functioning has concentrated on memory performance. However, it is also of interest to consider the effects of age and frontal lobe dysfunction on intelligence and problem-solving. Older adults are often reported to have impaired problem-solving ability (see reviews by Phillips and Forshaw, 1998; Salthouse, 1991), as also are patients with specific frontal lobe damage (Della Sala and Logie, 1998; McCarthy and Warrington, 1990; Shallice and Burgess, 1991).

Fifty years of research into the effects of age on intelligence and reasoning support the conclusion that increasing age results in preserved 'wisdom' or crystallized intelligence, but impaired 'wit' or fluid intelligence (e.g. Cattell, 1987). Fluid abilities involving abstract novel reasoning show relatively early and steep age-related declines (e.g. Phillips and Forshaw, 1998; Salthouse, 1993). Crystallized abilities reflecting acquired knowledge usually remain stable late into old age, particularly for socio-emotional material (e.g. Blanchard-Fields, 1996; Cornelius and Caspi, 1987).

The clinical picture presented by at least some patients with focal frontal lobe damage does not seem to match the effects of old age on fluid and crystallized intelligence. Consider the following description of the prototypical frontal lobe patient, Phineas Gage, in relation to the known pattern of age changes in cognition:

> After the accident he showed no respect for social convention; ethics, in the broad sense of the term were violated ... Another important aspect of Gage's story is the discrepancy between the degenerated character and the apparent intactness of the several instruments of mind — attention, perception, memory, language and intelligence.
>
> (Damasio, 1994, p. 11).

Other patients with substantial frontal lobe damage who show a similar dissociation between impaired wisdom and relatively spared intelligence have been described subsequently (Brazzelli, Colombo, Della Sala and Spinnler, 1994; Damasio and Van Hoesen, 1983; Eslinger and Damasio, 1985; Shallice and Burgess, 1991; Rolls, Hornak, Wade and McGrath, 1994). Reviews often emphasize the intact performance of patients with frontal lobe damage on intelligence and laboratory tests along with impaired social behavior, poor judgement, and inappropriate decisions taken in real-life (Benton, 1994; Damasio and Anderson, 1993; Milner, 1995; Parker and Crawford, 1992). This pattern suggests that precisely those problem-solving abilities which may be *spared* following frontal lobe lesions (e.g. abstract reasoning abilities, fluid intelligence) are *most affected* by aging; while the problem-solving abilities which are *impaired* after some frontal lobe lesions (e.g. social decision-making, using knowledge wisely) are *least affected* by aging. How can this discrepancy be addressed by a frontal lobe theory of cognitive changes with age?

One way of reconciling these apparent anomalies between performance of older adults and patients with frontal lobe damage is to consider the architecture of the frontal lobes in more detail. The frontal lobes occupy more than a third of the human cortex, and in order to make sense of such a large and diverse area many authors have proposed that the frontal lobes be subdivided. There has recently been much interest in the idea of localization of function within specific areas of the frontal lobes (see e.g. Beardsley, 1997; Darling, Della Sala, Gray and Trivelli, 1998; Eslinger, 1999; Sarazin, Pillon, Giannakopoulos et al., 1998). Relatively few studies of the frontal lobe theory of aging have explicitly examined the neuroanatomical and functional differences apparent within the prefrontal region. We will now outline in some detail the distinction between ventromedial and dorsolateral regions of prefrontal cortex, and the effects of age on these two regions, before returning to the implications that this has for cognitive function. We propose that adult aging differentially affects the dorsolateral prefrontal regions, in comparison to the ventromedial areas (see also Phillips and Della Sala, 1998).

Ventromedial and dorsolateral regions of the prefrontal cortex

Frontal theories of aging generally assume parallel decline across the various regions of the prefrontal cortex. However, the frontal lobes are a heterogeneous formation, and differentiable architectural and functional areas can be identified within this region. There is still considerable debate as to the number and function of distinct anatomical areas in the frontal lobes. For instance, there is some converging evidence of lateralization of function, at least in terms of left-frontal encoding and right-hemisphere retrieval mechanisms (e.g. Shallice, Fletcher, Frith et al., 1991; Tulving, Kapur, Craik et al., 1994), although this distinction may not apply to older adults (Cabeza, Grady, Nyberg et al., 1997). The distinction that might be particularly relevant to aging is between dorsolateral (DL) regions and ventromedial (VM) regions (see Fig. 1). The ventromedial region encompasses the lateral orbital gyrus, the middle orbital gyrus, the medial orbital gyrus and the gyrus rectus (Eslinger, 1999), and is sometimes also referred to as the orbitofrontal region.

The cerebral cortex does not have a homogeneous cellular structure and organization. Several different codes have been proposed to designate the various cortical areas (Crosby, Humphrey and Lauer, 1962). The most generally recognized terminology applied to the different areas of the cortex is that of Brodmann (1909), who used numbers to indicate distinct regions, simply following the order in which he studied them on one single case (Rajkowska and Goldman-Rakic, 1995). Using Brodmann's terminology, the DL region is centered on areas 9 and 46, while the VM region encompasses areas 10, 11, 12, 13, 14 and 47. It is interesting to note that in Brodmann's

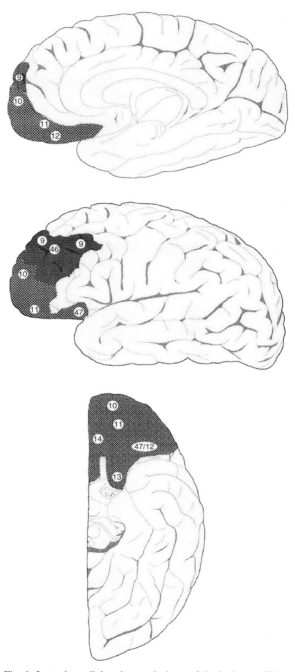

Fig. 1. Lateral, medial and ventral views of the brain specifying the location of the dorsolateral (light gray) and ventromedial (dark gray) prefrontal regions.

original map of the cerebral cortex the numbers 12, 13 and 14 are not mentioned (Braak, 1980; Gorman and Unutzer, 1993). Influenced by the work of his contemporary Vogt (1910), he later (Brodmann, 1910) inserted area 12 between area 10 and area 11 (Markowitsch, 1993). The location of areas 13 and 14, as currently agreed in the posterior orbitoventral surface of the human brain can be attributed to the work of Beck (1949) who mapped onto the human brain these two 'new' areas described by Walker (1940) in the orbital gyrus and gyrus rectus of the macaque brain. Several contemporary authors maintain that it would be more appropriate in humans to label Walker's (1940) area 12, which she described in the macaque, as area 47 (or 47/12). (For detailed discussions see: Barbas, 1995; Carmichael and Price, 1994; Eslinger, 1999; Pandya and Yeterian, 1996; Petrides and Pandya, 1994.)

The difference between the deficits following lesions in these two regions of the frontal lobes was recognized by early authors (here VM regions are referred to as basal, and DL regions as in the convexity of the frontal lobes). Cobb (1943, cited by Eslinger, 1999, p. 226) observed that " ... the basal cortex represents more the emotional integration and the convexity of the frontal lobe the intellectual integration". Years later, in reviewing the topic, Luria (1969) was unequivocal: "the syndrome arising in patients with ... convexity and basal lesions of the frontal regions are different" (p. 749). The fractionation between DL and VM areas is also seen ontogenetically: the cytoarchitectonic development of the orbital areas precedes that of the dorsolateral areas both in non-human primates and in man (Orzhekhovskaia, 1975, 1977). Indeed, studies carried out at the Moscow Brain Institute in the sixties (e.g. Kononova, 1962) showed that DL areas continue to develop after birth until the age of 12 years, and that their development finishes last in these regions compared to other areas of the brain, including orbital areas. Raz (1996) put forward the argument that those brain areas which are phylogenetically latest ("such as dorsolateral prefrontal and inferior parietal areas", p. 171) are also those which are most susceptible to the effects of aging: "the last to come (in evolution) is the first to go (in senium)" (p. 171).

VM and DL regions also differ in their cortico-subcortical connections (Petrides, 1994; Rolls,

1999a,b), indeed some authors clearly distinguish anatomical paths and functions of dorsolateral–subcortical and orbitofrontal–subcortical circuits (Fuster, 1996; Masterman and Cummings, 1997). Both VM and DL regions are richly connected to other parts of the brain (and each other). DL regions are intensely connected to primary sensory and motor regions, as well as the parietal cortex; while VM regions are heavily networked with the limbic system (Adolphs, Tranel, Bechara et al., 1996; Pandya and Yeterian, 1996; Rolls, 1996, 1999a,b). There is little doubt from studies of focal lesions in humans and monkeys, and from neuroimaging in normal adults, that these regions may have separable functions, although there is still debate as to what exactly these functions may be (Bechara, Damasio, Tranel and Anderson, 1998; Della Sala, Gray, Spinnler and Trivelli, 1998; Goldman-Rakic, 1996; Petrides, 1994; Rolls, 1996). However, there is some agreement that DL regions are involved in cognitive processes classified as 'executive functions', such as monitoring multiple events (Courtney, Ungerleider, Kell and Haxby, 1997) and abstract problem-solving (e.g., Prabhakaran, Smith, Desmond et al., 1997); while VM regions are involved in emotional processing (Rolls, 1996) and the regulation of social behavior (Anderson, Bechara, Damasio et al., 1999).

There is increasing recent evidence of the different functional domains of these regions. For example Sarazin et al. (1998) report that in patients with frontal lobe lesions cerebral glucose metabolism at rest in DL regions (but not VM regions) relates to executive test performance, while metabolism in VM regions (but not DL regions) relates to behavioral and emotional abnormalities.

The VM region of the brains receives inputs from object-processing visual areas and somatosensory areas; it is also involved in the processing of taste olfaction and autonomic regulation. However, for the sake of clarity, we will limit our discussion on the frontal theory of aging to its prominent function, that of an emotion relay. Within the DL and VM areas there may be further differentiation (Goldman-Rakic, 1996; Petrides, 1994; Rolls, 1996, 1999a,b) which is beyond the scope of the present chapter to discuss. Barbas, Gashghaei, Rempel-Clower and Xiao (2002, this volume) and Miller and Asaad (2002, this volume) discuss the properties and functions of the DL and VM regions more comprehensively.

Age effects on dorsolateral and ventromedial prefrontal regions

The majority of published studies on changes in brain neuroanatomy with age do not clearly distinguish between different regions within the frontal lobes. However, West (1996) has made the suggestion of regional differences within the prefrontal cortex in the effects of aging in passing:

> Within the prefrontal cortex, there also is some evidence suggestive of regional differences in the effects of increasing age, with the dorsolateral prefrontal region demonstrating a linear decline throughout adulthood, and the orbital prefrontal region showing evidence of a decline only during the late 7th decade and beyond.
>
> (West, 1996, p. 276).

Moreover, there are a number of studies which report the effects of age on neuronal density and size in one specific region of the frontal lobes, but differences in the methodology and sampling used in these studies makes it somewhat difficult to compare across studies.

Haug and co-workers have intensively studied age-related neuronal changes in area 11 (part of the VM prefrontal cortex) in comparison to other areas of the brain. Their findings show that in area 11 large neurons do tend to shrink with age, but that this process occurs later (after 65 years), less massively and less rapidly than in other areas, including prefrontal area 6 (Haug, 1985; Haug and Eggers, 1991; Haug, Barmwater, Eggers et al., 1983). Terry et al. (1987) report the effects of age on neurons in area 46 (DL prefrontal cortex). There was a substantial decrease in the number of counted large neurons (correlation between age and number of large neurons = −0.63), and a corresponding increase in the number of small neurons and glia (correlations with age = 0.33 for small neurons and 0.51 for glia). Terry et al. conclude that increasing age causes substantial shrinkage in the neurons of area 46. This decline appears to be linear from age 30 to 100 years, in contrast to the later and slower decline in area 11 (Haug and Eggers, 1991). At odds with these findings, a recent well-controlled MRI study (Raz, Gunning, Head et al., 1998) reported very similar decrements in the

volume of dorsolateral and orbital prefrontal regions from age 20 to 80 years (decrease = approximately 25%). However, these areas do not correspond well to our delineation of DL and VM regions: the area described as DL by Raz et al. (1998) includes Brodmann's areas 8, 10 and 45, as well as 9 and 46. Further, the orbital prefrontal region defined by Raz et al. comprises areas 11 and 47, and not the more ventral and medial regions (12, 13, 14) which are critical in determining social behavior (Rolls, 1996).

There are a large number of in vivo studies of regional cerebral activity with aging, although relatively few clearly distinguish between DL and VM areas (as defined in Fig. 1). Some evidence can be gleaned from the literature of a differential decrement of glucose uptake within the different regions of the prefrontal cortex. For example, Duara, Margolin, Roberston-Tchabo and London (1983) incidentally report (see their Table 3, p. 768) that age correlates negatively with weighted regional cerebral metabolic rates in the dorsolateral, but not in the orbitofrontal gyri. More recently, De Santi, de Leon, Convit et al. (1995) in a PET study of glucose metabolism in young and old adults conclude that "there is a stronger relationship between age and dorsal lateral frontal lobe metabolism than between age and orbital frontal lobe metabolism" (p. 367). Correspondingly, Marchal, Rioux, Petit-Taboué, et al. (1992), examining the changes in cerebral metabolic rate of oxygen in volunteers aged from 20 to 68 years, found age-related effects in nearly all the cortical gyri they analyzed, including DL prefrontal gyri, but did not find age-related changes in VM prefrontal regions. Garraux, Salmon, Degueldre et al. (1999) carried out a detailed study of frontal areas comparing PET activation during rest in 22 young and 21 old healthy adults. There were significant age effects on metabolism in dorsolateral areas 8, 9 and 46, and also other areas such as the anterior cingulate. There was no age effect on metabolism in ventral areas 10, 11 and 47. The more medial areas 12, 13, and 14 were not reported.

Therefore, there are at least some hints from studies of brain structure that aging may differentially affect the DL as opposed to VM prefrontal cortex. We will turn now to consider the effect of age on some of the psychological tests taxing the DL or the VM regions.

Cognitive–behavioral tests of dorsolateral and ventromedial function

There are few tasks for which evidence exists to indicate that they specifically tap DL or VM prefrontal function. This is partly because there have been few theoretically driven attempts to design tasks specifically testing either DL or VM prefrontal functioning in humans. Also, it is likely that these areas will often act in concert to influence performance on many complex tasks. If the functioning of these areas is taken crudely to reflect strategy generation and monitoring (DL) versus emotion and motivation (VM) it is clear that both of these functions will be present in a great many cognitive tests.

However, from the literature there is some evidence that some tests reflect the function of one area much more than the other. This evidence will be reviewed below, along with any literature on age differences on the tasks. Tests are only considered if they meet two criteria: (1) supporting evidence from lesion studies and neuroimaging of involvement of *either* DL *or* VM prefrontal regions in humans; and (2) some evidence to evaluate the effects of adult aging on the test. We do not discuss the considerable number of papers that indicate that the tests discussed below are generally sensitive to frontal lobe lesions but do not specify where in the frontal lobes such lesions occur. The prediction is that normal aging should cause poorer performance on tests sensitive to DL function but not on tests sensitive to VM functioning.

After this section, preliminary results from a study of aging and VM/DL functioning are reported looking at the effects of adult aging on six tests for which the literature provides evidence to suggest involvement of either DL or VM prefrontal function.

Functions of the DL prefrontal cortex

A number of the classic 'executive function tests' have been well validated as dependent on the functioning of the DL prefrontal cortex through neuroimaging experiments and studies of patients with focal lesions. Lesions localized to DL prefrontal regions are relatively rare, so there are comparatively few patient studies that clearly indicate the role of DL frontal regions. The tests considered here are the Wisconsin card sort task, letter fluency, Tower of London, delayed response, Stroop and recency judgement. Evidence for their selective dependence on DL rather than VM prefrontal areas is reviewed, followed by an evaluation of available evidence for the effects of adult aging on performance.

Wisconsin card sorting task

The WCST was developed to assess abstract reasoning, particularly the ability to identify abstract categories and shift cognitive set. The participant is presented with four stimulus cards with symbols that differ in shape, number and color and a pack of 128 response cards, and is asked to place each response card under one of the four stimulus cards. Only one of the dimensions is correct, color in the first instance. However, the participant is not told which dimension is correct and therefore must identify the matching card through trial and error, using the feedback provided. After 10 correct responses with color, the criterion is changed to shape without the participant's knowledge. When the sorting dimension is changed the participant must shift cognitive set to identify and attend to the new dimension.

Evidence of DL and VM involvement. Milner (1963) compared 18 patients with DL prefrontal lesions and a control group of patients with lesions elsewhere in the brain, including the VM prefrontal region, on the WCST. Patients with lesions in the DL prefrontal cortex made significantly more errors and achieved fewer sorting categories than the control group whereas, patients with VM prefrontal lesions performed as well as the control group. Furthermore, Drewe (1974) examined the performance of patients with orbital, medial or DL frontal lesions performing the WCST and found that patients with orbital damage achieved better performance on the WCST than those with frontal damage outside this area. Ahola, Vilkki and Servo (1996) did not find a significant difference in performance on the WCST between patients with frontal infarctions mainly involving the VM region and patients with nonfrontal lesions. However, group studies have indicated that DL damage does not necessarily affect WCST performance (e.g. Goldstein, Bernard, Fenwick et al., 1993). Anderson, Damasio, Jones and Tranel (1991) argue that site of lesion does not predict WCST performance, and provide evidence that some patients with severe bilateral DL damage perform relatively well on the WCST. In a recent paper Stuss, Levine, Alexander et al. (2000) argue that DL and VM lesions cause different patterns of deficits on the WCST, with DL lesions tending to cause a greater number of perseverative errors.

There is some support for the involvement of the DL prefrontal cortex in the WCST from neuroimaging studies. In a SPECT study, Rezai, Andreasen, Alliger et al. (1993) found that there was a significant increase in blood flow to the DL prefrontal cortex but not the VM prefrontal cortex during performance on the WCST. Barceló, Sanz, Molina and Rubia (1997) examined healthy volunteers performing a computerized version of the WCST and found significant event-related potential (ERP) changes in DL prefrontal cortex while participants performed the task. However, another neuroimaging study by Cantor-Graae, Warkentin, Franzén and Risberg (1993) investigating activation during the WCST in normal participants did not find significant DL activation. Other studies suggest that there is involvement of both DL and VM prefrontal cortex in the WCST. Berman, Ostrem, Randolph et al. (1995) obtained rCBF PET scans during the WCST. They found related activation in the DL, mesial, orbital and polar frontal cortex during the task.

Overall, the evidence does suggest that DL regions are involved in the WCST, particularly in determining perseverative responses. The evidence also suggests that VM regions are likely to be involved in other aspects of the task. The WCST is a multicomponent task, likely to recruit a complex network of brain areas.

Age effects. There are many studies that have demonstrated age-related decline in performance on the WCST. For example, Daigneault et al. (1992) found significant age declines from 40 years onwards on the WCST. Fristoe, Salthouse and Woodard (1997) compared the performance of young and older adults on the WCST, and found that older adults were impaired on all performance measures, even when years of education and health variables were partialled out. Experimental studies suggest that age differences on the WCST are due partially to less efficient use of feedback to determine future choices, as well as

to differences in working memory and processing speed (Fristoe et al., 1997). Nagahama, Fukuyama, Yamauchi et al. (1997) measured the rCBF activation in young and old participants while performing on the WCST. They found that both groups had significant activation in the left DL prefrontal cortex, however the activation in older participants was significantly lower than that of the younger participants. This supports the idea that any age-related deficit on the WCST may be linked to changes in the DL regions.

Letter fluency
Phonemic fluency tasks require the participant to generate words beginning with a particular letter in a short time period. Repeatedly retrieving words using a phonemic criterion is a relatively novel way of searching, so it has been argued that participants need to be able to generate and utilize effective retrieval strategies (e.g. Crowe, 1992).

Evidence of DL and VM involvement. There is evidence from neuroimaging and patient studies to support the involvement of DL prefrontal areas in letter fluency performance. Many neuroimaging studies such as those by Cantor-Graae et al. (1993); Frith et al. (1991) and Cuenod, Bookheimer, Hertz-Pannier et al. (1995); indicate significant activation in the left DL prefrontal cortex while normal volunteers perform letter fluency.

These findings are consistent with patient studies which suggest that there is a role for the DL prefrontal cortex in letter fluency. Stuss, Alexander, Hamer et al. (1998) compared the performance of a number of different patient groups on letter fluency. They found that patients with VM prefrontal damage were not impaired on letter fluency, but patients with left-sided DL prefrontal damage produced few words and made many task errors. Troyer, Moscovitch, Winocur et al. (1998) examined clustering and switching of word production on letter fluency in patients with focal brain lesions. They found that patients with left DL prefrontal lesions were impaired at switching from one retrieval strategy to another in comparison to healthy controls matched for age and gender. Evidence from patient studies suggests that lesions of VM regions do not substantially affect fluency performance (Hornak, Rolls and Wade, 1996;

Cicerone and Tanenbaum, 1997). Overall, there appears to be reasonably strong evidence of DL rather than VM frontal involvement in letter fluency tasks.

Age effects. Numerous studies have reported age deficits on tasks of letter fluency: older people tend to produce fewer words within the time limit (e.g. Daigneault et al., 1992; Tombaugh, Kozak and Rees, 1999). However, there are a number of different reasons to perform poorly on fluency tasks (Phillips, 1997), and other brain areas than DL prefrontal cortex are obviously involved in this task. Studies that have investigated qualitative production during fluency tasks indicate that age effects are not due to differences in switching strategies (Phillips, 1999; Troyer, Moscovitch and Winocur, 1997), unlike the production shown by patients with focal DL lesions. So, although there may be fluency deficits associated with DL frontal damage and age, these may not represent the qualitatively similar cognitive impairments.

Tower of London

The Tower of London task assesses the ability to plan the moves necessary to rearrange an array of colored disks on pegs from a starting position to match a predetermined position (Shallice, 1982). Some trials on the Tower of London involve making counterintuitive moves, which are necessary for solution, but do not place disks into their final goal position. Therefore, to perform the task successfully, the participant needs to look a few moves ahead and choose the appropriate sequence in which subgoals are to be reached (Gilhooly et al., 1999; Shallice, 1982; Ward and Allport, 1997).

Evidence of DL and VM involvement. There are many reports of the effects of frontal lobe damage on the Tower of London and related tasks (e.g. Shallice, 1982; Goel and Grafman, 1995). Morris, Miotto, Feigenbaum et al. (1997) report that patients with focal frontal lesions have particular difficulty in dealing with the first encounter of goal conflicts on planning tasks. Relatively few papers have reported in detail the lesions suffered by patients who perform poorly on the task. Owen, Downes, Sahakian et al. (1990) report the effects of frontal lobe lesions on planning in the TOL. They argue that there are no

differences between inferior, lateral and medial sites of lesion in terms of planning deficits. However, it is not clear how these locations relate to our current distinction between DL and VM. Morris et al. (1997) found no evidence of differences between patients with DL and VM lesions in performing the Tower of Hanoi task. However, some patients with lesions affecting VM rather than DL areas have been reported to perform normally on the TOL task (Cicerone and Tanenbaum, 1997; Hornak et al., 1996).

PET studies of TOL performance consistently indicate DL prefrontal activation. Morris, Ahmed, Syed and Toone (1993) measured rCBF of normal volunteers performing a computerized version of the TOL and found significant activation in the left DL prefrontal region and not the VM prefrontal cortex. Similarly, Owen, Doyon, Petrides and Evans (1996) and Baker, Rogers, Owen et al. (1996) found significant activation in area 9 of DL frontal cortex during TOL performance, but no significant activation of VM areas.

Patient studies are therefore equivocal about the relative roles of DL and VM regions in TOL performance, while neuroimaging studies provide a consistent picture of DL rather than VM activation while carrying out the TOL.

Age effects. Allamanno et al. (1987) found a significant deterioration in performance on the Tower of London task due to normal aging. Robbins, James, Owen et al. (1998) examined 341 participants aged 21–79 years performing the Tower of London and they found that the oldest groups solved fewer trials in the minimum moves possible. Gilhooly et al. (1999) report evidence from protocol analysis that older adults are less able to make accurate mental plans than younger adults.

The self-ordered pointing task

Petrides and Milner (1982) devised a task in which the subject has to organize, carry out and monitor a sequence of responses rather than reproduce a sequence directed by the experimenter: the self-ordered pointing task. For an example of the type of stimuli used in this task, see Fig. 2. Poor performance on the self-ordered pointing task may be due to poor organizational strategies and/or poor monitoring of responses. Patients with frontal lobe lesions perform

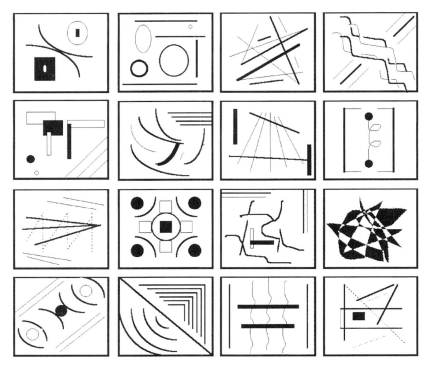

Fig. 2. An example of the type of stimuli used in the self-ordered pointing task. Participants see a series of cards with the same patterns, but on each card the patterns are in different spatial locations. The task is to choose a different pattern on each card.

worse on the task than those with temporal lobe lesions (Petrides and Milner, 1982).

Evidence of DL and VM involvement. There have been relatively few studies in humans that have specifically examined the role of DL and/or VM areas in performing the SOPT. De Zubicaray, Chalk, Rose et al. (1997) reported a patient with a DL lesion who performed poorly on the SOPT compared to 10 healthy controls. Petrides and colleagues (Petrides, Alivisatos, Meyer and Evans, 1993b; Petrides, Alivisatos, Evans and Meyer, 1993a) examined rCBF in normals while they performed a series of self-ordered pointing tasks. Petrides et al. (1993b) observed significant increases in rCBF within the right mid-DL frontal cortex when participants performed a task requiring the monitoring of self-generated responses from a set of abstract designs. They also found bilateral activation within the mid-DL frontal cortex when participants performed a task requiring the monitoring of verbal self-generated responses (Petrides et al., 1993a). The available

evidence therefore supports the idea that the SOPT involves the DL prefrontal regions, although there is little to rule out the involvement of VM regions also.

Age effects. A number of studies have found age effects on the SOPT (e.g. Daigneault et al., 1992; West, Ergis, Winocur and Saint-Cyr, 1998; Wiegersma and Meertse, 1990). There is evidence that age effects on the task begin quite early (before age 65 years, Daigneault and Braun, 1993) and are likely to be due to impaired monitoring of working memory (West et al., 1998).

Delayed response task

During the delayed response task, participants respond to a cue after a brief delay. The response is made on the basis of an internal representation maintained during the delay rather than directly in reaction to information present in the environment. The main neuropsychological processes examined by the delayed-response task are the representations of visuospatial stimuli, retention in short-term memory, retrieval, and the selection of an appropriate motor response (Teixeira-Ferreira, Vérin, Pillon et al., 1998).

Evidence of DL and VM involvement. Evidence from human patient populations supports the involvement of DL regions in the delayed response task, but also suggests that VM regions may be involved. Vérin, Partiot, Pillon et al. (1993) found that patients with DL prefrontal lesions had poor performance on delayed response tasks compared to patients with postcentral lesions and healthy controls. Freedman and Oscar-Berman (1986) found that patients with bilateral frontal lobe lesions of varying etiologies, including DL and VM were impaired on a delayed response task. Bechara et al. (1998), and Nies (1999) have also reported that patients with lesions involving only the VM prefrontal cortex are impaired on the delayed response task. Therefore, the involvement of the VM prefrontal cortex on the delayed response task cannot be excluded.

Neuroimaging studies in humans support the idea that DL prefrontal cortex is involved in the delayed response task. Goldberg, Berman, Randolph et al. (1996) and Honda, Barrett, Yoshimura et al. (1998) examined activation in the prefrontal cortex during delayed response tasks and found significant activation of the DL prefrontal cortex. Pascual-Leone and Hallett (1994) using transcranial magnetic cortical stimulation (TMS) found that DL prefrontal cortex is important for performance on the delayed response task.

Oscar-Berman (1975) found that lesions in the DL prefrontal cortex of non-human primates significantly impaired performance on delayed response tasks compared to lesions in the VM prefrontal cortex. However, animals with lesions in the VM prefrontal cortex also showed significantly impaired performance compared to intact controls. Neuronal activity has been observed in both the DL and VM prefrontal cortex of non-human primates during delayed response tasks (Hikosaka and Watanabe, 2000; Watanabe, 1996). This evidence suggests that VM activity is related to reward expectancy, while DL activity related to spatial working memory related in the delayed response task.

The evidence overall suggests a role for both DL and VM regions in delayed response tasks, although possibly for different aspects of task performance.

Age effects. We are not aware of any studies which have directly examined the role of adult aging in performance on delayed response tasks analogous to those used in patient populations. Some data relevant to this point are discussed below.

Stroop task
The aim of the Stroop task is to assess the performance of participants in a situation where stimuli can be classified according to conflicting categories. First, the participant is asked to read color names printed in black as quickly as possible. Then, the participant is asked to name the color of colored squares as quickly as possible. Finally, the color names are printed in an incompatible color of ink and the participant is asked to name the color of the ink rather than the color name. Longer latencies on the incompatible condition indicate inability to inhibit the habitual tendency to read the color name.

Evidence of DL and VM involvement. Perret (1974) found that patients with left frontal lesions demonstrated a greater progressive decrement in the Stroop test compared to the other patient groups and controls. There has subsequently been some disagreement in the literature as to whether DL or VM regions are most involved in the inhibitory functions necessary in the Stroop task. However, surprisingly, there has been little systematic attempt to look at the relationship between frontal lobe damage and Stroop test performance (Tranel, Anderson and Benton, 1994). Stuss, Benson, Kaplan et al. (1981) provide evidence that the VM prefrontal cortex is not important for performance on the Stroop task. In an article describing the Northampton Veterans Administration study, Stuss et al. (1981) compared the performance of patients with schizophrenia who had been 'treated' with prefrontal leukotomies and normal controls on the Stroop task. They found that these patients whose lesions involved the lower VM area of the frontal lobes performed similarly to normal controls on the Stroop task. Stuss (1991) also showed that patients with large bilateral VM lesions perform similarly to normal controls on tasks that include factors of interference such as the Stroop task.

Cabeza and Nyberg (1997) reviewed the PET studies of Stroop task performance up to 1995, and report rather mixed results: in some studies VM regions were activated during the color-ink naming, and in other studies DL regions were activated. Vendrell, Junqué, Pujol et al. (1995) investigated the areas of the brain involved in performance on the Stroop task using fMRI in patients with brain lesions. They found significant activation in the right DL region when participants performed on the Stroop task. Peterson, Skudlarski, Gatenby et al. (1999) carried out an fMRI study of Stroop performance and found substantial activation in very many brain areas, including both DL (areas 9, 45, 46) and VM (areas 10, 12) regions of the frontal lobes. Peterson et al. summarize their own along with a number of other Stroop activation studies, and conclude that the most consistent brain area activated is the anterior cingulate, with additional activation in dorsolateral area 46.

There are a number of methodological issues concerning how best to assess inhibition in the Stroop task, in particular, which baseline condition to compare to Stroop performance. Taylor, Kornblum, Lauber et al. (1997) compared PET activation during Stroop color-ink naming to a number of different baseline conditions, and found that the only region that consistently showed activation was the left inferior frontal gyrus (BA 44/45).

Age effects. A large number of studies have reported age differences in performance on the inhibition component of the Stroop task (e.g. Daigneault et al., 1992; Lepage, Stuss and Richer, 1999). However, there are problems in interpreting age-related slowing on this task, and some authors have argued that the major cause of age differences in performance is cognitive slowing rather than inhibitory failure (e.g. Boone et al., 1990; Uttl and Graf, 1997; Verhaeghen and De Meersman, 1998). Rabbitt (1997) outlines some of the problems associated with trying to assess 'inhibition' through tasks like the Stroop.

Recency judgement

In recency judgement tasks, participants are presented with a series of stimulus items and then have to judge which item of a pair was presented most recently. The participant is required to search back through memory for the previous occurrence of the items to compare their recency (Milner, Corsi and Leonard, 1991). Milner claimed that deficits on recency judgement may be the result of the inability to structure and segregate events in memory (Milner and Teuber, 1968). Shimamura, Janowsky and Squire (1990) found that patients with frontal lobe lesions are impaired at organizing information within a temporal context.

Evidence for DL and VM involvement. Milner et al. (1991) examined the areas of the frontal lobes important for making recency judgements. They found that patients with lesions involving the mid-DL prefrontal cortex were significantly impaired on verbal recency judgements, more so when the lesion involved the left hemisphere. In contrast, patients with lesions that did not encroach upon the DL prefrontal cortex were not impaired. More recently, Kopelman, Stanhope and Kingsley (1997) compared the performance of patients with focal frontal lesions on measures of temporal context memory. They found that patients with frontal lesions involving the DL region were significantly impaired on a recency task compared with patients with temporal lobe lesions. Patients with frontal lesions that did not involve DL regions did not show any impairment in recency judgement. However, Butters, Kaszniak, Glisky et al. (1994) were unable to identify any specific relationship between temporal context memory deficits and the frontal lesion site. Zorrilla, Aguirre, Zarahn et al. (1996) examined the neural basis of recency judgement in healthy individuals, using fMRI. They found that there was significant activation in bilateral DL prefrontal cortex and not VM prefrontal cortex during a recency task. Therefore, neuroimaging provides further support for the claim that the DL prefrontal cortex is involved in making recency judgements.

Age effects. Mittenberg et al. (1989) found that both recency for words and recency for pictures were affected by age. Fabiani and Friedman (1997) also found that younger participants performed significantly better than the older participants on both word sequencing and picture sequencing judgements.

Functions of the VM prefrontal cortex

There are relatively few specific tests for which there is evidence from patient or lesion studies of involvement of the VM prefrontal cortex. We will therefore review in general terms the types of disorder associated with VM prefrontal decline, before proceeding to the few paradigms for which there is specific empirical evidence. Unfortunately, many neuroimaging studies do not cover the areas 11, 12, 13 and 14, which it makes it more difficult to draw conclusions about the role of these areas in psychological functions.

In general terms, damage to VM regions is associated with impaired social behavior (e.g. Brazzelli et al., 1994; Saver and Damasio, 1991) poor financial management (e.g. Damasio, Tranel and Damasio, 1991), inability to maintain gainful employment (e.g. Eslinger and Damasio, 1985) and personality change (e.g Meyer and McLardy, 1948). The specific association between personality disorders and lesions in the VM regions of the frontal lobes was observed at the turn of the century by German neurologists (Knörlein, 1865; Jastrowitz, 1888; Oppenheim, 1890; Voegelin, 1897) and subsequently put to the test with large group studies (Kleist, 1934; Zangwill, 1966). Welt (1888), quoted by Markowitsch (1992) reported the case of a patient who showed dramatic changes in his character after an injury confined to the VM aspects of the frontal lobes.[1] She compared this case with several others taken from earlier literature, including the case of Phineas Gage. By comparing the available evidence she concluded that changes in character followed damage to the VM regions of the frontal lobes, in particular the right VM area. More recently, Rolls et al. (1994) reported a group of patients with VM lesions who show impaired social behavior, yet intact cognitive performance. In contrast, they report two patients with DL damage who show the opposite pattern of results: impaired cognitive performance but acceptable social behavior. The problem behaviors shown by VM patients included disinhibition, sexually inappropriate advances, boastfulness, mis-

interpretation of others' mood states, impulsivity, and aggressive behavior. Duffy and Campbell (1994) reviewed the effects of lesions in specific regions of the frontal lobes and conclude that VM lesions result in impulsivity, aggressiveness, lewdness, and lack of empathy.

Behavior and personality change such as that described in VM patients is rarely associated with normal adult aging. There are no age changes in traits such as aggressiveness (Renaud and Murray, 1996), and age does not relate to either experimental or questionnaire measures of impulsivity (Phillips and Rabbitt, 1995). Furthermore, the incidence of personality disorders decreases in the course of normal aging (Ames and Molinari, 1994).

There are some tasks involving aspects of emotional functioning and social cognition for which there is evidence available to evaluate the role of VM and DL functioning, and these are now reviewed. The tasks described are the gambling task, emotion identification, and theory of mind.

The gambling task

Patients with VM lesions, but not those with DL damage, often make disastrous financial decisions in real life (Eslinger and Damasio, 1985). Bechara, Damasio, Damasio and Anderson (1994) therefore designed a laboratory task that aims to assess decision-making: the gambling task (see Fig. 3). The aim of the gambling task is to try to win as much money as possible. Participants are presented with four decks of cards and $2000 of fake money. The participant is asked to choose one card at a time from any of the four decks, and initially, the participant wins money no matter which pile is chosen. However, as the task continues, the participant will also pay penalties, which vary with the deck and the position of the card in the deck. Two of the decks of cards are considered high risk because they have both high rewards and high penalties, whereas two decks of cards are considered low risk because they have low rewards but even lower penalties. Picking cards from the low risk decks is more profitable in the long run. In order to perform well on the task, the participant must develop a hunch that the high paying decks are 'bad' and the low paying decks are 'good'.

---

[1] The example that Welt gives of the preposterous character and outlandish personality of her patient is his lack of concern in annoying his fellow inmates by vividly complaining that in the hospital he was getting 'sour staff' to drink rather than the smooth French wines he was used to (Markowitsch, 1992). Of course, this was well before the advent of the modern public health services.

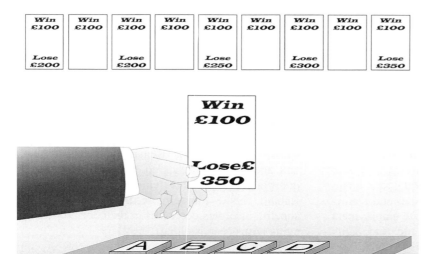

Fig. 3. The Bechara et al. (1994) gambling task. Participants must choose one card at a time from any deck (A, B, C or D) that they want, changing deck whenever they wish. Each card has printed on it a reward and penalty, which is transacted in fake money. The sequence shown at the top of this figure is the rewards and punishments associated with cards from deck A, a high-risk deck. Picking cards from decks A or B will eventually result in losses, while picking cards from decks C or D will result in gains.

Evidence of DL and VM involvement. Bechara et al. (1994) found that patients with VM frontal damage select fewer cards from the good decks and choose more from the bad decks, whereas healthy controls and patients with lesions in the occipital, temporal and DL frontal regions select more cards from the good decks and tend to avoid the bad decks. In a later study Bechara et al. (1998) showed again that patients with VM damage are impaired on the gambling task and yet patients with DL damage perform normally on the task. Furthermore, Levine, Freedman, Dawson et al. (1999) describe a patient with a right ventral frontal lesion (areas 47, 10 and 45) who performed poorly on the gambling task.

Further support for the involvement of the VM prefrontal cortex on the gambling task is apparent in studies measuring the skin conductance responses (SCRs) of participants during the gambling task (Bechara, Tranel, Damasio and Damasio, 1996; Bechara, Damasio, Tranel and Damasio, 1997). Bechara et al. (1996) claimed that somatic state activation is necessary to make the distinction between good and bad decks in the gambling task, therefore participants should show somatic state activation when they are choosing between good and bad

decks. Bechara et al. (1996) found that patients with VM prefrontal damage, as well as a healthy control group, generated reward and punishment SCRs. However, as the control group became experienced at the task, they also began to generate SCRs prior to choosing some cards, whereas the patient group did not. These anticipatory SCRs were generally higher in relation to the disadvantageous decks than the advantageous decks.

Overall, the findings suggest that performance on the gambling task depends upon the VM prefrontal cortex and not the DL prefrontal cortex.

Age effects. There is not yet any published work that we are aware of on the effects of age on the gambling task. However, on related topics, there are reported to be no age differences in efficacy of decisions on investing money (Walsh and Hershey, 1993) or decision-making in a task involving choice about which insurance policy to select (Hartley, 1989). Relevant data are also presented below.

Emotion identification tasks
In contrast to patients with DL lesions, patients with VM damage show lower affective responses to emo-

tionally loaded stimuli, have less control over their emotions, and poorer ability to interpret and empathize with the emotional states of others (Hornak et al., 1996; Bechara et al., 1996). In the facial expression version of the emotion identification task, the participant is presented with a photograph of a face, underneath which are a series of adjectives describing different emotions. The participant is instructed to choose the adjective that best describes the emotion displayed by the face in the photograph. In the vocal version of the emotion identification task, the participant is presented with a tape of emotional sounds and is instructed to choose from a list of emotions, the emotion that is best expressed by that sound.

Evidence of DL and VM involvement. Hornak et al. (1996) compared the effects of VM prefrontal damage with lesions elsewhere in the brain on the identification of facial and vocal expressions of emotion. Hornak et al. (1996) found that patients with lesions in the VM prefrontal cortex were significantly impaired on emotion identification tests compared with patients with lesions elsewhere in the brain. VM damage caused poorer ability to recognize emotional expressions, while damage to the DL prefrontal regions did not. Patients with VM lesions who performed poorly on emotion identification tasks also reported alterations in the experience of emotions.

Relatively few studies have looked at the performance of patients with focal frontal lobe lesions on emotion identification tasks, but there are a large number of neuroimaging studies. Most of these indicate activation related to emotion identification in a number of different brain regions usually including the frontal lobes. In a neuroimaging study using fMRI, George, Ketter, Gill et al. (1993) attempted to identify the areas of the brain involved in recognizing emotion in faces, and found highly significant activation in ventral areas of the prefrontal cortex during an emotion matching task, as well as less intense activation in DL areas. Other studies have found that both VM and DL areas of prefrontal cortex are involved in identifying facial and vocal expressions of emotion (Imaizumi, Mori, Kiritani et al., 1997; Sprengelmeyer, Rausch, Eysel and Przuntek, 1998). There is also evidence that VM prefrontal

cortex may be differentially depending on which basic emotion is viewed (Blair, Morris, Frith et al., 1999; Phillips, Young, Senior et al., 1997). Blair et al. (1999) found that VM prefrontal regions are involved particularly in processing expressions of anger, rather than sadness.

Age effects. There has been relatively little research on changes in perceiving emotions with age (Carstensen, Isaacowitz and Charles, 1999; Schaie and Willis, 1996; Weiner and Graham, 1989). Some theories suggest that increasing age should be associated with improvements, or at least stability, in the ability to interpret others emotions (e.g. Carstensen et al., 1999). Some studies suggest that increasing age may be associated with better ability to interpret and regulate emotion (Blanchard-Fields, Jahnke and Camp, 1995). Montepare, Koff, Zaitchik and Albert (1999) argue that age differences are generally not found in interpreting emotions from static pictures of faces, while age differences are more likely to be found in interpreting emotions from dynamic cues such as bodily movements.

Theory of mind

Theory of mind tests aim to assess the ability to accurately attribute mental states to other people (Premack and Woodruff, 1978). Impairments of 'theory of mind' have been argued to underlie the social and communicative difficulty experienced by individuals with autism (Baron-Cohen, Jolliffe, Mortimore and Robertson, 1997). In most theory of mind tests, participants are given a story to read and have to assess the state of mind of a protagonist in the story. For example, in the 'Faux Pas' test (Stone, Baron-Cohen and Knight, 1998) participants must judge whether someone described in a story has done or said something socially inappropriate, whether the person making the faux pas is aware of their error, and whether the person hearing it would feel hurt or insulted by what was said.

Evidence of DL and VM involvement. Baron-Cohen and Ring (1994) claim that the VM cortex is involved in theory of mind. Stone et al. (1998) examined five patients with damage to the left lateral frontal cortex including the dorsal areas and more ventral areas performing on the Faux Pas task. They also tested patients with bilateral damage to the VM prefrontal cortex from head trauma. Stone et al. (1998) found that both control participants and patients with DL prefrontal lesions detected and understood the faux pas. However, most of the patients with ventral prefrontal lesions made errors in detecting the faux pas.

Studies of neuroimaging provide further evidence that the frontal lobes are associated with theory of mind. A PET study by Fletcher, Happé, Frith et al. (1995) demonstrated that the left medial prefrontal cortex is activated when normal volunteers attempt to attribute mental states to others. In a later study, Happé, Ehlers, Fletcher et al. (1996) found VM activation in individuals with Asperger's syndrome, a mild variant of autism, performing the same mental state attribution task. Baron-Cohen, Ring, Moriarty et al. (1994) found that there was significant activation in the VM, but not the DL, prefrontal cortex while normal participants performed a mental state recognition task. More recently, Gallagher, Happé, Brunswick et al. (2000) found activation in the medial prefrontal cortex when normal volunteers were presented with theory of mind stories and cartoons using fMRI.

Age effects. Few studies have looked at the effects of adult aging on theory of mind abilities. Happé, Winner and Brownell (1998) presented adults of various ages with a number of short passages then asked a question in which they had to attribute mental states to the protagonist of the story. They found that the older group actually performed better than the younger adults on the theory of mind stories, which would suggest that theory of mind ability is well-preserved in old age.

Summary of dorsolateral and ventromedial test findings

Tasks that are dependent on DL prefrontal cortex could generally be construed as assessing executive function. Such tests are in the main sensitive to normal adult aging. In contrast, tasks with high VM prefrontal involvement tend to require emotionally salient decision-making, and the sparse evidence that is available suggests that such tasks are little affected by age. We therefore propose that cognitive changes

with age are better described in terms of deterioration of DL prefrontal cortex than in terms of general frontal lobe decline (see also Phillips and Della Sala, 1998). Next, we describe a pilot study to examine the effects of adult age on a selection of the tasks described above that are selectively sensitive to DL or VM prefrontal functioning.

The effect of adult aging on six tests of DL or VM prefrontal functioning: some preliminary data

To date, no studies have directly assessed the effects of adult aging on a range of DL and VM prefrontal tasks. We have some preliminary data to report on age differences on six of the tests outlined above for which there is some evidence for relatively specific involvement of either DL or VM frontal regions: (1) the Wisconsin card sort test (Grant and Berg, 1948); (2) a version of the self-ordered pointing task using abstract designs (Petrides and Milner, 1982); (3) a visuospatial delayed response task (Teixeira-Ferreira et al., 1998); (4) the gambling task (Bechara et al., 1994); (5) identifying facial expression of emotion identification using the color Ekman faces (Hornak et al., 1996); and (6) a faux-pas test of theory of mind (Stone et al., 1998). It is predicted that there will be age effects on the first three (DL) tests but not of the latter three (VM) tests.

Ninety participants were recruited: 30 aged between 20 and 39 years (mean age = 28.80, SD = 6.00, range = 20–38 years), 30 aged between 40 and 59 years (mean age = 50.27, SD = 5.65, range = 40–59 years) and 30 aged between 60 and 80 years (mean age = 69.93, SD = 5.47, range = 61–80 years). The groups were matched for vocabulary scores, and all participants met the WAIS-III UK and WMR-III UK selection criterion and did not have any history of medical or neurological problems. Graphs of the basic results are shown in Fig. 4.

A preliminary analysis of the data revealed significant age effects on all three tests sensitive to dorsolateral prefrontal dysfunction. There was a significant effect of age group on: the number of perseverative errors made on the WCST $F(2, 85) = 4.880$, $P < 0.05$; the number of errors made during the self-ordered pointing task, $F(2, 87) = 17.779$, $P < 0.05$; and also the number of errors made during the delayed response task $F(2, 87) = 6.522$, $P < 0.05$.

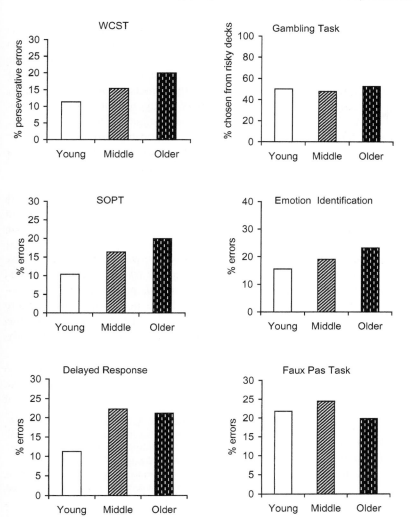

Fig. 4. Performance of young, middle and old participants on the six tests of dorsolateral and ventromedial prefrontal functioning. WCST = Wisconsin card sort test; SOPT = self-ordered pointing task.

In contrast, there were no age effects found on two of the tests sensitive to ventromedial prefrontal dysfunction: number of cards chosen from risky decks in the gambling task $F(2, 87) = 0.916$, NS; and number of errors made in identifying socially inappropriate responses in the faux pas task $F(2, 86) = 0.961$, NS. However, there was an effect of age on overall accuracy on the emotion identification task $F(2, 87) = 3.414$, $P < 0.05$. Inspection of the effects of age on identifying particular types of emotion showed that out of the seven types of emotion portrayed, an age impairment was only found on

labeling 'sadness', with no age differences in identifying the other six emotion states. This may be due to different interpretation of the verbal labels used for some of the emotions by young and old adults.

These results suggest that there may be a distinction that can be made in terms of the effects of normal adult aging on tests of DL and VM prefrontal function. More detailed analysis of test performance, and in particular an attempt to understand the reason underlying any age differences on a particular test, will allow a richer picture of the nature of age deficits on the tasks.

Conclusions

The distinction between the effects of dorsolateral and ventromedial prefrontal lesions on human behavior has often been made in the clinical literature. However, this distinction has rarely been applied to the process of normal adult aging, despite intense interest in the effects of age on the frontal lobes. There are probably no tests which specifically tap only DL or VM functioning: the involvement of DL regions in cognitive control processes and VM regions in emotional and motivational processes mean that both are likely to be involved in a range of different cognitive tasks. Distinctions can, however, be made between some tests that are largely sensitive to DL but not VM damage (e.g. the self-ordered pointing test) and others that are largely sensitive to VM but not DL damage (e.g. the gambling task). Also, future analyses may help to distinguish qualitative patterns of performance indicative of VM or DL involvement in particular tasks (e.g. working memory versus reward anticipation in the delayed response task). Most 'frontal lobe tests' are multicomponential, and so it is only by detailed analysis of performance that underlying reasons for poor performance can be worked out (Phillips, 1997, 1999).

In the current chapter we argue that:

(1) Neuroimaging and histological evidence supports the hypothesis that DL prefrontal regions are impacted by normal aging earlier, and more severely, than VM prefrontal regions.

(2) The clinical picture seen in normal aging matches more closely the type of deficits seen in patients with DL lesions (i.e. executive and problem-solving dysfunction) rather than the deficits seen in VM patients (i.e. social and emotional dysfunction).

(3) Evidence from the literature and the current reported pilot study indicates substantial age-related decline on most tasks sensitive to DL function, i.e. the Wisconsin card sort test, self-ordered pointing, Stroop, fluency, Tower of London, recency judgement. However, there must be some caution in interpreting these results because the reasons for age-related changes and DL frontal involvement may be different. All of these executive tests are relatively complex, involving a number of different information processing components, and there are correspondingly many different ways to fail these tests.

(4) Evidence from the literature is much more scant on paradigms which might assess VM prefrontal functions, and the effects of age on such tests. The preliminary study reported here suggests that there are no age effects on tests of gambling, theory of mind (supporting a similar finding by Happé et al., 1998), and identification of all basic emotions with the exception of sadness. Further studies will explore in more detail the nature of such deficits in patients with VM lesions, and qualitative similarities or differences between the effects of VM lesions and the effects of normal aging.

From the current review, we conclude that there is growing empirical support for the differential functional roles of VM and DL prefrontal areas in cognition and emotion. There is also tentative support from brain imaging and behavioral studies for the hypothesis that normal adult aging affects primarily DL but not VM regions of the prefrontal cortex.

References

Adolphs R, Tranel D, Bechara A, Damasio H, Damasio AR: Neuropsychological approaches to reasoning and decision-making. In Damasio AR (Ed), Neurobiology of Decision-Making. Berlin: Springer Verlag, pp. 157–179, 1996.

Ahola K, Vilkki J, Servo A: Frontal tests do not detect frontal infarctions after ruptured intracranial aneurysm. Brain and Cognition: 31; 1–16, 1996.

Albert ML: Clinical Neurology of Aging. New York: Oxford University Press, 1984.

Allamanno N, Della Sala S, Laiacona M, Pasetti C, Spinnler H: Problem solving ability in aging and dementia: normative data on a non-verbal test. Italian Journal of Neurological Sciences: 8; 111–120, 1987.

Ames A, Molinari V: Prevalence of personality disorders in community living elderly. Journal of Geriatric Psychiatry and Neurology: 7; 189–194, 1994.

Anderson SW, Bechara A, Damasio H, Tranel D, Damasio AR: Impairment of social and moral behavior related to early damage in human prefrontal cortex. Nature Neuroscience: 2; 1032–1037, 1999.

Anderson SW, Damasio H, Jones RD, Tranel D: Wisconsin Card Sorting Test performance as a measure of frontal lobe damage. Journal of Clinical and Experimental Neuropsychology: 13; 909–922, 1991.

Axelrod BN, Henry RR: Age-related performance on the Wisconsin Card Sorting, Similarities and Controlled Oral Word

Association tests. The Clinical Neuropsychologist: 6; 16–26, 1992.

Baddeley A, Della Sala S: Working memory and executive control. Philosophical Transactions of the Royal Society of London Series B: Biological Sciences: 351; 1397–1401, 1996.

Baker SC, Rogers RD, Owen AM, Frith CD, Dolan RJ, Frackowiak RSJ, Robbins TW: Neural systems engaged by planning — a Pet study of the Tower-of-London task. Neuropsychologia: 34; 515–526, 1996.

Barbas H: Anatomic basis of cognitive–emotional interactions in the primate prefrontal cortex. Neuroscience and Biobehavioral Reviews: 19; 499–510, 1995.

Barbas H, Gashghaei HT, Rempel-Clower NL, Xiao D: Anatomic basis of functional specialization in prefrontal cortices in primates. In Boller F, Grafman J (Eds), Handbook of Neuropsychology, 2nd edn, Vol. 7. Amsterdam: Elsevier, pp. 1–27, 2002.

Barceló F, Sanz M, Molina V, Rubia FJ: The Wisconsin Card Sorting Test and the assessment of frontal function: a validation study with event-related potentials. Neuropsychologia: 35; 399–408, 1997.

Baron-Cohen S, Jolliffe T, Mortimore C, Robertson M: Another advanced test of theory of mind: evidence from very high functioning adults with autism or asperger syndrome. Journal of Child Psychology and Psychiatry: 38; 813–822, 1997.

Baron-Cohen S, Ring H: A model of the mindreading system: neuropsychological and neurobiological perspectives. In Lewis C, Mitchell P (Eds), Children's Early Understanding of Mind. Hove: Lawrence Erlbaum, pp. 183–207, 1994.

Baron-Cohen S, Ring H, Moriarty J, Schmitz B, Costa D, Ell P: Recognition of mental state terms. Clinical findings in children with autism and a functional neuroimaging study of normal adults. British Journal of Psychiatry: 165; 640–649, 1994.

Beardsley T: The machinery of thought. Scientific American, August, pp. 58–63, 1997.

Bechara A, Damasio AR, Damasio H, Anderson SW: Insensitivity to future consequences following damage to human prefrontal cortex. Cognition: 50; 7–15, 1994.

Bechara A, Damasio H, Tranel D, Damasio AR: Deciding advantageously before knowing the advantageous strategy. Science: 275; 1293–1295, 1997.

Bechara A, Damasio H, Tranel D, Anderson SW: Dissociation of working memory from decision making within the human prefrontal cortex. The Journal of Neuroscience: 18; 428–437, 1998.

Bechara A, Tranel D, Damasio H, Damasio AR: Failure to respond autonomically to anticipated future outcomes following damage to the prefrontal cortex. Cerebral Cortex: 6; 215–225, 1996.

Beck E: A cytoarchitectural investigation into the boundaries of cortical areas 13 and 14 in the human brain. Journal of Anatomy: 83; 147–155, 1949.

Benton A: The frontal lobes: A historical sketch. In Boller F, Spinnler H, Hendler JA (Eds), Handbook of Neuropsychology. Vol 9. The Frontal Lobes. Amsterdam: Elsevier, pp. 3–16, 1994.

Berman, Ostrem, Randolph, Gold, Goldberg, Coppola: Carson,

Herscovitch and Weinberger: Physiological activation of a cortical network during performance of the Wisconsin Card Sorting Test: A positron emission tomography study. Neuropsychologia: 33; 1027–1046, 1995.

Blair RJR, Morris JS, Frith CD, Perrett DI, Dolan RJ: Dissociable neural responses to facial expressions of sadness and anger. Brain: 122; 883–893, 1999.

Blanchard-Fields F: Social cognitive development in adulthood and aging. In Blanchard-Fields F, Hess TM (Eds), Perspectives on Cognitive Change in Adulthood and Aging. New York: McGraw Hill, pp. 454–487, 1996.

Blanchard-Fields F, Jahnke HC, Camp C: Age differences in problem-solving style: The role of emotional salience. Psychology and Aging: 10; 173–180, 1995.

Boone KB, Miller BL, Lesser IM, Hill E, D'Elia L: Performance on frontal lobe tests in healthy older adults. Developmental Neuropsychology: 6; 215–223, 1990.

Braak H: Architectonics of the Human Telencephalic Cortex. New York: Springer-Verlag, 1980.

Brazzelli M, Colombo N, Della Sala S, Spinnler H: Spared and impaired cognitive abilities after bilateral frontal damage. Cortex: 30; 27–51, 1994.

Brodmann K: Vergleichende Lokalisationslehre der Grobhirnrinde in Ihren Prinzipien Dargestellt auf Grund des Zellenbaues. Leipzig: Barth, 1909.

Brodmann K: Feinere anatomie des grosshirns. In Lewandowsky M (Ed), Handbuch der Neurologie, Vol 1: Allgemeine Neurologie (Part 1) Berlin: Springer, pp. 206–307, 1910.

Butters MA, Kaszniak AW, Glisky EL, Eslinger PJ, Schacter DL: Recency discrimination deficits in frontal lobe patients. Neuropsychology: 8; 343–353, 1994.

Cabeza R, Grady CL, Nyberg L, McIntosh AR, Tulving E, Kapur S, Jennings JM, Houle S, Craik FIM: Age-related differences in neural activity during memory encoding and retrieval: a positron emission tomography study. The Journal of Neuroscience: 17; 391–400, 1997.

Cabeza R, Nyberg L: Imaging cognition: an empirical review of PET studies with normal subjects. Journal of Cognitive Neuroscience: 9; 1–26, 1997.

Cantor-Graae E, Warkentin S, Franzén G, Risberg J: Frontal lobe challenge: a comparison of activation procedures during rCBF measurements in normal subjects. Neuropsychiatry, Neuropsychology, and Behavioral Neurology: 6; 83–92, 1993.

Carmichael ST, Price JL: Architectonic subdivision of the orbital and medial prefrontal cortex in the macaque monkey. Journal of Comparative Neurology: 346; 366–402, 1994.

Carstensen LL, Isaacowitz DM, Charles ST: Taking time seriously: A theory of socioemotional selectivity. American Psychologist: 54; 165–181, 1999.

Cattell RB: Intelligence: Its Structure, Growth and Action. Amsterdam: North-Holland, 1987.

Chao LL, Knight RT: Age-related prefrontal alterations during auditory memory. Neurobiology of Aging: 18; 87–95, 1997.

Cicerone KD, Tanenbaum LN: Disturbance of social cognition after traumatic orbitofrontal brain injury. Archives of Clinical Neuropsychology: 12; 173–188, 1997.

Coffey CE, Wilkinson WE, Parashos IA, Soady SAR, Sullivan

RJ, Patterson IJ, Figiel GS, Webb MC, Spritzer CE, Djang WT: Quantitative cerebral anatomy of the aging human brain: a cross-sectional study using magnetic resonance imaging. Neurology: 42; 527–536, 1992.

Coleman PD, Flood DG: Neuron numbers and dendritic extent in normal aging and Alzheimer's disease. Neurobiology of Aging: 8; 521–545, 1987.

Cornelius SW, Caspi A: Everyday problem-solving in adulthood and old age. Psychology and Aging: 2; 144–153, 1987.

Courtney SM, Ungerleider LG, Kell K, Haxby JV: Transient and sustained activity in a distributed neural system for human working memory. Nature: 386; 608–611, 1997.

Crosby EC, Humphrey T, Lauer EW: Correlative Anatomy of the Nervous System. New York: Macmillan, pp. 434–441, 1962.

Crowe SF: Dissociation of two frontal lobe syndromes by a test of verbal fluency. Journal of Clinical and Experimental Neuropsychology: 14; 327–339, 1992.

Cuenod CA, Bookheimer SY, Hertz-Pannier L, Zeffiro TA, Theodore WH, Le Bihan D: Functional MRI during word generation, using conventional equipment: a potential tool for language localization in the clinical environment. Neurology: 45; 1821–1827, 1995.

Daigneault S, Braun CMJ: Working memory and the self-ordered pointing task: further evidence of early prefrontal decline in normal aging. Journal of Clinical and Experimental Neuropsychology: 15; 881–895, 1993.

Daigneault S, Braun CMJ, Whitaker HA: Early effects of normal aging on perseverative and non-perseverative prefrontal measures. Developmental Neuropsychology: 8; 99–114, 1992.

Damasio AR: Descartes' Error: Emotion, Reason and the Human Brain. New York: Grosset/Putnam, 1994.

Damasio AR, Anderson SW: The frontal lobes. In Heilman KM, Valenstein E (Eds), Clinical Neuropsychology. New York: Oxford University Press, pp. 409–459, 1993.

Damasio AR, Tranel D, Damasio H: Somatic markers and the guidance of behavior. In Levin HS, Eisenberg HM, Benton AL (Eds), Clinical Neuropsychology, 2nd ed. New York: Oxford University Press, pp. 339–376, 1991.

Damasio AR, Van Hoesen GW: Emotional disturbances associated with focal lesions of the frontal lobe. In Satz P (Ed), Neuropsychology of Human Emotion. New York: Guilford Press, pp. 85–110, 1983.

Damasio H, Grabowski T, Frank R, Galaburda AM, Damasio AR: The return of Phineas Gage: clues about the brain from the skull of a famous patient. Science: 264; 1102–1105, 1994.

Darling S, Della Sala S, Gray C, Trivelli C: Putative functions of the prefrontal cortex: historical perspectives and new horizons. In Mazzoni G, Nelson TO (Eds), Metacognition and Cognitive Neuropsychology. Hove: Lawrence Erlbaum, 1998.

De Santi S, de Leon MJ, Convit A, Tarshish C, Rusinek H, Tsui WH: Sinaiko E, Wang, G.-J., Bartlet E, Volkow N: Age-related changes in brain: II. Positron emission tomography of frontal and temporal lobe glucose metabolism in normal subjects. Psychiatric Quarterly: 66; 357–370, 1995.

De Zubicaray GI, Chalk JB, Rose SE, Semple J, Smith GA: Deficits on self ordered tasks associated with hyperostosis frontalis interna. Journal of Neurology, Neurosurgery, and Psychiatry: 63; 309–314, 1997.

Della Sala S, Gray C, Spinnler H, Trivelli C: Frontal lobe functioning in man: the riddle revisited. Archives of Clinical Neuropsychology: 13; 1–19, 1998.

Della Sala S, Logie RH: Dualism down the drain: thinking in the brain. In Logie RH, Gilhooly KJ (Eds), Working Memory and Thinking. Hove, Psychology Press, pp. 45–66, 1998.

Drewe EA: The effect of type and area of brain lesion on Wisconsin Card Sorting Test performance. Cortex: 10; 159–170, 1974.

Duara R, Margolin RA, Roberston-Tchabo EA, London ED: Schwartz M, Renfrew JW et al.: Cerebral glucose utilization, as measured with positron emission tomography in 21 resting healthy men between the ages of 21 and 83. Brain: 106; 761–775, 1983.

Duffy JD, Campbell JJ: The regional prefrontal syndromes: a theoretical and clinical overview. Journal of Neuropsychiatry and Clinical Neurosciences: 6; 379–387, 1994.

Eslinger PJ, Damasio AR: Severe impairment of higher cognition after bilateral ablation: patient EVR. Neurology: 35: 1731–1741, 1985.

Eslinger PJ: Orbital frontal cortex: historical and contemporary views about its behavioral and physiological significance. An introduction to special topic papers: part 1. Neurocase: 5; 225–229, 1999.

Fabiani M, Friedman D: Dissociations between memory for temporal order and recognition memory in aging. Neuropsychologia: 35; 129–141, 1997.

Fletcher P, Happé F, Frith U, Baker SC, Dolan RJ, Frackowiak RSJ, Frith CD: Other minds in the brain: a functional imaging study of 'theory of mind' in story comprehension. Cognition: 57; 109–128, 1995.

Freedman M, Oscar-Berman M: Bilateral frontal lobe disease and selective delayed-response deficits in humans. Behavioral Neuroscience: 100; 337–342, 1986.

Fristoe NM, Salthouse TA, Woodard JL: Examination of age-related deficits on the Wisconsin Card Sorting Test. Neuropsychology: 11; 428–436, 1997.

Frith CD, Friston KJ, Liddle PF, Frackowiak RSJ: A PET study of word finding. Neuropsychologia: 29; 1137–1148, 1991.

Fuster JM: The Prefrontal Cortex. Raven Press: New York, 1996.

Gallagher HL, Happé F, Brunswick N, Fletcher PC, Frith U, Frith CD: Reading the mind in cartoons and stories: an fMRI study of 'theory of mind' in verbal and nonverbal tasks. Neuropsychologia: 38; 11–21, 2000.

Garraux C, Salmon E, Degueldre C, Lemaire C, Laureys S, Franck G: Comparison of impaired subcortico-frontal metabolic networks in normal aging, subcortico-frontal dementia, and cortical frontal dementia. NeuroImage: 10; 149–162, 1999.

George MS, Ketter TA, Gill DS, Haxby JV, Ungerleider LG, Herscovitch P, Post RM: Brain regions involved in recognising facial emotion or identity: an oxygen-15 PET study. Journal of Neuropsychiatry: 5; 384–394, 1993.

Gilhooly KJ, Phillips LH, Wynn VE, Logie RH, Della Sala S:

Planning processes and age in the 5 disc Tower of London task. Thinking and Reasoning: 5; 339–361, 1999.

Glisky EL, Polster MR, Routhieaux BRC: Double dissociation between item and source memory. Neuropsychology: 9; 229–235, 1995.

Goel V, Grafman J: Are the frontal lobes implicated in 'planning' functions? Interpreting data from the Tower of Hanoi. Neuropsychologia: 33; 623–642, 1995.

Goldberg TE, Berman KF, Randolph C, Gold JM, Weinberger DR: Isolating the mnemonic component in spatial delayed response: a controlled PET ^{15}O-labeled water regional cerebral blood flow study in normal humans. NeuroImage: 3; 69–78, 1996.

Goldman-Rakic PS: The prefrontal landscape: implications of functional architecture for understanding human mentation. Philosophical transactions of the Royal Society of London Series B: Biological Sciences: 351; 1445–1453, 1996.

Goldstein LH, Bernard S, Fenwick PBC, Burgess PW, McNeil J: Unilateral frontal lobectomy can produce strategy application disorder. Journal of Neurology, Neurosurgery, and Psychiatry: 56; 274–277, 1993.

Gorman DG, Unutzer J: Brodmann's 'missing' numbers. Neurology: 43; 226–227, 1993.

Grafman J: Alternative frameworks for the conceptualization of prefrontal lobe functions. In Boller F, Spinnler H, Hendler JA (Eds), Handbook of Neuropsychology, Vol 9. The Frontal Lobes. Amsterdam: Elsevier, pp. 187–202, 1994.

Grant DA, Berg EA: A behavioral analysis of degree of reinforcement and ease of shifting to new responses in a Weigl-type card sorting problem. Journal of Experimental Psychology: 38; 404–411, 1948.

Gur RC, Gur RE, Obrist WD, Skolnick BE, Reivich M: Age and regional cerebral blood flow at rest and during cognitive activity. Archives of General Psychiatry: 44; 617–621, 1987.

Happé FGE, Ehlers S, Fletcher P, Frith U, Johansson M, Gillberg C: Dolan R, Frackowiak R, Frith C: 'Theory of mind' in the brain. Evidence from a PET scan study of Asperger syndrome. NeuroReport: 8; 197–201, 1996.

Happé FGE, Winner E, Brownell H: The getting of wisdom: theory of mind in old age. Developmental Psychology: 34; 358–362, 1998.

Harlow JM: Recovery from the passage of an iron bar through the head. Publications of the Massachusetts Medical Society: 2; 327–347. 1868.

Hartley AA: The cognitive ecology of problem-solving. In Poon L, Rubin D, Wilson B (Eds), Everyday Cognition in Adulthood and Late Life. New York: Cambridge University Press, pp. 300–329, 1989.

Haug H: Are neurons of the human cerebral cortex really lost during aging? A morphometric examination. In Traber J, Gispen WH (Eds), Senile Dementia of the Alzheimer Type. Berlin: Springer-Verlag, pp. 150–163, 1985.

Haug H, Barmwater U, Eggers R, Fischer D, Kühl S, Sass, N.-L.: Anatomical changes in the aging brain: Morphometric analysis of the human prosencephalon. In Cervós-Navarro J, Sarkander H-I (Eds), Brain Aging: Neuropathology and Neuropharmacology. New York: Raven Press, pp. 1–12, 1983.

Haug H, Eggers R: Morphometry of the human cortex cerebri and corpus striatum during aging. Neurobiology of Aging: 12; 336–338, 1991.

Hikosaka K, Watanabe M: Delay activity of orbital and lateral prefrontal neurons of the monkey varying with different rewards. Cerebral Cortex: 10; 263–271, 2000.

Honda M, Barrett G, Yoshimura N, Sadato N, Yonekura Y, Shibasaki H: Comparative study of event-related potentials and positron emission tomography activation during a paired-associate memory paradigm. Experimental Brain Research: 119; 103–115, 1998.

Hornak J, Rolls ET, Wade D: Face and voice expression identification in patients with emotional and behavioural changes following ventral frontal lobe damage. Neuropsychologia: 34; 247–261, 1996.

Imaizumi S, Mori K, Kiritani S, Kawashima R, Sugiura M, Fukuda H, Itoh K, Kato T, Nakamura A, Hatano K, Kojima S, Nakamura K: Vocal identification of speaker and emotion activates different brain regions. NeuroReport: 8; 2809–2812, 1997.

Jastrowitz M: Beiträge zur localisation im grosshirn und über deren praktische verwerthung [Contributions to localisation in the cerebrum and their practical evaluation]. Deutsche Medicinische Wochenschrift: 14; 81–83, 108–112, 1888.

Kleist K: Gehirnpathologie. Vornehmlich auf Grund der Kriegserfahrungen [Brain pathology. Based on war experiences]. Leipzig: Barth, 1934.

Knörlein (n.n.g.): Krankengeschichte und sectionsbefund eines basalen hirntumors [History and post-mortem findings in a case of tumor in the basal brain]. Allgemeine Wiener Medicinische Zeitschrift: 15; 250–252, 1865.

Kononova EP: The Frontal Region of the Brain. Leningrad: Medgiz Press, 1962.

Kopelman MD, Stanhope N, Kingsley D: Temporal and spatial context memory in patients with focal frontal, temporal lobe, and diencephalic lesions. Neuropsychologia: 35; 1533–1545, 1997.

Lepage M, Stuss DT, Richer F: Attention and the frontal lobes: a comparison between normal aging and lesion effect. Brain and Cognition: 39; 63–66, 1999.

Levine B, Freedman M, Dawson D, Black S, Stuss DT: Ventral frontal contribution to self-regulation: convergence of episodic memory and inhibition. Neurocase: 5; 263–275, 1999.

Levine B, Stuss DT, Milberg WP: Effects of aging on conditional associative learning: process analyses and comparison with focal frontal lesions. Neuropsychology: 11; 367–381, 1997.

Lezak MD: Neuropsychological Assessment. 3rd ed. New York: Oxford University Press, 1995.

Libon DJ, Glosser G, Malamut BL, Kaplan E, Goldberg E, Swanson R, Sands LP: Age, executive functions and visuospatial functioning in healthy older adults. Neuropsychology: 8; 38–43, 1994.

Luria AR: Frontal lobe syndrome. In Vinken PJ, Bruyn GW (Eds), Localization in Clinical Neurology. Amsterdam: North-Holland, pp. 725–757, 1969.

Markowitsch HJ: Intellectual Functions and the Brain. An Historical Perspective. Göttingen: Hogrefe and Huber Publishers, 1992.

Markowitsch HJ: Brodmann's numbers. Letters, Neurology: 43; 1863–1864, 1993.

Marchal G, Rioux P, Petit-Taboué, M.-C., Sette G, Travère, J.-M., Le Poec C, Courtheoux P, Derlon, J.-M., Baron, J.-C.: Regional cerebral oxygen consumption, blood flow, and blood volume in heathy human aging. Archives of Neurology: 49; 1013–1020, 1992.

Masliah E, Mallory M, Hansen L, DeTeresa R, Terry R: Quantitative synaptic alterations in the human neocortex during normal aging. Neurology: 43; 192–197, 1993.

Masterman DL, Cummings JL: Frontal–subcortical circuits: the anatomic basis of executive, social and motivated behaviors. Journal of Psychopharmacology: 11; 107–114, 1997.

Mayes AR, Daum I: How specific are the memory and other cognitive deficits caused by frontal lobe lesions. In Rabbitt PMA (Ed), Methodology of Frontal and Executive Function. Hove: Psychology Press, pp. 155–176, 1997.

McCarthy RA, Warrington EK: Cognitive Neuropsychology. London: Academic Press, 1990.

Meyer A, McLardy T: Posterior cuts in prefrontal leucotomy: a clinico-pathological study. The Journal of Medical Science: 94; 555–564, 1948.

Miller EK, Asaad WF: The prefrontal cortex: conjunction and cognition. In Boller F, Grafman J (Eds), Handbook of Neuropsychology, 2nd edn, Vol. 7. Amsterdam: Elsevier, pp. 29–54, 2002.

Milner B: Effects of different brain lesions on card sorting. Archives of Neurology: 9; 90–100, 1963.

Milner B: Aspects of human frontal lobe function. In Jasper HH, Riggio S, Goldman-Rakic PS (Eds), Epilepsy and the Functional Anatomy of the Frontal Lobe. New York: Raven Press, pp. 67–84, 1995.

Milner B, Corsi P, Leonard G: Frontal-lobe contribution to recency judgements. Neuropsychologia: 29; 601–618, 1991.

Milner B, Teuber HL: Alteration of perception and memory in man: reflections on methods. In Weiskrantz L (Ed), Analysis of Behavioural Change. Harper and Row, New York, pp. 268–375, 1968.

Mittenberg W, Seidenburg M, O'Leary DS, DiGiulio DV: Changes in cerebral functioning associated with normal aging. Journal of Clinical and Experimental Neuropsychology: 11; 918–932, 1989.

Montepare J, Koff E, Zaitchik D, Albert M: The use of body movements and gestures as cues to emotions in younger and older adults. Journal of Nonverbal Behavior: 23; 133–152, 1999.

Morris RG, Ahmed S, Syed GM, Toone BK: Neural correlates of planning ability: frontal lobe activation during the Tower of London test. Neuropsychologia: 31; 1367–1378, 1993.

Morris RG, Miotto EC, Feigenbaum JD, Bullock P, Polkey CE: The effect of goal–subgoal conflict on planning ability after frontal- and temporal-lobe lesions in humans. Neuropsychologia: 35; 1147–1157, 1997.

Morrison JH, Hof PR: Life and death of neurons in the aging brain. Science: 278; 412–418, 1997.

Moscovitch M, Winocur G: The neuropsychology of memory and aging. In Craik FIM, Salthouse TA (Eds), The Handbook of Aging and Cognition. Hillsdale, NJ: Erlbaum, pp. 315–372, 1992.

Moscovitch M, Winocur G: Frontal Lobes, memory, and aging. Annals of the New York Academy of Sciences: 769; 119–150, 1995.

Mukaetova-Ladinska EB, Hurt J, Wischik CM: Biological determinants of cognitive change in normal aging and dementia. International Review of Psychiatry: 7; 399–417, 1995.

Nagahama Y, Fukuyama H, Yamauchi H, Katsumi Y, Magata Y, Shibasaki H, Kimura J: Age-related changes in cerebral blood flow activation during a card sorting test. Experimental Brain Research: 114; 571–577, 1997.

Nies KJ: Cognitive and social-emotional changes associated with mesial orbitofrontal damage: assessment and implications for treatment. Neurocase: 5; 313–324, 1999.

Oppenheim H: Zur pathologie der grosshirngeschwülste [On the pathology of cerebral tumors]. Archiv für Psychiatrie und Nervenkrankheiten: 21; 560–587, 705–745, 1890.

Oscar-Berman M: The effects of dorsolateral-frontal and ventrolateral-orbitofrontal lesions on spatial discrimination learning and delayed response in two modalities. Neuropsychologia: 13; 237–246, 1975.

Orzhekhovskaia NS: Comparative study of formation of the frontal cortex of the brain of the monkeys and man in ontogenesis. Arkhiv Anatomii Gistologii i Embriologii: 68; 43–49, 1975.

Orzhekhovskaia NS: Comparison of the field formation in the frontal area during prenatal period in macaca (macacus rhesus, s. Macaca mulatta) and man. Arkhiv Anatomii Gistologii i Embriologii: 72; 32–38, 1977.

Owen AM, Downes JJ, Sahakian BJ, Polkey CE, Robbins TW: Planning and spatial working memory following frontal lobe lesions in man. Neuropsychologia: 28; 1021–1034, 1990.

Owen AM, Doyon J, Petrides M, Evans AC: Planning and spatial working memory: a positron emission tomography study in humans. European Journal of Neuroscience: 8; 353–364, 1996.

Pandya DN, Yeterian EH: Comparison of prefrontal architecture and connections. The orbitofrontal cortex. Philosophical Transactions of the Royal Society of London Series B: Biological Sciences: 351; 1423–1432, 1996.

Parker DM, Crawford JR: Assessment of frontal lobe dysfunction. In Crawford JR, Parker DM, McKinlay WM (Eds), A Handbook of Neuropsychological Assessment. Hove: Erlbaum, pp. 267–291, 1992.

Parkin AJ: Normal age-related memory loss and its relation to frontal lobe dysfunction. In Rabbitt PMA (Ed), Methodology of Frontal and Executive Function. Hove: Psychology Press, pp. 177–190, 1997.

Pascual-Leone A, Hallett M: Induction of errors in a delayed response task by repetitive transcranial magnetic stimulation of the dorsolateral prefrontal cortex. NeuroReport: 5; 2517–2520, 1994.

Perret E: The left frontal lobe of man and the suppression of habitual responses in verbal categorical behavior. Neuropsychologia: 12; 323–330, 1974.

Peterson BS, Skudlarski P, Gatenby C, Zhang H, Anderson AW, Gore JC: An fMRI study of Stroop word-color interference: evidence for cingulate subregions subserving multiple distributed attentional systems. Biological Psychiatry: 45; 1237–1258, 1999.

Petrides M: Frontal lobes and working memory: evidence from investigations of the effects of cortical excisions in non-human primates. In Boller F, Spinnler H, Hendler JA (Eds), Handbook of Neuropsychology, Vol 9. The Frontal Lobes. Amsterdam: Elsevier, pp. 59–82, 1994.

Petrides M, Alivisatos B, Evans AC, Meyer E: Dissociation of human mid-dorsolateral from posterior dorsolateral frontal cortex in memory processing. Proceedings of the National Academy of Science of the United States of America: 90; 873–877, 1993a.

Petrides M, Alivisatos B, Meyer E, Evans AC: Functional activation of the human frontal cortex during the performance of verbal working memory tasks. Proceedings from the National Academy of Sciences of the United States of America: 90; 878–882, 1993b.

Petrides M, Milner B: Deficits on subject-ordered tasks after frontal- and temporal-lobe lesions in man. Neuropsychologia: 20; 249–262, 1982.

Petrides M, Pandya DN: Comparative architectonic analysis of the human and macaque frontal cortex. In Grafman J, Boller F (Eds), Handbook of Neuropsychology, Vol 9. Amsterdam: Elsevier, pp. 17–58, 1994.

Phillips LH: Do 'frontal tests' measure executive function? Issues of assessment and evidence from fluency tests. In Rabbitt PMA (Ed), Methodology of Frontal and Executive Function. Hove: Psychology Press, pp. 191–214, 1997.

Phillips LH: Age and individual differences in letter fluency. Developmental Neuropsychology: 15; 249–267, 1999.

Phillips LH, Della Sala S: Aging, intelligence and anatomical segregation in the frontal lobes. Learning and Individual Differences: 10; 217–243, 1998.

Phillips LH, Forshaw MJ: The role of working memory in age differences in reasoning. In Logie RH, Gilhooly KJ (Eds), Working Memory and Thinking. Hove: Psychology Press, pp. 23–43, 1998.

Phillips LH, Rabbitt PMA: Impulsivity and speed-accuracy strategies in intelligence test performance. Intelligence: 21; 13–29, 1995.

Phillips ML, Young AW, Senior C, Brammer M, Andrew C, Calder AJ: Bullmore ET, Perrett DI, Rowland D, Williams SCR, Gray JA, David AS: A specific neural substrate for perceiving facial expressions of disgust. Nature: 389; 495–498, 1997.

Prabhakaran V, Smith JAL, Desmond JE, Glover GH, Gabrieli JDE: Neural substrates of fluid reasoning: an fMRI study of neocortical activation during performance of the Raven's Progressive Matrices test. Cognitive Psychology: 33; 43–63, 1997.

Premack D, Woodruff G: Does the chimpanzee have a 'theory of mind'? Behaviour and Brain Sciences: 4; 515–526, 1978.

Rabbitt PMA: Introduction: methodoliges and models in the study of executive function. In Rabbitt PMA (Ed), Method-

ology of Frontal and Executive Function. Hove: Psychology Press, pp. 1–38, 1997.

Rajkowska G, Goldman-Rakic PS: Cytoarchitectonic definition of prefrontal areas in the normal human cortex: 1. Remapping of areas 9 and 46 using quantitative criteria. Cerebral Cortex: 5; 307–322, 1995.

Raz N: Neuroanatomy of the aging brain observed in vivo: a review of structural MRI findings. In Bigler ED (Ed), Neuroimaging II: Clinical Applications. New York: Plenum Press, pp. 153–184, 1996.

Raz N, Gunning FM, Head D, Dupuis JH, Acker JD: Neuroanatomical correlates of cognitive aging: evidence from structural magnetic resonance imaging. Neuropsychology: 12; 95–114, 1998.

Renaud RD, Murray HG: Aging, personality, and teaching effectiveness in academic psychologists. Research in Higher Education: 37; 323–340, 1996.

Rezai K, Andreasen NC, Alliger R, Cohen G, Swayze V II, O'Leary DS: The neuropsychology of the prefrontal cortex. Archives of Neurology: 50; 636–642, 1993.

Robbins TW, James M, Owen AM, Sahakian BJ, Lawrence AD, McInnes L, Rabbitt PMA: A study of performance on tests from the CANTAB battery sensitive to frontal lobe dysfunction in a large sample of normal volunteers: implications for theories of executive functioning and cognitive aging. Journal of the International Neuropsychological Society: 4; 474–490, 1998.

Rolls ET: The orbitofrontal cortex. Philosophical Transactions of the Royal Society of London Series B: Biological Sciences: 351; 1433–1444, 1996.

Rolls ET: The Brain and Emotion. Oxford University Press: New York, 1999a.

Rolls ET: The functions of the orbitofrontal cortex. Neurocase: 5; 301–312, 1999b.

Rolls ET, Hornak J, Wade D, McGrath J: Emotion-related learning in patients with social and emotional change associated with frontal lobe damage. Journal of Neurology, Neurosurgery and Psychiatry: 57; 1518–1524, 1994.

Salthouse TA: Theoretical Perspectives on Cognitive Aging. Hillsdale, NJ: Erlbaum, 1991.

Salthouse TA: Influence of working memory on adult age differences in matrix reasoning. British Journal of Psychology: 84; 171–199, 1993.

Sarazin M, Pillon B, Giannakopoulos P, Rancurel G, Samson Y, Dubois B: Clinicometabolic dissociation of cognitive functions and social behavior in frontal lobe lesions. Neurology: 51; 142–148, 1998.

Saver JL, Damasio AR: Preserved access and processing of social knowledge in a patient with acquired sociopathy due to ventromedial frontal damage. Neuropsychologia: 29; 1241–1249, 1991.

Schaie KW, Willis SL: Adult Development and Aging, 4th ed. New York: Harper Collins, 1996.

Shallice T: Specific impairments of planning. Philosophical Transactions of the Royal Society of London Series B: Biological Sciences: 298; 199–209, 1982.

Shallice T: From Neuropsychology to Mental Structure. Cambridge: Cambridge University Press, 1988.

Shallice T, Burgess P: Deficits in strategy application following frontal lobe damage in man. Brain: 114; 727–741, 1991.

Shallice T, Fletcher F, Frith CD, Grasby P, Frackowiak RSJ, Dolan RJ: Brain regions associated with acquisition and retrieval of verbal episodic memory. Nature: 368; 633–635, 1991.

Shaw TG, Mortel KF, Stirling Meyer J, Rogers RL, Hardenberg J, Cutaia MM: Cerebral blood flow changes in benign aging and cerebrovascular disease. Neurology: 34; 855–862, 1984.

Shimamura AP: Neuropsychological perspectives on memory and cognitive decline in normal human aging. Seminars in the Neurosciences: 6; 387–394, 1994.

Shimamura AP, Janowsky JS, Squire LR: Memory for the temporal-order of events in patients with frontal-lobe lesions and amnesic patients. Neuropsychologia: 28; 803–813, 1990.

Sprengelmeyer R, Rausch M, Eysel UT, Przuntek H: Neural structures associated with recognition of facial expressions of basic emotions. Philosophical Transactions of the Royal Society of London Series B: Biological Sciences: 265; 1927–1931, 1998.

Stone VE, Baron-Cohen S, Knight RT: Frontal lobe contributions to theory of mind. Journal of Cognitive Neuroscience: 10; 640–656, 1998.

Struble RG, Price DL, Cork LA, Price DL: Senile plaques in cortex of aged normal monkeys. Brain Research: 361; 267–275, 1985.

Stuss DT: Interference effects on memory functions in postleukotomy patients: An attentional perspective. In Levin HS, Eisenberg HM, Benton AL (Eds), Frontal Lobe Dysfunction and Dysfunction. New York: Oxford University Press, pp. 157–172, 1991.

Stuss DT, Alexander MP, Hamer L, Palumbo C, Dempster R, Binns M: Levine B, Izukawa D: The effects of focal anterior and posterior brain lesions on verbal fluency. Journal of the International Neuropsychological Society: 4; 265–278, 1998.

Stuss DT, Benson DF, Kaplan EF, Weir WS, Della Malva C: Leucotomised and nonleucotomized schizophrenics: comparison on tests of attention. Biological Psychiatry: 16; 1085–1100, 1981.

Stuss DT, Levine B, Alexander MP, Hong J, Palumbo C, Hamer L: Murphy KJ, Izukawa D: Wisconsin Card Sorting Test performance in patients with focal frontal and posterior brain damage: effects of lesion location and test structure on separable cognitive processes. Neuropsychologia: 38; 388–402, 2000.

Taylor SF, Kornblum S, Lauber EJ, Minoshima S, Koeppe RA: Isolation of specific interference processing in the Stroop task: PET activation studies. NeuroImage: 6; 81–92, 1997.

Teixeira-Ferreira CT, Vérin M, Pillon B, Levy R, Dubois B, Agid Y: Spatio-temporal working memory and frontal lesions in man. Cortex: 34; 83–98, 1998.

Terry RD, De Teresa R, Hansen LA: Neocortical cells counts in normal adult aging. Annals of Neurology: 21; 530–539, 1987.

Tombaugh TN, Kozak J, Rees L: Normative data stratified by age and education for two measures of verbal fluency: FAS and animal naming. Archives of Clinical Neuropsychology: 14; 167–177, 1999.

Tranel D, Anderson SW, Benton A: Development of the concept of 'executive function' and its relationship to the frontal lobes. In Boller F, Grafman J (Eds), Handbook of Neuropsychology, Vol 9. Amsterdam: Elsevier, pp. 125–148, 1994.

Troyer AK, Moscovitch M, Winocur G: Clustering and switching as two components of verbal fluency: evidence from younger and older healthy adults. Neuropsychology: 11; 138–146, 1997.

Troyer AK, Moscovitch M, Winocur G, Alexander MP, Stuss D: Clustering and switching on verbal fluency: the effects of focal frontal- and temporal-lobe lesions. Neuropsychologia: 36; 499–504, 1998.

Tulving E, Kapur S, Craik FIM, Moscovitch M, Houle S: Hemispheric encoding/retrieval asymmetry in episodic memory: positron emission tomography findings. Proceedings of the National Academy of Sciences of the United States of America: 91; 2016–2020, 1994.

Uttl B, Graf P: Color-word stroop test performance across the adult life span. Journal of Clinical and Experimental Neuropsychology: 19; 405–420, 1997.

Vendrell P, Junqué C, Pujol J, Jurado MA, Molet J, Grafman J: The role of prefrontal regions in the stroop task. Neuropsychologia: 33; 341–352, 1995.

Verhaeghen P, De Meersman L: Aging and the Stroop effect: a meta-analysis. Psychology and Aging: 13; 120–126, 1998.

Vérin M, Partiot A, Pillon B, Malapani C, Agid Y, Dubois B: Delayed response tasks and prefrontal lesions in man — evidence for self generated patterns of behaviour with poor environmental modulation. Neuropsychologia: 31; 1379–1396, 1993.

Voegelin H: Beitrag zur kenntnis der stirnhirn-erkrankungen [Contribution to the study of frontal lobe diseases]. Allgemeine Zeitschrift fur Psychiatrie und ihre Grenzgebiete: 54; 588–599, 1897.

Vogt O: Die myeloarchitektonische feldrung des menschlichen stirnhirns. Journal of Psychology and Neurology: 15; 221–232, 1910.

Walker AE: A cytoarchitectural study of the prefrontal area of the macaque monkey. Journal of Comparative Neurology: 73; 59–86, 1940.

Walsh DA, Hershey D: Mental models and the maintenance of complex problem-solving skills into old age. In Cerella J, Hoyer W (Eds), Adult information processing: limits on loss. New York: Academic Press, pp. 553–584, 1993.

Ward G, Allport A: Planning and problem-solving using the 5-disc Tower of London task. Quarterly Journal of Experimental Psychology: 50; 49–78, 1997.

Watanabe M: Reward expectancy in primate prefrontal neurons. Nature: 382; 629–932, 1996.

Weiner B, Graham S: Understanding the motivational role of affect: lifespan research from a attributional perspective. Cognition and Emotion: 3; 401–419, 1989.

Welt L: Über charakterveränderungen des menschen infolge der läsionen des stirnhirns [On character changes of man as a consequence of lesions of the frontal lobe]. Deutsches Archiv für Klinische Medicin: 42; 339–390, 1888.

West RL: An application of prefrontal cortex function theory to cognitive aging. Psychological Bulletin: 120; 272–292, 1996.

West R, Ergis AM, Winocur G, Saint-Cyr J: The contribution of impaired working memory monitoring to performance of the Self-Ordered Pointing task in normal aging and Parkinson's disease. Neuropsychology: 12; 546–554, 1998.

Whelihan WM, Lesher EL: Neuropsychological changes in frontal functions with aging. Developmental Neuropsychology: 1; 371–380, 1985.

Wiegersma S, Meertse K: Subjective ordering, working memory, and aging. Experimental Aging Research: 16; 73–77, 1990.

Woodruff-Pak DS: The Neuropsychology of Aging. Malden, MA: Blackwell, 1997.

Zangwill OL: Psychological deficits associated with frontal lobe lesions. International Journal of Neurology: 5; 395–402, 1966.

Zorrilla LTE, Aguirre GK, Zarahn E, Canon TD, D'Esposito M: Activation of the prefrontal cortex during judgments of recency: a functional MRI study. NeuroReport: 7; 15–17, 1996.

Handbook of Neuropsychology, 2nd Edition, Vol. 7
J. Grafman (Ed)

CHAPTER 5

The frontal lobes and frontal–subcortical circuits in neuropsychiatric disorders

Susan McPherson [a] and Jeffrey L. Cummings [a,b,*]

[a] *Department of Neurology, School of Medicine, University of California at Los Angeles, Los Angeles, CA 90095-176919, USA*
[b] *Department of Psychiatry and Biobehavioral Sciences, School of Medicine, University of California at Los Angeles, 2-232 RNRC, Los Angeles, CA 90095-176919, USA*

Introduction

Of the numerous structures in the human brain, the frontal lobes provide the richest example of the convergence of neurology, psychiatry and neuropsychology in understanding human behavior. The impact of frontal lobe damage in modulating human behavior was dramatically documented in the early 19th century with the case of Phineas Gage, who manifested marked personality changes after sustaining a massive frontal lobe injury. Recent advances in neuroimaging have led to improved understanding of the neuroanatomy of the frontal lobes and the relationship between behavior and frontal dysfunction while providing evidence for the role of frontal–subcortical circuits in a variety of neuropsychiatric disorders.

This chapter provides an outline of the anatomy and role of the frontal–subcortical circuits in human behavior and presents evidence for the role of the frontal lobes using data from both neuroimaging and neuropsychology. Neuropsychiatric syndromes associated with frontal–subcortical circuit dysfunction are described and the occurrence of these syndromes in common neurologic and psychiatric disorders is described. Finally, techniques for assessing behavioral changes in specific frontal lobe disorders are discussed.

*Corresponding author. Tel.: +1 (310) 206-5238; Fax: +1 (310) 206-5287; E-mail: cummings@ucla.edu

Frontal–subcortical circuits

Descriptions of several parallel frontal–subcortical circuits linking regions of the frontal lobe to subcortical structures (Alexander and Crutcher, 1990; Alexander, Crutcher and DeLong, 1990; Alexander, DeLong and Strick, 1986) provide a framework for understanding the similarity between behavioral changes produced by frontal lobe damage and those produced by lesions in other anatomic regions. Five frontal–subcortical circuits have been identified: (1) a motor circuit, with origins in the supplementary motor area, (2) an oculomotor circuit, which originates in the frontal eye fields, and three circuits with origins in the prefrontal cortex ([3] dorsolateral prefrontal cortex, [4] lateral orbital cortex, [5] anterior cingulate cortex) (Alexander and Crutcher, 1990; G.E. Alexander et al., 1986; Alexander et al., 1990). The circuits share the following similarities: (1) all involve the same basic anatomic structures including the striatum, globus pallidus, substantia nigra, and thalamus; (2) all circuits are contiguous, but remain largely segregated throughout; (3) all circuit structures receive input from other brain regions which are functionally related to the specific circuit to which they project (Cummings, 1993); (4) each circuit has a direct and indirect pathway, both of which project to the thalamus. The direct pathway connects the striatum with the globus pallidus interna/substantia nigra complex (Alexander et al., 1990). The indirect pathway connects the striatum with the globus pallidus externa followed by projection to the subthalamic nucleus, and to the globus pallidus interna/substantia nigra (Alexander et al., 1990).

Behavioral disorders similar to those described with frontal lobe lesions result from injury to these frontal–subcortical structures. The three frontal–subcortical circuits originating in the prefrontal cortex are most associated with cognitive and behavioral functions.

Dorsolateral prefrontal circuit

The dorsolateral prefrontal circuit projects primarily to the dorsolateral portion of the head of the caudate nucleus (see Fig. 1) (G.E. Alexander et al., 1986; Alexander et al., 1990). This circuit originates in Broadmann's areas 9 and 10, projects to the caudate, and connects the caudate to the dorsomedial globus pallidus interna and rostral substantia nigra through the direct pathway. The indirect pathway links the

globus pallidus externa to the subthalamic nucleus, and back to the globus pallidus interna and substantia nigra. Neurons of the globus pallidus interna and substantia nigra project to the ventral anterior and medial dorsal thalamic nuclei which in turn connect with the dorsolateral prefrontal region. Damage to the dorsolateral prefrontal circuit results primarily in neuropsychological deficits marked by the dysexecutive syndrome and abnormalities in motor programming.

Lateral orbitofrontal circuit

Originating in the inferolateral prefrontal cortex (Broadmann's area 10), the lateral orbitofrontal circuit projects primarily to the ventromedial caudate nucleus (see Fig. 2) (G.E. Alexander et al., 1986; Alexander et al., 1990). The circuit begins in the caudate region and projects to the dorsomedial pallidum and the rostromedial substantia nigra, medial to the area receiving projections from the dorsolat-

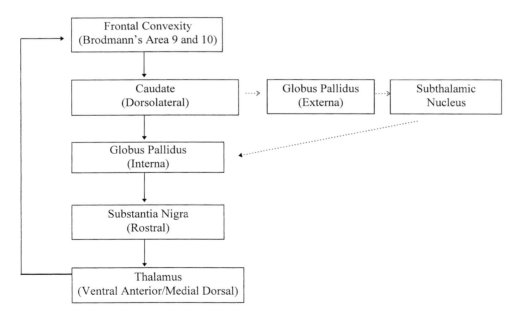

Indirect pathway is shown by dotted lines.

Fig. 1. Schematic anatomy of the dorsolateral prefrontal–subcortical circuit.

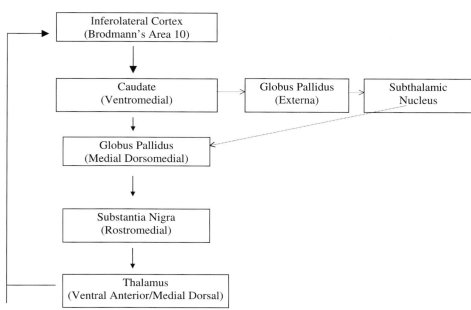

Indirect pathway is shown by dotted lines.

Fig. 2. Schematic anatomy of the orbitofrontal–subcortical circuit.

eral caudate, forming the direct pathway. The indirect pathway connects the globus pallidus externa and subthalamic nucleus with the caudate and in turn projects to the globus pallidus interna/substantia nigra. Projection back to the orbitofrontal cortex occurs via the medial portions of the ventral anterior and medial dorsal thalamic nuclei receiving fibers from the pallidum and nigra.

The primary behavioral feature of orbitofrontal dysfunction is marked change in personality including loss of tact, irritability, and elevated mood. Imitation and utilization behaviors have been observed in patients with anterior orbitofrontal lobe lesions (Lhermitte, Pillon and Serdaru, 1986), reflecting enslavement to environmental cues, enforced utilization of objects in the environment, or automatic imitation of the gestures and actions of others. The stimulus-bound behavior associated with disruption to this circuit may provide explanations for the etiology of obsessive–compulsive disorders.

Anterior cingulate circuit

The anterior cingulate circuit begins in the cortex of the anterior cingulate gyrus (Brodmann's area 24) and projects to the ventral striatum which includes the nucleus accumbens, olfactory tubercle, and the ventromedial portions of the caudate and putamen (see Fig. 3) (G.E. Alexander et al., 1986; Alexander et al., 1990; Parent, 1990). The ventral striatal area also receives projections from the hippocampus, amygdala, and entorhinal cortices. Efferent connections exist between ventral striatum and ventral and rostrolateral globus pallidus and rostrodorsal substantia nigra. Specific regions of the medial dorsal nucleus of the thalamus, the ventral tegmental area, habenula, hypothalamus, and amygdala receive efferents from the globus pallidus and substantia nigra. Projections from medial dorsal thalamic neurons back to the anterior cingulate cortex complete the circuit. Direct and indirect pathways

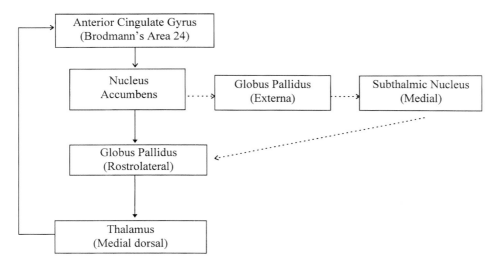

Indirect pathway is shown by dotted lines.

Fig. 3. Schematic anatomy of the anterior cingulate-subcortical circuit.

have not been identified but are thought likely to exist (Alexander et al., 1990). Anterior cingulate injury has been associated with profound apathy and akinetic mutism.

Each of the circuits described above provides important clues explaining the etiology of behaviors in a variety of neuropsychiatric and neurologic disorders. The following section will present the evidence from neuropsychology and neurophysiology for the role of frontal–subcortical circuits in several different conditions.

The frontal lobes in neuropsychiatric and neurologic diseases

Neuropsychiatric syndromes

Apathy

Apathy occurs in a variety of neuropsychiatric diseases and neurological disorders. Apathy has been defined as diminished motivation not attributable to decreased level of consciousness, cognitive impairment, or emotional distress (Marin, 1990). Apathy

has been documented as one of the most frequent behavioral changes associated with Alzheimer's disease (AD) (Mega, Cummings, Fiorello and Gornbein, 1996) and is one of the most personally distressing behavioral changes for caregivers of patients with AD (Greene, Smith, Gardiner and Timbury, 1982). Apathy commonly occurs in patients with Parkinson's disease, Huntington's disease, and thalamic lesions, disrupting the subcortical connections of the anterior cingulate-subcortical circuit (Burns, Folstein, Brandt and Folstein, 1990; Starkstein, Mayberg, Preziosi et al., 1992b; Stuss, Guberman, Nelson and Larochelle, 1988). Apathy can be distinguished from depression in patients with neurodegenerative diseases (Levy, Cummings, Fairbanks et al., 1998). A few investigations address the underlying neurobiology of apathy. Craig, Cummings, Fairbanks et al. (1996) studied the regional cerebral blood flow (rCBF) of 31 patients with Alzheimer's disease (AD), rated on severity of apathy from none to severe using the neuropsychiatric inventory (NPI; Cummings, Mega, Gray et al., 1994). Apathy was associated with decreased hypoperfusion in the prefrontal (medial and lateral) and anterior temporal regions.

Akinetic mutism

Patients with akinetic mutism are profoundly apathetic, do not speak spontaneously and answer questions in monosyllables, if at all. These patients are often incontinent, eat and drink only if fed and move very little. They are indifferent to dire circumstances and display no emotion, even when in pain (Barris and Schuman, 1953; Fesenmeier, Kuzniecky and Garcia, 1990; Nielsen and Jacobs, 1951). Bilateral occlusion of the anterior communicating artery, which supplies the medial frontal cortex, has been associated with akinetic mutism, as has damage to the thalamus and basal ganglia (Chui and Willis, 1999). Unilateral lesions of the cingulate can produce transient akinetic mutism (Damasio and Damasio, 1989).

Hypo/hyperkinetic states

Hypokinesia is defined as a slowness of voluntary movements and a reduction in automatic movements. Hyperkinetic syndromes are marked by increased muscular movement.

Examples of hyperkinetic disorders include Huntington's disease (HD) and Gilles de la Tourette syndrome (GTS), while Parkinson's disease (PD) and progressive supranuclear palsy (PSP) are examples of hypokinetic disorders. In addition to the motor abnormalities, studies have shown that patients with hyper/hypokinetic disorders manifest neuropsychiatric disturbances such as depression, apathy or irritability (Albert, Feldman and Willis, 1974; Chui, 1995; Cummings, Diaz, Levy et al., 1996; Litvan, Mega, Cummings and Fairbanks, 1996; Starkstein, Mayberg, Leiguarda et al., 1992a; Starkstein et al., 1992b). In a recent investigation, Litvan, Paulsen, Mega and Cummings (1998) compared scores on the NPI, a measure used to assess the frequency and severity of ten behaviors commonly reported in patients with cognitive disturbances, in patients with hyperkinetic (HD) versus hypokinetic movement disorders (PSP). Patients with HD produced higher irritability, anxiety, disinhibition and agitation scores while patients with PSP had higher scores on apathy. Litvan and colleagues suggest that the hyperkinetic syndrome associated with HD is the result of

excitatory subcortical output through the medial and orbitofrontal circuits to the pallidum, thalamus and cortex, thus producing more behaviors associated with hyperactivity. In contrast, the apathy commonly associated with PSP is hypothesized to result from hypostimulation of the frontal–subcortical circuits resulting from damage to the substantia nigra, striatum and pallidum.

Disinhibition syndromes

Disinhibition is marked by a lack of restraint manifested in several ways including disregard for social conventions, impulsivity and poor risk assessment. Patients exhibit undue familiarity, talking to strangers and touching others without permission. Conversation is marked by a lack of tact and uncivil or lewd remarks. There is a lack of concern regarding behavior and a lack of conscientiousness regarding completion of assigned tasks. According to Starkstein and Robinson (1997), the disinhibition syndrome includes motor, instinctual, emotional, cognitive and perceptual aspects with signs and symptoms similar to the diagnostic criteria for mania. Motor disinhibition involves pressured speech, hyperactivity and decreased need for sleep. Cognitive disinhibition is marked by grandiose or paranoid delusions and flight of ideas (Starkstein and Robinson, 1997). Hypersexuality, hyperphagia, and aggressive outbursts are indicative of disinhibited instinctual drives.

Disinhibition syndromes have been linked to the right frontal lobe, but have also been attributed to dysfunction of the orbitofrontal and basal temporal cortex (Cummings, 1993; Starkstein and Robinson, 1997). Nuclear imaging studies have provided support for the role of the frontal lobes in disinhibition syndrome. Investigators have reported syndromes characterized by sexual disinhibition, irritability, bursts of laughter and loud speech, in patients with decreased metabolic activity in the orbitofrontal regions as shown on functional imaging as measured by PET (Kumar, Schapiro, Haxby et al., 1990). Starkstein and Robinson (1997) also found decreased perfusion in the orbitofrontal, dorsolateral frontal, and basotemporal cortices and basal ganglia in patients with dementia who exhibited loss of insight, hyperphagia, hypersexuality, euphoria and irritability. Social disintegration in primates has been observed with orbitofrontal lesions (Kling and Stek-lis, 1976).

Cognitive syndromes

Executive dysfunction

Deficits in executive functions and abnormalities in motor programming, associated with damage to the dorsolateral subcortical circuit, characterize the dysexecutive syndrome, which has been associated with damage to the dorsolateral subcortical circuit. Patients with dysexecutive dysfunctions have difficulty in organizing behavioral responses to solve complex problems, such as learning new information, copying complex figures or systematically searching memory (Chow and Cummings, 1999). Other abnormalities include a dependence on environmental cues, poor activation of remote memories, inability to shift and maintain cognitive and behavioral sets, poor generation of motor programs, and an inability to use verbal skills to guide behavior (Chow and Cummings, 1999).

Studies of neuropsychological function have provided some of the best evidence for the connection between frontal–subcortical circuits and a dysexecutive syndrome. Research has shown that patients with a dysexecutive syndrome are unable to shift set required by changing task demands, have reduced verbal fluency and poor organizational strategies on tests of learning, and exhibit poor constructional strategies for copying complex designs (Benton, 1968; Jones-Gotman and Milner, 1977; Shimamura, Gershberg, Jurica et al., 1992). Poor performance is evident on motor measures, particularly on reciprocal and sequential motor tests (Cummings, 1985). Although performance on standard tests of new learning capacity such as word recognition, cued recall, and paired-associate learning may remain intact (Janowsky, Shimamura, Kritchevsky and Squire, 1989a), performance on tests of free recall, temporal order, source memory, and metamemory is impaired (Janowsky et al., 1989a; Janowsky, Shimamura and Squire, 1989b; Janowsky, Shimamura and Squire, 1989c; Milner, 1971; Shimamura, Janowsky and Squire, 1991).

Dysexecutive syndromes are described in patients with frontal and frontal–subcortical disorders including frontotemporal dementias, Huntington's disease, and Parkinson's disease.

Neurologic disorders

Gilles de la Tourette syndrome

Gilles de la Tourette syndrome (GTS) is a chronic disorder characterized by multiple motor and one or more verbal tics occurring many times per day over the course of one year. Onset is before age 18 and generally begins in early childhood, with mean age of onset at age 7 (American Psychiatric Association, 1994). The course of the disease is one of remission and relapse. The syndrome begins with motor tics in 80% of patients, commonly involving the face in the form of eye blinking, rolling of the eyes, nose twitching, or forced eye closure (Greenberg, Aminoff and Simon, 1993). Patients generally progress to a number of motor tics. Verbal tics are present and include grunts, barks, sniffing, hisses, throat clearing, or coughing. Coprolalia (vulgar or obscene speech), echolalia (parroting the speech of others), echopraxia (imitating the movements of others) and palilalia (repetition of words or phrases) occurs in some cases (Greenberg et al., 1993).

Neuropsychiatric symptoms are common in GTS. Attention deficit–hyperactive disorder (ADHD) occurs in approximately 50% of children with GTS (Towbin and Riddle, 1991). Fifty to sixty percent of patients have a comorbid OCD (Frankel, Cummings, Robertson et al., 1986). As compared with patients with OCD alone, patients with GTS and OCD have more violent, sexual and symmetrical obsessions. Compulsions in these patients consist of more touching, blinking, counting and self-damaging behaviors.

No structural basis for the clinical disorder has been found. However, structural imaging studies have reported smaller volumes of basal ganglia (putamen or caudate) in children with GTS as compared with normal controls (Peterson, Riddle, Cohen et al., 1993; Singer, Reiss, Brown et al., 1993). Positron emission tomography studies reveal decreased metabolism in the caudate and thalamic nuclei with increased metabolism in the lateral premotor and supplementary motor regions (Eidelberg, Moeller, Antonini et al., 1997). Diminished blood flow in the caudate and anterior cingulate of GTS patients has been reported (Moriarty, Costa, Schmitz et al., 1995). Children with GTS show greater binding to D2 dopamine receptors in the caudate, as compared with their discordant twin (Wolf, Jones, Knable et al., 1996). Studies using ligand-based SPECT reveal increased dopamine transporter activity (Malison, McDougle, van Dyck et al., 1995).

The pathophysiology of GTS is not well known. Hyperactivity of dopaminergic systems in the basal ganglia, including the caudate and putamen has been suggested. Genetic studies indicate that dopaminergic hyperactivity is inherited as an autosomal dominant condition which may manifest as OCD, GTS, tic syndromes not meeting criteria for GTS, or GTS plus OCD. The putamen, involved primarily in the motor circuitry of the frontal–subcortical circuits, may mediate tic symptoms while obsessions and compulsions would be mediated primarily through the ventral striatum (Alexander, Baker and Kaplan, 1986; Cummings, 1993).

Huntington's disease

Huntington's disease is a hereditary degenerative disorder characterized by the clinical triad of chorea, dementia and a history of familial occurrence (Cummings and Benson, 1992). It is an autosomal dominant disorder affecting 50% of the offspring of the affected individual and resulting from a triplet repeat mutation on chromosome 4. The disease affects males and females equally. Average age at onset is between the ages of 35 and 40 years with an average duration of 14 years. There is no treatment that can change the course of Huntington's disease, although medication such as major tranquilizers can help limit the choreic movements.

The neuropsychological profile of HD is similar to that exhibited by patients with dorsolateral prefrontal dysfunction. Impaired performance is documented on card-sorting tests (Weinberger, Berman and Ladorola et al., 1988b). Patients with HD exhibit deficits in concentration and judgment, are unable to initiate problem solving behavior (Cummings and Benson, 1992), and have difficulty with tasks that require organization, planning, and sequential arrangement of information and perform badly on some visuospatial tasks (Brandt, Folstein and Folstein, 1988; Caine, Hunt, Weingartner and Ebert, 1978). Impaired verbal fluency, one of the earliest manifestations, precedes decline in memory and other cognitive functions (Butters, Sax, Montgomery and Tarlow, 1978). Deficits in confrontation naming and language tests requiring organization and se-

quencing of information are documented in patients with middle and advanced stages of HD (Podoll, Caspary, Lange and Noth, 1988), although other language disturbances such as paraphasias do not occur. Dysarthria is a prominent feature.

Deficits in sequential motor programming, such as alternating hand sequences, have been documented and are in excess of what can be ascribed to the movement disorder (Cummings and Benson, 1992). These impairments reflect involvement of the heads of the caudate nuclei which receive lateral prefrontostriatal projections, providing the best example of cognitive and behavioral symptoms associated with damage to the caudate nuclei.

Behavioral and personality alterations occur early in HD and may include apathy, mood disorder, irritability, explosive disorder, and obsessive–compulsive disorder (Cummings and Cunningham, 1992; Folstein, 1989). Such alterations correspond to initial involvement of medial caudate regions, particularly those which receive projections from the orbitofrontal and anterior cingulate circuits. These circuits mediate limbic system functions and emotion (Vonsattel, Myers, Stevens et al., 1985).

Taken together the above observations implicate dysfunction of frontal–subcortical circuits in HD. Impairments in executive functions impugn the dorsolateral prefrontal circuit. Involvement of the orbitofrontal circuit is indicated by disinhibition and inappropriate behavior. Finally, damage to the anterior cingulate circuit accounts for apathy and lack of initiative.

Psychiatric disorders

Schizophrenia

Schizophrenia is a disorder marked by characteristic symptoms including delusions, hallucinations, disorganized speech, grossly disorganized or catatonic behavior, or negative symptoms (affective flattening, alogia or avolition) (American Psychiatric Association, 1994). The symptoms must be significant enough to interfere with daily functioning, continuous signs of the disturbance must be present for 6 months, and this period must include one month in which at least two of the symptoms listed above are present. Reported prevalence rates vary from 0.2% to 2% of the population. The onset may be abrupt or insidious and usually occurs in the mid-20s for men and the late 20s for women. Treatment involves the use of a variety of neuroleptic medications.

Significant advances have been made in recent years in understanding the psychophysiological processes underlying schizophrenia, particularly with regard to the involvement of the frontal lobe and frontal–subcortical circuits. Research from neuropsychological investigations suggests significant frontal lobe dysfunction in patients with schizophrenia. The following section focuses on the frontal and executive dysfunction observed schizophrenia as well as the underlying physiological mechanisms implicated.

Neuropsychological research suggests that schizophrenic patients exhibit difficulties on tests of planning, organizing, and manipulating complex, novel and abstract information. An extensive amount of research has been conducted using the Wisconsin Card Sorting Test (WCST, Milner, 1963), which assesses the frontal systems functions of set shifting, abstraction and response to feedback. These investigations showed chronic schizophrenic patients to be impaired (Braff, Heaton, Kuck et al., 1991; Fey, 1951; Stuss, Benson, Kaplan et al., 1983). More specifically, schizophrenic patients have difficulty abstracting the concept to be applied and perseverated incorrect responses, despite being provided with potentially corrective feedback. Studies also revealed that schizophrenic patients did not activate the dorsolateral prefrontal cortex (DLPFC) during the WCST as compared with normal controls (Weinberger, Berman and Zec, 1986). Furthermore, the reduction in blood flow correlated with the number of perseverative errors made by the schizophrenic patients. Other investigators suggest that disordered performance on the WCST may not be directly related to a decline in DLPFC, but may be a function of poor attention, lack of motivation or poor cooperation (Goldberg, Weinberger, Berman et al., 1987). Although studies have shown that schizophrenic patient's performance on the WCST improves when provided with explicit instructions about the underlying principles of the test and reinforcement (Bellack, Mueser, Morrison et al., 1990; Green, Ganzell, Satz and Vaclav, 1990), they continue to exhibit more perseverations than normal controls (Goldberg and Weinberger, 1994). Medication does not account

for the poorer performance on measures that assess frontal lobe abilities (Andreasen, Rezai, Alliger et al., 1992; Berman, Zec and Weinberger, 1986).

Schizophrenic patients who present with more negative symptoms have been reported to have more difficulty on neuropsychological tasks that assess frontal lobe ability (Kolb and Wishaw, 1983; Liddle and Morris, 1991; Palmer, Heaton, Paulson et al., 1997). The negative symptoms of schizophrenia, which include alogia, apathy, avolition, and inattention, have been compared to the same behavioral deficits observed in patients with frontal lobe lesions and many similarities noted thereby providing additional support for the role of frontal lobe dysfunction in schizophrenia (Perry, Swerdlow, McDowell and Braff, 1999).

Studies on working memory provide additional support implicating frontal lobe dysfunction in schizophrenia. Working memory has been defined as the ability "to maintain information over short delays while that information is being transformed or coordinated with other ongoing mental operations in the service of a response" (Goldberg and Gold, 1995, p. 152). Working memory is a higher-level operation associated with the functional substrate of the prefrontal cortex (Baddeley and Wilson, 1988). Schizophrenic patients perform poorly on working memory tasks such as digit repetition, mental arithmetic reasoning (Gold, Hermann, Wyler et al., 1994), and on variants of the Brown-Peterson task which require subjects to "hold" information while simultaneously processing additional information such as counting backward while holding a word in memory (Fleming, Goldberg, Gold and Weinberger, 1995). Goldberg, Bigelow, Weinberger et al., 1991) discovered that the introduction of the dopamine agonist dextroamphetamine to schizophrenic patients affected working memory. Taken together, the above neuropsychological data lend substantial support to the relationship between schizophrenia and the frontal lobe.

The field of psychophysiology has provided information in support of a frontal systems deficit in schizophrenia. Research by Swerdlow and Koob (1987) suggests that dysfunction at any one of the levels of cortico-striato-pallido-thalamic circuitry may be responsible for the deficits in information processing. In addition, investigators discovered that diminished prefrontal activation was correlated with reduced hippocampal volume (Weinberger, Berman, Sudduth and Torrey, 1992), reductions in dopaminergic activity (Weinberger, Berman and Illowsky, 1988a), and may be reversed by amphetamine, a dopamine agonist (Daniel, Weinberger, Jones et al., 1991). Additional studies suggest that schizophrenic patients who present with the core negative symptom profile produce patterns of hypofrontality on functional imaging. Severity of negative symptoms and hypofrontality has been associated with both reduction in rCBF (Suziki, Kurachi, Kawasaki et al., 1992) and decreased frontal glucose metabolism (Tamminga, Thaker, Buchanan et al., 1992; Wolkin, Sanfilipo, Wolf et al., 1992).

The research cited above provides support for the notion that the deficits in cognition and behavioral changes seen in schizophrenia can be explained primarily by a lack of activation of the frontal lobes on frontal–subcortical circuits. The negative symptom patterns observed in some schizophrenics mirror those of patients with frontal lesions or frontal damage as the result of closed head injury. Perseveration, observed in many schizophrenics, can be explained by the inability of the patient to inhibit behavior, a task ascribed to the orbitofrontal–subcortical circuit.

Depression
The essential feature of a major depressive episode is a two-week period of continuous symptoms including either depressed mood or the loss of interest or pleasure in nearly all activities. In addition to depressed mood, the patient must experience at least four additional symptoms including: (1) changes in appetite or weight, sleep and psychomotor activity; (2) decreased energy; (3) feelings of worthlessness or guilt; (4) difficulty in thinking, concentrating or making decisions; or (5) recurrent thoughts of death or suicide ideation, plans or attempts (American Psychiatric Association, 1994). Symptoms usually develop over days to weeks and duration is variable with an untreated episode lasting six months or longer regardless of age at onset. Treatment may include the use of antidepressant medication, psychotherapy or, in treatment-resistant cases, electroconvulsive therapy (ECT).

Depression has been identified as one of the more frequent emotional disorders occurring after stroke (Robinson, 1998). Prior to the Diagnostics and Statistical Manual of Mental Disorders, fourth edition (DSM-IV; American Psychiatric Association, 1994) depression related to stroke was categorized as an organic mood disorder. According to DSM-IV, patients who meet criteria for depression following a stroke would be diagnosed as mood disorder, due to stroke. The relationship between brain dysfunction and mood disorders is increasingly recognized. The rates of depression following stroke provide strong evidence for the role of frontal lobes and frontal–subcortical circuits in the etiology of mood disorders. The frequency of post-stroke depression in a community sample ranges from 15% (Burvill, Johnson, Jamrozik et al., 1995) to 22% (Wade, Legh-Smith and Hewer, 1987) with a mean of 13% when all studies are considered (Robinson, 1998). In hospitalized patients, the prevalence has been reported at around 25% with an overall mean of 23% for rehabilitation hospital settings (Robinson, 1998).

Most of the research conducted to date has focused on the role of the left hemisphere in mediating depression following a stroke. Robinson and Starkstein (1990) found that in the acute post-stroke period both cortical and subcortical left frontal lobe lesions are associated with depressive mood changes, as compared with lesions in posterior and right hemisphere structures. Robinson (1998) recently suggested that laterality and interhemispheric lesions may only play a role in determining depression in the acute stroke phases as several studies reported depression in patients with right hemispheric lesions several weeks following a stroke (Eastwood, Rifat, Nobbs and Ruderman, 1990; House, Dennis, Warlow et al., 1990; Morris, Robinson and Raphael, 1990). Several investigators have reported finding a significant inverse correlation between severity of depression and the distance of the lesion from the left frontal pole (Eastwood et al., 1990; Morris, Robinson and Raphael, 1992; Robinson, Starr, Lipsey et al., 1984), although this correlation dissipates over time (Robinson, 1998). Depression is frequent in patients with a stroke to the left caudate and/or putamen (Starkstein, Robinson, Berthier et al., 1988), lacunar state and Binswanger disease, providing additional evidence for the influence of the frontal–subcortical circuits in affective disorders (Ishii, Nishara and Imamura, 1986).

Bipolar disorder: mania

A manic episode is characterized by a period of at least one week during which there is an abnormally and persistently elevated, expansive or irritable mood. The mood must be accompanied by at least three of the following symptoms: inflated self-esteem or grandiosity; decreased need for sleep; pressured speech; flight of ideas or racing thoughts; distractibility; increase in goal-directed activity or psychomotor agitation; and excessive involvement in pleasurable activities that have a high potential for painful consequences such as sexual indiscretions, buying sprees, or foolish business investments (American Psychiatric Association, 1994). Age of onset for a first manic episode is usually in the early 20s with some cases starting in adolescence and others beginning after age 50. Episodes have an acute onset with escalation of symptoms occurring over a few days. In most cases, the episodes are briefer than depressive episodes and resolve more quickly. In general, mania is treated with lithium, although carbamazapine and valproate are frequently used.

Relatively fewer studies have been conducted on primary mania regarding frontal and executive dysfunction and the existing studies have produced inconsistent findings. Studies utilizing diagnostic imaging have found no differences in rCBF between medicated manic patients and depressed patients (Silfverskiold and Risberg, 1989), increased global CBF in small samples of patients with mixed or manic states (Baxter, Phelps, Mazziotta et al., 1985) or increased hemispheric blood flows in rapid cycling-bipolar patients (Mukherjee, Prohovnik, Sackheim et al., 1984). Unfortunately, many studies included patients treated with a variety of medications which may have masked important findings. However, relevant studies suggest, albeit weakly, that mania, relative to major depression my be associated with increased CBF and cerebral metabolism ratio (Rubin, Sackeim, Prohovnik et al., 1995). At least one study suggested that minimally medicated patients who were in a manic episode and patients with major depression both presented with patterns of reduced CBF in selective frontal and anterior temporal regions. Other studies have reported a relative decrease in CBF in the right basal temporal cortex of patients with primary mania (Migliorelli, Starkstein,

Teson et al., 1993). Thus, functional imaging has produced inconsistent results.

Using magnetic resonance spectroscopy to study the brain biochemistry, Deicken, Fein and Weiner (1995) found that non-medicated, euthymic bipolar patients have abnormal phosphorous metabolism in the frontal lobes as compared with normal controls, suggesting that abnormal frontal lobe phospholipid metabolism may play a role in bipolar disorder.

Secondary mania is a circuit-related behavior and can result from strokes, head injury or degenerative conditions that affect the orbitofrontal circuit, caudate nucleus and perithalamic areas (Chow and Cummings, 1999). Whereas a classic bipolar type of mood disorder, with distinct periods of mania and depression, has been reported following lesions to the thalamus and caudate nucleus, lesions to the cortex have generally resulted in mania not associated with a cyclic pattern (Starkstein, Fedoroff, Berthier and Robinson, 1991). Mania also is produced with lesions to non-frontal circuit-related areas such as the amygdala, temporal stem (Berthier, Starkstein, Robinson and Leiguarda, 1993; Starkstein, Pearlson, Boston and Robinson, 1987) and right hemisphere (Cummings, 1995).

Obsessive–compulsive disorder

Obsessive–compulsive disorder (OCD) is a psychiatric disorder characterized by recurrent obsessions and/or compulsions severe enough to be time consuming (more than one hour a day) and that cause significant distress or interfere with occupational or social functioning (American Psychiatric Association, 1994). A diagnosis can be made if the patient experiences either obsessions, defined as persistent thoughts, ideas, impulses or images that are experienced as intrusive or inappropriate and cause marked anxiety or distress, or compulsions, which are defined as repetitive behaviors, the goal of which is to prevent anxiety. The disorder cannot be attributed to an underlying medical condition or the direct physiological effects of a substance. Although the underlying physiological mechanism of OCD is unknown, symptom amelioration occurs in about 50% of patients placed on serotonin reuptake inhibitors (i.e., fluoxetine, paroxetine, sertraline, fluvoxamine, clomipramine) (Rubin and Harris, 1999).

Early neuropsychological studies of OCD produced mixed results with some studies reporting deficits consistent with frontal dysfunction and others which found visuospatial and memory deficits without frontal findings (Otto, 1992). More recently, Cavedini, Ferri, Scarone and Bellodi (1998) reported finding no differences between OCD and depressed patients on tests of set shifting or word fluency, but did find that OCD patients made significantly more perseverative errors on an object alternation test (OAT) used to assess orbitofrontal cortex dysfunction in both humans and non-human primates. Some investigators have found that patients with OCD are slower than controls in performing frontally mediated tasks and have treated speed of performance as a confounding variable (Christensen, Kim, Dysken and Hoover, 1992; Insel, Donnelly, Lalakea et al., 1983; Zielinsky, Taylor and Juzwin, 1991). Galderisi, Mucci, Catapano et al. (1995) hypothesized that patients with OCD perform more slowly on measures of frontal abilities secondary to disruption of frontal–subcortical circuits and not as a result of meticulous concern for correct execution or intrusion of obsessive thoughts during the task. These investigators found that patients with OCD required more time than controls to provide correct responses on a self-ordered pointing task requiring initiation, organization and monitoring of behavior, but they did not require more time on other measures sensitive to left and right temporo-hippocampal functions such as Corsi block tapping or digit sequence repetition. Veale, Sahakian, Owen and Marks (1996) found that OCD patients had greater difficulty on a test requiring shifting of attention implicating involvement of frontostriatal systems. They suggested that OCD patients are easily distracted by competing stimuli, require excessive monitoring and checking of responses to ensure that mistakes do not occur and are more rigid at setting aside the main goal and planning necessary subgoals.

Neuroimaging studies of blood flow by single photon emission computed tomography (SPECT) or glucose metabolism by positron emission tomography (PET) have provided the strongest evidence implicated dysfunction of the frontal lobes and frontal–subcortical circuits in the pathophysiology of OCD. Studies of patients before treatment reported a higher ratio of medial–frontal cortex to whole cortex perfusion as compared with controls (Machlin, Harris, Pearlson et al., 1991), and significant positive correlations between symptom severity and increased rCBF in bilateral dorsal parietal cortex, left posterofrontal cortex, and bilateral orbitofrontal cortex but decreased flow in the head of the caudate bilaterally (Rubin, Villanueva-Meyer, Ananth et al., 1992). Lucey, Costa, Adshead et al. (1997) reported significant lower blood flow in the superior frontal cortex and right caudate of patients with OCD and posttraumatic stress disorder as compared with normal controls. Blood flow studies of OCD patients after treatment have reported decreased medial–frontal perfusion in relationship to whole cortex perfusion (Hoehn-Saric, Pearlson, Harris et al., 1991), and decreased flow in cortical areas and bilateral caudate (Rubin, Ananth, Villanueva-Meyer et al., 1995).

Studies using glucose to assess brain metabolism in OCD report significantly elevated absolute glucose metabolic rates for the entire cortex, caudate nuclei, and orbital gyri of patients with concurrent diagnoses of major depression (Baxter, Phelps, Mazziotta et al., 1987), as well as patients with OCD only (Baxter, Schwartz, Mazziotta et al., 1988). Similar studies have reported increased glucose metabolism in the orbitofrontal regions (Nordahl, Benkelfat, Semple et al., 1989; Perani, Columbo, Bressi et al., 1995; Sawle, Hymas, Lees and Frackowiak, 1991; Swedo, Shapiro, Grady et al., 1989), prefrontal and anterior cingulate regions (Swedo et al., 1989), premotor, and midfrontal cortex (Sawle et al., 1991), and cingulate, lenticular and thalamic regions (Perani et al., 1995). Studies which have investigated the effects of treatment, either with medications or behavior therapy, have documented regional reductions in glucose metabolism. Decreased metabolism in the orbitofrontal areas has been documented in at least two studies (Benkelfat, Nordahl, Semple et al., 1990; Swedo, Pietrini, Leonard et al., 1992). Additional studies have reported decreased metabolism to the left caudate (Benkelfat et al., 1990), right caudate (Baxter, Schwartz, Bergman et al., 1992; Saxena, Brody, Colgan et al., 1999), both caudates (Schwartz, Stoessel, Baxter et al., 1996) and cingulate cortex (Perani et al., 1995).

Saxena, Brody, Schwartz and Baxter (1998) hypothesized that the symptoms in OCD may be the result of a 'captured' signal in the direct pathway of the orbitofrontal–subcortical circuit. They suggested that there is an imbalance between the direct versus indirect pathways, with the direct pathway having greater influence than the indirect pathway. In the case of OCD, these authors suggest that greater inhibition of the globus pallidus interna leads to more thalamo-cortico activation. Excessive activation of this direct circuit, in relationship to the indirect basal ganglia pathway may result in an inability to switch the obsessive thoughts and compulsive behaviors to other behaviors (Saxena et al., 1998).

Attention deficit–hyperactivity disorder

Attention deficit–hyperactivity disorder is marked by a persistent pattern of inattention and/or hyper-activity–impulsivity that is more frequent and severe than can be expected for the child's level of development (American Psychiatric Association, 1994). Hyperactivity is manifested by fidgetiness, squirming or by an inability to remain seated when expected to do so. Additional symptoms include difficulty engaging quietly in leisure activities, excessive talking and appearing to be 'on the go'. Impulsivity is manifested as impatience, difficulty in delaying responses, blurting out answers before questions have been completed, difficulty waiting one's term, and frequent intruding or interrupting of others to the point of causing difficulties in social, academic or occupational settings (American Psychiatric Association, 1994). According to DSM-IV (American Psychiatric Association, 1994) the disorder manifests in approximately 3–5% of school-age children, is more common in males, and is typically diagnosed during the elementary school years.

Several investigators have suggested that impulsivity, or the failure to inhibit or delay behavioral responses, is the core deficit in ADHD (Barkley, 1994; Trommer, Hoeppner and Zecker, 1991; Van der Merre and Sergeant, 1988), implicating frontal striatal circuits and prefrontal cortex (Baxter et al., 1988; Zametkin, Nordahl, Gross et al., 1990). Neuroimaging studies have reported significantly smaller volumetric measures of caudate nucleus and globus pallidus (but not the putamen) predominantly in the right hemisphere in boys with ADHD versus normal controls (Castellanos, Giedd, Eckburg et al., 1994; Castellanos, Giedd, Marsh et al., 1996). In a study of 26 males with ADHD, Casey, Castellanos, Giedd et

al. (1997) found that performance on measures of response selection and response execution, correlated with prefrontal cortex, caudate nucleus and globus pallidus volumes, predominantly in the right hemisphere. More specifically, performance on tasks of sensory selection correlated with right prefrontal and right caudate measures while performance on tasks of response selection and response execution, correlated with caudate symmetry and left globus pallidus measures. These authors suggest that their findings implicate a role for the right prefrontal cortex in suppressing responses to salient events, while the basal ganglia are involved in executing behavioral responses.

Antisocial behavior

Studies of sociopathic individuals provide information regarding the role of the frontal lobes, social behavior and the disinhibition syndrome. According to Lapierre, Braun and Hodgins (1995), sociopathic individuals and patients with orbitofrontal lesions are similar in that both may exhibit a disinhibition syndrome characterized by promiscuous and maladaptive behaviors, lack of social and ethical judgment, lack of foresight, irritability and antisocial tendencies. Lapierre et al. (1995) conducted a study of 30 convicted criminals diagnosed as "psychopaths" versus 30 convicted criminals who did not meet criteria for psychopathy, using neuropsychological measures sensitive to both orbitofrontal and dorsolateral frontal dysfunction. Subjects diagnosed with psychopathy performed more poorly on the orbitofrontal measures (go/no-go discrimination, mazes, smell identification), than non-psychopathic criminals. However, no differences were found on the dorsolateral frontal measure (card sorting test). Research by Raine, Buchsbaum, Stanley et al. (1994) also supports the role of the frontal lobes in sociopathic behavior. Using functional imaging (PET), Raine and colleagues found that individuals convicted of murder had significantly lower glucose metabolism in both lateral and medial prefrontal cortex compared with controls, but no differences in posterior frontal, temporal or parietal glucose metabolism. These investigators suggested that prefrontal dysfunction could result in a loss of inhibition on subcortical structures that may act as a predisposition to violence.

Summary

The evidence from neuropsychological and neuroimaging studies helps to bridge the gap between those disorders formally thought to be purely psychiatric and behavioral manifestations seen as part of neurologic syndromes. These studies provide compelling evidence regarding the relationship of frontal dysfunction and disruption of the frontal–subcortical circuits in the cognitive and behavioral changes of a variety of neuropsychiatric disorders.

Measurement of behavioral changes

Although several scales exist to assess behavior of psychiatric patients, few scales assess the behaviors common to both neuropsychiatric and neurological disorders. As described above, damage to the anterior cingulate circuit results in apathy, with irritability and disinhibition produced by orbitofrontal circuit damage. Apathy is one of the behaviors common to schizophrenia, and disinhibition is seen in disorders such as ADHD and obsessive–compulsive disorder. Recently, Cummings et al. (1994) developed the NPI, a scale that assesses behaviors which are common in both neuropsychiatric and neurological disorders, but are rarely measured, including apathy, irritability and disinhibition.

Neuropsychiatric inventory (NPI)

The NPI was originally developed for assessment of psychopathology in dementia. Information is gathered from an informant who has daily contact with the patient and is familiar with the patient's behavior. Ten behavioral domains are assessed by the NPI: delusions, hallucinations, agitation/aggression, dysphoria, anxiety, euphoria, apathy, disinhibition, irritability/lability, and aberrant motor behavior. The instrument utilizes screening questions which provide an overview of each behavioral domain. It uses scripted questions which determine if the patient's behavior has changed since the onset of the disease. Behavior is rated on both frequency and severity. The NPI has been shown to have acceptable validity and reliability (Cummings et al., 1994).

The NPI has been used to document the neuropsychiatric changes in dementing disorders such as Alzheimer's disease (Mega et al., 1996), frontotemporal dementia (Levy, Miller, Cummings et al., 1996), HD and PSP (Litvan et al., 1998). A study of normal elderly (Cummings et al., 1994) indicated that NPI scores were not influenced by normal aging. Conversely, Cummings and colleagues suggest that any elevation of an NPI scale should be regarded as important evidence for the presence of psychopathology.

Given the ability of this measure to detect apathy and disinhibition, the NPI may be a useful tool in assessing behavioral changes related to disorders the affect the frontal lobes and frontal–subcortical circuits.

Summary

Frontal–subcortical circuitry represents an anatomic framework relevant to the pathophysiology of neuropsychiatric as well as neurobehavioral disorders. Although the three behaviorally relevant frontal–subcortical circuits described involve identical anatomic structures, each circuit remains discrete and has unique functions. Identifiable circuit-specific behaviors exist for each circuit and are analogous to the cognitive and behavioral changes exhibited in patients with frontal lobe dysfunction. The dorsolateral prefrontal–subcortical circuit is responsible for the dysexecutive syndrome characteristic of several disorders including the cognitive deficits associated with schizophrenia. Irritability, impulsivity, agitation, mania, obsessions and compulsions characterize damage to the orbitofrontal–subcortical circuit. Damage to the anterior cingulate–subcortical circuit manifests in apathy and akinetic mutism.

Further study and validation of the role of frontal circuits in neuropsychiatric disorders is needed. The framework and research provided above facilitates hypothesis generation which can help guide future research exploring the relationship between cognitive and behavioral disorders and frontal–subcortical circuits.

Acknowledgements

This project was supported by a National Institute on Aging Alzheimer's Disease Core Center grant (AG 10123), an Alzheimer's Disease Research Center of California grant, and the Sidell-Kagan Foundation.

References

Albert ML, Feldman RG, Willis AL: The 'subcortical dementia' of progressive supranuclear palsy. Journal of Neurology, Neurosurgery and Psychiatry: 37; 121–130, 1974.

Alexander GE, Crutcher MD: Functional architecture of basal ganglia circuits: Neural substrates of parallel processing. Trends in Neurosciences: 13; 26–271, 1990.

Alexander GE, Crutcher MD, DeLong MR: Basal ganglia–thalamocortical circuits: Parallel substrates for motor, oculomotor, prefrontal and limbic functions. Progress in Brain Research: 85; 119–146, 1990.

Alexander GE, DeLong MR, Strick PL: Parallel organization of functionally segregated circuits linking basal ganglia and cortex. Annual Review of Neuroscience: 9; 357–381, 1986.

Alexander MP, Baker E, Kaplan E: Dimensions of performance in patients with ideomotor apraxia. Neurology: 36; 345, 1986.

American Psychiatric Association: Diagnostic and Statistical Manual of Mental Disorders, 4th ed. (DSM-IV). Washington, DC: American Psychiatric Press, 1994.

Andreasen NC, Rezai K, Alliger R, Swayze VW II, Flaum M, Kirchner P, Cohen G, O'Leary DS: Hypofrontality in neuroleptic-naive patients and inpatients with chronic schizophrenia. Assessment with Xenon 133 single photon emission computed tomography and the Tower of London. Archives of General Psychiatry: 49; 943–958, 1992.

Baddeley A, Wilson BA: Frontal amnesia and the dysexecutive syndrome. Brain and Cognition: 7; 212–224, 1988.

Barkley RA: Impaired delayed responding: a unified theory of attention deficit hyperactivity disorder. In Routh DK (Ed), Disruptive Behavior Disorders in Childhood. New York: Plenum, pp. 11–57, 1994.

Barris RW, Schuman HR: Bilateral anterior cingulate gyrus lesions. Neurology: 3; 44–52, 1953.

Baxter LR, Phelps ME, Mazziotta JC, Guze BH, Schwartz JM, Selin CE: Local cerebral glucose metabolic rates in obsessive–compulsive disorder — a comparison with rates in unipolar depression and in normal controls. Archives of General Psychiatry: 44; 211–218, 1987.

Baxter LR, Phelps ME, Mazziotta JC, Schwartz JM, Gerner R, Selin C, Sumida R: Cerebral metabolic rates for glucose in mood disorders. Archives of General Psychiatry: 42; 441–447, 1985.

Baxter LR, Schwartz JM, Bergman KS, Szuba MP, Guze BH, Mazziotta JC, Alazraki A, Selin CE, Ferng HK, Munford P et al.: Caudate glucose metabolic rate changes with both drug and behavior therapy for obsessive–compulsive disorder. Archives of General Psychiatry: 49; 681–689, 1992.

Baxter LR, Schwartz JM, Mazziotta JC, Phelps ME, Pahl JJ, Guze BH, Fairbanks L: Cerebral glucose metabolic rates in non-depressed obsessive–compulsives. American Journal of Psychiatry: 145; 1560–1563, 1988.

Bellack AS, Mueser KT, Morrison RL, Tierney A, Podell K: Remediation of cognitive deficits in schizophrenia. American Journal of Psychiatry: 147; 1650–1655, 1990.

Benkelfat C, Nordahl TE, Semple WE, King AC, Murphy LD, Cohen RM: Local cerebral glucose metabolic rates in obsessive–compulsive disorder: Patients treated with clomipramine. Archives of General Psychiatry: 47; 840–848, 1990.

Benton AL: Differential behavioral effects in frontal lobe disease. Neuropsychologia: 6; 53–60, 1968.

Berman KF, Zec RF, Weinberger DR: Physiologic dysfunction of dorsolateral prefrontal cortex in schizophrenia: Role of neuroleptic treatment, attention and mental effort. Archives of General Psychiatry: 42; 814–821, 1986.

Berthier ML, Starkstein SE, Robinson RG, Leiguarda R: Limbic lesions in a patient with recurrent mania [letter]. Journal of Neuropsychiatry and Clinical Neurosciences: 2; 235–236, 1993.

Braff DL, Heaton R, Kuck J, Cullum M, Moranville J, Grant I, Zisook S: The generalized pattern of neuropsychological deficits in outpatients with chronic schizophrenia with heterogeneous Wisconsin Card Sorting Test results. Archives of General Psychiatry: 48; 891–898, 1991.

Brandt J, Folstein ME, Folstein MF: Differential cognitive impairment in Alzheimer's disease and Huntington's disease. Annals of Neurology: 23; 555–556, 1988.

Burns A, Folstein S, Brandt J, Folstein M: Clinical assessment of irritability, aggression, and apathy in Huntington and Alzheimer disease. Journal of Nervous and Mental Disease: 178; 20–26, 1990.

Burvill PW, Johnson GA, Jamrozik KD, Anderson CS, Stewart-Synne EG, Chakera TMH: Prevalence of depression after stroke: The Perth community stroke study. British Journal of Psychiatry: 166; 320–327, 1995.

Butters N, Sax D, Montgomery K, Tarlow S: Comparison of the neuropsychological deficits associated with early and advanced Huntington's disease. Archives of Neurology: 35; 585–589, 1978.

Caine ED, Hunt RD, Weingartner H, Ebert MH: Huntington's dementia: Clinical and neuropsychological features. Archives of Neurology: 35; 377–384, 1978.

Casey BJ, Castellanos FX, Giedd JN, Marsh WL, Hamburger SD, Schubert AB, Vauss YC, Vaituzis AC, Dickstein DP, Sarfatti SE et al.: Implication of right frontostriatal circuitry in response inhibition and attention-deficit/hyperactivity disorder. Journal of the American Academy of Child and Adolescent Psychiatry: 36; 374–383, 1997.

Castellanos FX, Giedd JN, Eckburg P, Marsh WL, Vaituzis AC, Kaysen D, Hamburger SD, Rapoport JL: Quantitative morphology of the caudate nucleus in attention deficit hyperactivity disorder. American Journal of Psychiatry: 151; 1791–1796, 1994.

Castellanos FX, Giedd JN, Marsh WL, Hamburger SD, Baituzis AC, Dickstein DP, Sarfatti SE, Vauss YC, Snell JW, Lange N et al.: Quantitative brain magnetic resonance imaging in attention-deficit/hyperactivity disorder. Archives of General Psychiatry: 53; 607–626, 1996.

Cavedini P, Ferri S, Scarone S, Bellodi L: Frontal lobe dysfunction in obsessive–compulsive disorder and major depression: A clinical–neuropsychological study. Psychiatry Research: 78; 21–28, 1998.

Chow TW, Cummings JL: Frontal–subcortical circuits. In Miller

BL, Cummings JL (Eds), The Human Frontal Lobes. New York: Guilford Press, Ch. 1, pp. 3–26, 1999.

Christensen KJ, Kim SW, Dysken MW, Hoover KM: Neuropsychological performance in obsessive–compulsive disorder. Biological Psychiatry: 31; 4–18, 1992.

Chui HF: Psychiatric aspects of progressive supranuclear palsy. General Hospital Psychiatry: 17; 135–143, 1995.

Chui H, Willis L: Vascular diseases of the frontal lobes. In Miller BL, Cummings JL (Eds), The Human Frontal Lobes. New York: Guilford Press, pp. 370–401, 1999.

Craig AH, Cummings JL, Fairbanks L, Itti L, Miller BL, Li J, Mena I: Cerebral blood flow correlates of apathy in Alzheimer disease. Archives of Neurology: 53; 1116–1120, 1996.

Cummings JL: Clinical Neuropsychiatry. New York, NY: Grune and Stratton, Inc., 1985.

Cummings JL: Frontal–subcortical circuits and human behavior. Archives of Neurology: 50; 873–880, 1993.

Cummings JL: Anatomic and behavioral aspects of frontal–subcortical circuits. In Grafman J, Holyoak KJ, Boller F (Eds), Structure and Functions of the Human Prefrontal Cortex. New York: New York Academy of Sciences, pp. 1–13, 1995.

Cummings JL, Benson DF: Dementia: A Clinical Approach. Boston, MA: Butterworths–Heinemann, 1992.

Cummings JL, Cunningham K: Obsessive–compulsive disorder in Huntington's disease. Biological Psychiatry: 31; 263–270, 1992.

Cummings JL, Diaz C, Levy M, Binetti G, Litvan L: Neuropsychiatric syndromes in neurodegenerative diseases: frequency and significance. Seminars in Clinical Neuropsychiatry: 1; 241–247, 1996.

Cummings JL, Mega M, Gray K, Rosenberg-Thompson S, Carusi DA, Gornbein J. The Neuropsychiatric Inventory: Comprehensive assessment of psychopathology in dementia. Neurology: 44; 2308–2314, 1994.

Damasio H, Damasio AR: Lesion Analysis in Neuropsychology. New York: Oxford University Press, 1989.

Daniel DG, Weinberger DR, Jones DW, Zigun JR, Coppola R, Handel S, Bigelow LB, Goldberg TE, Berman KF, Kleinman JE: The effect of amphetamine on regional cerebral blood flow during cognitive activation in schizophrenia. Journal of Neuroscience: 11; 1907–1917, 1991.

Deicken RF, Fein G, Weiner MW: Abnormal frontal lobe phosphorous metabolism in bipolar disorder. American Journal of Psychiatry: 152; 915–918, 1995.

Eastwood MR, Rifat SL, Nobbs H, Ruderman J: Mood disorder following cerebrovascular accident. British Journal of Psychiatry: 154; 195–200, 1990.

Eidelberg D, Moeller JR, Antonini A, Kazumata K, Dhawan V, Budman C, Feigin A: The metabolic anatomy of Tourette's syndrome. Neurology: 48; 927–934, 1997.

Fesenmeier JT, Kuzniecky R, Garcia JH: Akinetic mutism caused by bilateral anterior cerebral tuberculous obliterative arteritis. Neurology: 30; 1005–1006, 1990.

Fey ET: The performance of young schizophrenics and young normals on the Wisconsin Card Sorting Test. Consulting Psychology: 15; 311–319, 1951.

Fleming K, Goldberg TE, Gold JM, Weinberger DR: Brown Pe-

terson performance in patients with schizophrenia. Psychiatry Research: 56; 155–161, 1995.

Folstein SE: Huntington's Disease: A Disorder of Families. Baltimore, MD: Johns Hopkins University Press, 1989.

Frankel M, Cummings JL, Robertson MM, Trimble MR, Hill MA, Benson DF: Obsessions and compulsions in Gilles de la Tourette's syndrome. Neurology: 36; 378–382, 1986.

Galderisi S, Mucci A, Catapano F, D'Amato AC, Maj M: Neuropsychological slowness in obsessive–compulsive patients. Is it confined to tests involving the fronto-subcortical systems? British Journal of Psychiatry: 167; 394–398, 1995.

Gold JM, Hermann BP, Wyler A, Randolf C, Goldberg TE, Weinberger DR: Schizophrenia and temporal lobe epilepsy: A neuropsychological study. Archives of General Psychiatry: 51; 265–272, 1994.

Goldberg TE, Bigelow LB, Weinberger DR, Daniel DG, Kleinman JR: Cognitive and behavioral effects of coadministration of dextroamphetamine and haloperidol in schizophrenia. American Journal of Psychiatry: 148; 78–84, 1991.

Goldberg TE, Gold JM: Neurocognitive deficits in schizophrenia. In Hirsch SR, Weinberger DR (Eds), Schizophrenia. Oxford: Blackwell Science, Ltd., Ch. 10, pp. 146–162, 1995.

Goldberg TE, Weinberger DR: Schizophrenia training paradigms, and the Wisconsin Card Sorting Test redux. Schizophrenia Research: 11; 291–296, 1994.

Goldberg TE, Weinberger DR, Berman KF, Plishkin NH, Podd MH: Further evidence of dementia of the prefrontal type in schizophrenia. Archives of General Psychiatry: 44; 1008–1014, 1987.

Greenberg DA, Aminoff MJ, Simon RP: Clinical Neurology, 2nd ed. Norwalk: Appleton and Lange, 1993.

Green MF, Ganzell S, Satz P, Vaclav J: Teaching the Wisconsin Card Sorting Test to schizophrenic patients [letter to the editor]. Archives of General Psychiatry: 47; 91–92, 1990.

Greene JG, Smith R, Gardiner M, Timbury GC: Measuring behavioral disturbance of elderly demented patients in the community and its effects on relatives: A factor analytic study. Age and Aging: 11; 121–126, 1982.

Hoehn-Saric R, Pearlson GD, Harris GJ, Machlin SR, Camargo EE: Effects of fluoxetine on regional cerebral blood flow in obsessive–compulsive patients. American Journal of Psychiatry: 148; 1243–1245, 1991.

House A, Dennis M, Warlow C, Hawton K, Molyneux K: Mood disorders after stroke and their relation to lesion location. A CT scan study. Brain: 113; 1113–1130, 1990.

Insel TR, Donnelly EF, Lalakea ML, Alterman IS, Murphy DL: Neurological and neuropsychological studies of patients with obsessive–compulsive disorder. Biological Psychiatry: 18; 741–751, 1983.

Ishii N, Nishara Y, Imamura T: Why do frontal lobe symptoms predominate in vascular dementia with lacunes? Neurology: 36; 340–345, 1986.

Janowsky JS, Shimamura AP, Kritchevsky M, Squire LR: Cognitive impairment following frontal lobe damage and its relevance to human amnesia. Behavioral Neuroscience: 103; 548–560, 1989a.

Janowsky JS, Shimamura AP, Squire LR: Memory and metamemory: Comparisons between patients with frontal lobe and amnesic patients. Psychobiology: 17; 3–11, 1989b.

Janowsky JS, Shimamura AP, Squire LR: Source memory impairment in patients with frontal lobe lesions. Neuropsychologia: 27; 1043–1056, 1989c.

Jones-Gotman M, Milner B: Design fluency: The invention of nonsense drawings after focal cortical lesions. Neuropsychologia: 15; 653–674, 1977.

Kling A, Steklis HD: A neural substrate for affiliative behavior in non-human primates. Brain, Behavior and Evolution: 13; 216–238, 1976.

Kolb B, Wishaw IQ: Performance of schizophrenic patients on tests sensitive to left or right frontal, temporal or parietal function in neurologic patients. Journal of Nervous and Mental Disorders: 171; 435–443, 1983.

Kumar A, Schapiro MD, Haxby JV, Grady CL, Friedland RP: Cerebral metabolic and cognitive studies in dementia with frontal lobe behavioral features. Journal of Psychiatric Research: 24; 97–109, 1990.

Lapierre D, Braun CMJ, Hodgins S: Ventral frontal deficits in psychopathy: neuropsychological test findings. Neuropsychology: 33; 139–151, 1995.

Levy ML, Cummings JL, Fairbanks L, Masterman D, Miller BL, Craig AH, Paulsen JS, Litvan I: Apathy is not depression. The Journal of Neuropsychiatry and Clinical Neurosciences: 10; 314–319, 1998.

Levy ML, Miller BL, Cummings JL, Fairbanks L, Craig AH: Alzheimer disease and frontotemporal dementias. Archives of Neurology: 53; 687–690, 1996.

Lhermitte F, Pillon B, Serdaru M: Human anatomy and the frontal lobes. Part I: Imitation and utilization behavior: a neuropsychological study of 75 patients. Annals of Neurology: 19; 326–334, 1986.

Liddle PF, Morris DL: Schizophrenic syndromes and frontal lobe performance. British Journal of Psychiatry: 158; 340–345, 1991.

Litvan I, Mega MS, Cummings JL, Fairbanks L: Neuropsychiatric aspects of progressive supranuclear palsy. Neurology: 47; 1184–1189, 1996.

Litvan I, Paulsen JS, Mega MS, Cummings JL: Neuropsychiatric assessment of patients with hyperkinetic and hypokinetic movement disorders. Archives of Neurology: 55; 1313–1319, 1998.

Lucey JV, Costa DC, Adshead G, Deahl M, Busatto G, Gacinovic S, Travis M, Pilowsky L, Ell PJ, Marks IM: et al. Brain blood flow in anxiety disorders. OCD, panic disorder with agoraphobia, and post-traumatic stress disorder on 99mTcHM-PAO single photon emission computed tomography (SPECT). British Journal of Psychiatry: 171; 346–350, 1997.

Machlin SR, Harris GJ, Pearlson GD, Hoehn-Saric R, Jeffery P, Camargo EE: Elevated medial–frontal cerebral blood flow in obsessive–compulsive patients: A SPECT study. American Journal of Psychiatry: 148; 1240–1242, 1991.

Malison RT, McDougle CJ, van Dyck CH, Scahill L, Baldwin RM, Seibyl JP, Price LH, Leckman JF, Innis RB: [123I]b-CIT SPECT imaging of striatal dopamine transporter binding in

Tourette's disorder. American Journal of Psychiatry: 152; 1359–1361, 1995.

Marin RS: Differential diagnosis and classification of apathy. American Journal of Psychiatry: 147; 22–30, 1990.

Mega MS, Cummings JL, Fiorello T, Gornbein J: The spectrum of behavioral changes in Alzheimer's disease. Neurology: 46; 130–135, 1996.

Migliorelli R, Starkstein SE, Teson A, de Quiros G, Vazquez S, Leiguardia RG: SPECT findings in patients with primary mania. Journal of Neuropsychiatry: 5; 379–383, 1993.

Milner B: Effect of different brain lesions in card sorting: The role of the frontal lobes. Archives of Neurology: 9; 100–110, 1963.

Milner B: Interhemispheric differences in the location of psychological processes in man. British Medical Bulletin: 27; 272–277, 1971.

Moriarty J, Costa DC, Schmitz B, Trimble MR, Ell PJ, Robertson MM: Brain perfusion abnormalities in Gilles de la Tourette's syndrome. British Journal of Psychiatry: 167; 249–254, 1995.

Morris PLP, Robinson RG, Raphael B: Prevalence and course of depressive disorders in hospitalized stroke patients. International Journal of Psychiatric Medicine: 20; 349–364, 1990.

Morris PLP, Robinson RG, Raphael B: Lesion location and depression in hospitalized stroke patients: evidence supporting a specific relationship in the left hemisphere. Neuropsychiatry, Neuropsychology and Behavioral Neurology: 3; 75–82, 1992.

Mukherjee S, Prohovnik I, Sackheim HA, Decina P, Lee C: Regional cerebral blood flow in a rapid cycling patient (Abstract). Presented at the Annual Meeting of American Psychiatric Association, Los Angeles, CA, 1984.

Nielsen JM, Jacobs LL: Bilateral lesions of the anterior cingulate gyri. Bulletin of the Los Angeles Neurological Society: 16; 231–234, 1951.

Nordahl TE, Benkelfat C, Semple WE, Gross M, King AC, Cohen RM: Cerebral glucose metabolic rates in obsessive–compulsive disorder. Neuropsychopharmacology: 2; 23–28, 1989.

Otto MW: Normal and abnormal information processing: A neuropsychological perspective on obsessive–compulsive disorder. Psychiatric Clinics of North America: 15; 825–848, 1992.

Palmer BW, Heaton RK, Paulson JS, Kuck J, Braff D, Harris MJ, Zisook S, Jeste DV: Is it possible to be schizophrenic yet neuropsychologically normal? Neuropsychology: 11; 437–436, 1997.

Parent A: Extrinsic connections of the basal ganglia. Trends in Neuroscience: 13; 254–258, 1990.

Perani D, Columbo C, Bressi S, Bonfanti A, Grassi F, Scarone S, Bellodi L, Smeraldi E, Fazio F: [18]FDG PET Study in obsessive–compulsive disorder: A clinical/metabolic correlation study after treatment. British Journal of Psychiatry: 166; 244–250, 1995.

Perry W, Swerdlow NR, McDowell JE, Braff DL: Schizophrenia and frontal lobe functioning: Evidence from neuropsychology, cognitive neuroscience, and psychophysiology. In Miller BL, Cummings JL (Eds), The Human Frontal Lobes. New York: Guilford Press, pp. 509–521, 1999.

Peterson B, Riddle MA, Cohen DJ, Katz LD, Smith JC, Hardin

MT, Leckman JF: Reduced basal ganglia volumes in Tourette's syndrome using three-dimensional reconstruction techniques from magnetic resonance images. Neurology: 43; 941–949, 1993.

Podoll K, Caspary P, Lange HW, Noth J: Language functions in Huntington's disease. Brain: 11; 1475–1503, 1988.

Raine A, Buchsbaum MS, Stanley J, Lottenberg S, Abel L, Stoddard J: Selective reductions in prefrontal glucose metabolism in murderers. Biological Psychiatry: 36; 365–373, 1994.

Robinson RG: The Clinical Neuropsychiatry of Stroke. Cambridge: Cambridge University Press, 1998.

Robinson RG, Starkstein SE: Current research in affective disorders following stroke. Journal of Neuropsychiatry and Clinical Neurosciences: 2; 1–14, 1990.

Robinson RG, Starr LB, Lipsey JR, Rao K, Price TR: A two-year longitudinal study of post-stroke mood disorders: dynamic changes in associated variables over the first six months of follow-up. Stroke: 15; 510–517, 1984.

Rubin E, Sackeim HA, Prohovnik I, Moeller JR, Schnur DB, Mukherjee S: Regional cerebral blood flow in mood disorders: IV. Comparison of mania and depression. Psychiatry Research: 61; 1–10, 1995.

Rubin RT, Ananth J, Villanueva-Meyer J, Trajmar PG, Mena I: Regional ^{133}Xenon cerebral blood flow and cerebral Tc-HMPAO uptake in patients with obsessive–compulsive disorder before and after treatment. Biological Psychiatry: 38; 429–437, 1995.

Rubin RT, Harris GJ: Obsessive–compulsive disorder and the frontal lobes. In Miller BL, Cummings JL (Eds), The Human Frontal Lobes. New York: Guilford Press, pp. 522–536, 1999.

Rubin RT, Villanueva-Meyer J, Ananth J, Trajmar PG, Mena I: Regional ^{133}Xe cerebral blood flow and cerebral 99m-HMPAO uptake in unmedicated obsessive–compulsive disorder patients and matched normal control subjects: Determination by high-resolution single-photon emission computed tomography. Archives of General Psychiatry: 49; 695–702, 1992.

Sawle GV, Hymas NF, Lees AJ, Frackowiak RS: Obsessional slowness: Functional studies with positron emission tomography. Brain: 114; 2191–2202, 1991.

Saxena S, Brody AL, Maidment KM, Dunkin JJ, Colgan ME, Alborzian S, Phelps ME, Baxter LR Jr: Localized orbitofrontal and subcortical metabolic changes and predictors of response to paroxetine treatment in obsessive–compulsive disorder. Neuropsychopharmacology: 21; 683–693, 1999.

Saxena S, Brody AL, Schwartz JM, Baxter LR: Neuroimaging and frontal subcortical circuitry in obsessive–compulsive disorder. British Journal of Psychiatry: 173(Suppl. 35); 26–37, 1998.

Schwartz JM, Stoessel PW, Baxter LR, Martin KM, Phelps ME: Systematic changes in cerebral glucose metabolic rate after successful behavior modification treatment of obsessive–compulsive disorder. Archives of General Psychiatry: 53; 109–113, 1996.

Shimamura AP, Gershberg FB, Jurica PJ, Mangels JA, Knight RT: Intact implicit memory in patients with frontal lobe lesions. Neuropsychologia: 30; 931–937, 1992.

Shimamura AP, Janowsky JS, Squire LR: What is the role of

frontal lobe damage in memory disorders? In: Levin HS, Eisenberg HM, Benton AL (Eds), Frontal Lobe Function and Dysfunction. New York: Oxford University Press, pp. 173–195, 1991.

Silfverskiold P, Risberg J: Regional cerebral blood flow in depression and mania. Archives of General Psychiatry: 46; 253–259, 1989.

Singer HS, Reiss AL, Brown JE, Aylward EH, Shih B, Chee E, Harris EL, Reader MJ, Chase GA, Bryan RN et al.: Volumetric MRI changes in basal ganglia of children with Tourette's syndrome. Neurology: 43; 950–956, 1993.

Starkstein SE, Fedoroff P, Berthier ML, Robinson RG: Manic–depressive and pure manic states after brain lesions. Biological Psychiatry: 29; 149–158, 1991.

Starkstein SE, Mayberg HS, Leiguarda R, Preziosi TJ, Robinson RG: A prospective longitudinal study of depression, cognitive decline, and physical impairment in patients with Parkinson's disease. Journal of Neurology, Neurosurgery and Psychiatry: 55; 377–382, 1992a.

Starkstein SE, Mayberg HS, Preziosi TJ, Andrezejewski P, Leiguarda R, Robinson R: Reliability, validity and clinical correlates of apathy in Parkinson's disease. Journal of Neuropsychiatry and Clinical Neurosciences: 4; 134–139, 1992b.

Starkstein SE, Pearlson GD, Boston J, Robinson RG: Mania after brain injury. A controlled study of causative factors. Archives of Neurology: 44; 1069–1073, 1987.

Starkstein SE, Robinson RG: Mechanism of disinhibition after brain lesions. The Journal of Nervous and Mental Disease: 185; 108–114, 1997.

Starkstein SE, Robinson RG, Berthier ML, Parikh RM, Price TR: Differential mood changes following basal ganglia versus thalamic lesions. Archives of Neurology: 45; 723–730, 1988.

Stuss DT, Benson DF, Kaplan EF, Weir WS, Naeser MA, Lieberman I, Ferrill D: The involvement of orbitofrontal cerebrum in cognitive tasks. Neuropsychologia: 21; 235–248, 1983.

Stuss DT, Guberman A, Nelson R, Larochelle S: The neuropsychology of paramedian thalamic infarction. Brain and Cognition: 8; 348–378, 1988.

Suziki M, Kurachi M, Kawasaki Y, Kiba KB, Yamaguchi N: Left hypofrontality correlates with blunted affect in schizophrenia. Japanese Journal of Psychiatry and Neurology: 46; 653–657, 1992.

Swedo SE, Pietrini P, Leonard HL, Schapiro MB, Rettew DC, Goldberger EL, Rapoport SI, Rapoport JL, Grady CL: Cerebral glucose metabolism in childhood onset obsessive–compulsive disorder: Revisualization during pharmacotherapy. Archives of General Psychiatry: 49; 690–694, 1992.

Swedo SE, Shapiro MG, Grady CL, Cheslow DL, Leonard HL, Kumar A, Friedland R, Rapoport SE, Rapoport JL: Cerebral glucose metabolism in childhood onset obsessive–compulsive disorder. Archives of General Psychiatry: 46; 518–523, 1989.

Swerdlow NR, Koob GF: Dopamine, schizophrenia, mania and depression: Toward a unified hypothesis of cortico-striato-pallido-thalamic function. Behavioural and Brain Sciences: 10; 197–245, 1987.

Tamminga CA, Thaker GK, Buchanan R, Kirkpatrick B, Alphs LD, Chase TN, Carpenter WT: Limbic system abnormalities identified in schizophrenia using positron emission tomography with fluorodeoxyglucose and neocortical alterations with deficit syndrome. Archives of General Psychiatry: 49; 522–530, 1992.

Towbin KE, Riddle MA: Obsessive–compulsive disorder. In Lewis M (Ed), Child and Adolescent Psychiatry: a Comprehensive Textbook. Baltimore, MD: Williams and Wilkins, pp. 685–697, 1991.

Trommer BL, Hoeppner JA, Zecker SG: The go no-go test in attention deficit disorder is sensitive to methylphenidate. Journal of Child Neurology: 6; 128–131, 1991.

Van der Merre J, Sergeant J: Acquisition of attention skill in pervasively hyperactive children. Journal of Child Psychology and Psychiatry: 29; 301–310, 1988.

Veale DM, Sahakian BJ, Owen AM, Marks IM: Specific cognitive deficits in tests sensitive to frontal lobe dysfunction in obsessive–compulsive disorder. Psychological Medicine: 26; 1261–1269, 1996.

Vonsattel JP, Myers RH, Stevens TJ, Ferrante RJ, Bird ED, Richardson EP, Jr.: Neuropathological classification of Huntington's disease. Journal of Neuropathology and Experimental Neurology: 44; 559–577, 1985.

Wade DT, Legh-Smith J, Hewer RA: Depressed mood after stroke, a community study of its frequency. British Journal of Psychiatry: 151; 200–205, 1987.

Weinberger DR, Berman KF, Illowsky BP: Physiological dysfunction of the dorsolateral prefrontal cortex in schizophrenia. III. A new cohort of evidence for a monoaminergic mechanism. Archives of General Psychiatry: 45; 609–615, 1988a.

Weinberger DR, Berman KF, Ladorola M et al.: Prefrontal cortical blood flow and cognitive function in Huntington's disease. Journal of Neurology, Neurosurgery, and Psychiatry: 51; 94–104, 1988b.

Weinberger DR, Berman KF, Sudduth R, Torrey EF: Evidence of a dysfunction of a prefrontal–limbic network in schizophrenia: a magnetic resonance imaging and regional cerebral blood flow study of discordant monozygotic twins. American Journal of Psychiatry: 149; 890–897, 1992.

Weinberger DR, Berman KF, Zec RF: Physiologic dysfunction of dorsolateral prefrontal cortex in schizophrenia: I. Regional cerebral blood flow evidence. Archives of General Psychiatry: 4; 114–124, 1986.

Wolf SS, Jones DW, Knable MB, Gorey JG, Sam Lee K, Hyde TM, Coppola R, Weinberger DR: Tourette syndrome: prediction of phenotypic variation in monozygotic twins by caudate nucleus D2 receptor binding. Science: 273; 1225–1227, 1996.

Wolkin A, Sanfilipo M, Wolf AP, Angrist B, Brodie JD, Rotrosen J: Negative symptoms and hypofrontality in chronic schizophrenia. Archives of General Psychiatry: 49; 959–965, 1992.

Zametkin AJ, Nordahl TE, Gross M, King AC, Semple WE, Rumsey J, Hamburger S, Cohen RM: Cerebral glucose metabolism in adults with hyperactivity of childhood onset. New England Journal of Medicine: 323; 1361–1366, 1990.

Zielinsky CM, Taylor MA, Juzwin KR: Neuropsychological deficits in obsessive–compulsive disorder. Neuropsychology, Neuropsychiatry and Behavioral Neurology: 4; 110–126, 1991.

CHAPTER 6

The somatic marker hypothesis and decision-making

Antoine Bechara *, Daniel Tranel and Antonio R. Damasio

Department of Neurology, Division of Behavioral Neurology and Cognitive Neuroscience, University of Iowa College of Medicine, Iowa City, IA 52242, USA

Introduction

Most current theories of choice are cognitive in perspective, with the exception of a few theories that addressed emotion (Janis and Mann, 1977; Mann, 1992). These theories of choice assume that decisions derive from an assessment of the future outcomes of various options and alternatives through some type of cost–benefit analyses. Some of these theories have addressed emotion as a factor in decision-making. However, they addressed emotions that are the consequence of some decision (e.g., the disappointment or regret experienced after some risky decision that worked out badly), rather than the affective reactions arising directly from the decision itself at the time of deliberation. Some theorists have proposed that people make judgements not only by evaluating the consequences and their probability of occurring, but also and even sometimes primarily at a gut or emotional level (Damasio, 1994; Schwartz and Clore, 1983; Zajonc, 1984). Our work with patients with bilateral lesions of the ventromedial (VM) prefrontal cortex provides evidence that supports this proposal. Lesions of the VM prefrontal cortex interfere with the normal processing of somatic or emotional signals, while sparing most basic cognitive functions. Such damage leads to impairments in the decision-making process, which seriously compromise the quality of decisions in daily life. In this chapter, we review evidence in support of the view that the process of decision-making depends in many important ways on neural substrates that regulate homeostasis, emotion, and feeling.

One of the first and most famous cases of the so-called 'frontal lobe syndrome' was the patient Phineas Gage, described by Harlow (Harlow, 1848, 1868). Phineas Gage was a railroad construction worker, and survived an explosion that blasted an iron tamping bar through the front of his head. Before the accident, Gage was a man of normal intelligence, energetic and persistent in executing his plans of operation. He was responsible, sociable, and popular among peers and friends. After the accident, his medical recovery was remarkable. He survived the accident with normal intelligence, memory, speech, sensation, and movement. However, his behavior changed completely. He became irresponsible, untrustworthy, and impatient of restraint or advice when it conflicted with his desires. Using modern neuroimaging techniques, Damasio and colleagues have reconstituted the accident by relying on measurements taken from Gage's skull (Damasio, Grabowski, Frank et al., 1994). The key finding of this neuroimaging study was that the most likely placement of Gage's lesion included the VM region of the prefrontal cortex, bilaterally. The ventromedial sector includes both the gyrus rectus and mesial half of the orbital gyri, as well as the inferior half of the medial prefrontal surface, from its most caudal aspect to its most rostral in the frontal pole. Mesial sectors of areas 10 and 11, areas 12, 13, and 25, and subgenual sectors of areas 24 and 32 of Brodmann are included in this sector, as is the white matter subjacent to all of these areas (Fig. 1; also see pp. 24 and 25 of Damasio, 1995).

---

* Corresponding author. Fax: +1 (319) 356-4505;
E-mail: antoine-bechara@uiowa.edu

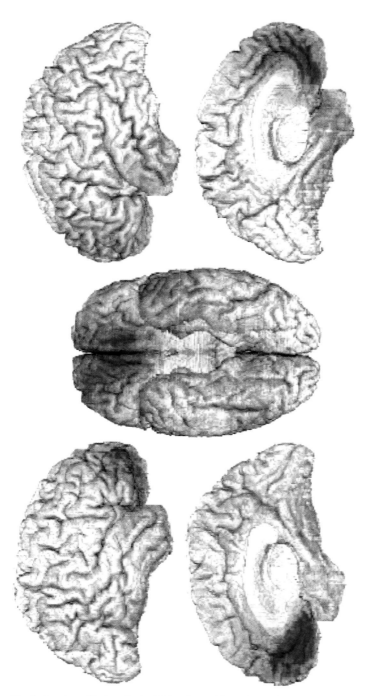

Fig. 1. Overlap of lesions in the VM patients. The red color indicates an overlap of four or more patients.

The case of Phineas Gage paved the way for the notion that the frontal lobes were linked to social conduct, judgement, decision-making, and personality. A number of instances similar to the case of Phineas Gage have since appeared in the literature (Ackerly and Benton, 1948; Brickner, 1932; Welt, 1888). Interestingly, all these cases received little attention for many years. The revival of interest in various aspects of the 'frontal lobe syndrome' was triggered in part by the patient described by Eslinger and Damasio (1985), a modern counterpart to Phineas Gage. Over the years, we have studied numerous patients with this type of lesion. Such patients with damage to the VM prefrontal cortex develop severe impairments in personal and social decision-making, in spite of otherwise largely preserved intellectual abilities. These patients had normal intelligence and creativity before their brain damage. After the damage, they begin to have difficulties planning their workday and future, and difficulties in choosing friends, partners, and activities. The actions they elect to pursue often lead to losses of diverse order, e.g., financial losses, losses in social standing, losses of family and friends. The choices they make are no longer advantageous, and are remarkably different from the kinds of choices they were known to make in the pre-morbid period. These patients often decide against their best interests. They are unable to learn from previous mistakes as reflected by repeated engagement in decisions that lead to negative consequences. In striking contrast to this real-life decision-making impairment, problem-solving abilities in laboratory settings remain largely normal. As noted, the patients have normal intellect, as measured by a variety of conventional neuropsychological tests (Bechara, Damasio, Tranel and Anderson, 1998a; Damasio, 1994; Damasio, Tranel and Damasio, 1990; Eslinger and Damasio, 1985).

The somatic marker hypothesis

This particular class of patients presents a puzzling defect. It is difficult to explain their disturbance in terms of defects in knowledge pertinent to the situation (Anderson, Bechara, Damasio et al., 1999; Saver and Damasio, 1991) or general intellectual compromise (Damasio et al., 1990). The disturbance also cannot be explained in terms of defects in language

comprehension or expression (Anderson, Damasio, Jones and Tranel, 1991), or in working memory or attention (Anderson et al., 1991; Bechara et al., 1998a). For many years, the condition of these patients has posed a double challenge. First, although the decision-making impairment is obvious in the real life of these patients, there had been no laboratory probe to detect and measure this impairment. Second, there had been no satisfactory account of the neural and cognitive mechanisms underlying the impairment. While these VM patients were intact on nearly all neuropsychological tests, the patients did have a compromised ability to express emotion and to experience feelings in appropriate situations. In other words, despite normal intellect, there were abnormalities in emotion and feeling, along with the abnormalities in decision-making. Based on these observations, the *somatic marker hypothesis* (Damasio, 1994; Damasio, Tranel and Damasio, 1991) was proposed.

Background assumptions of the hypothesis

In addition to an operative self and consciousness (Damasio, 1999), the somatic marker hypothesis is based on four main assumptions. The first is that human reasoning and decision-making depend on many levels of neurobiological operation, some of which are conscious and overt, and some of which are not. Conscious, overt cognitive operations depend on sensory images that are based on the activity of early sensory cortices. Second, those cognitive operations, regardless of the content of images, depend on support processes such as attention and working memory. Third, reasoning and decision-making depend on the availability of knowledge about *situations*, *actors*, *options for action*, and *outcomes*. Such knowledge is stored in 'dispositional' form throughout higher-order cortices and some subcortical nuclei. (The term dispositional is synonymous with implicit, and non-topographically organized; see Damasio (1989a,b, 1994) and Damasio and Damasio (1994) for details on dispositional knowledge and the convergence zone framework.) Dispositional knowledge can be made explicit in (a) motor responses of varied types and complexity (some combinations of which can constitute emotions), and in (b) images. The results of all motor responses, including those that are not generated consciously, can be represented in images. Finally, knowledge can be classified as follows. (a) Innate and acquired knowledge concerning bioregulatory processes and body states and actions, including those which are made explicit as emotions. (b) Knowledge about entities, facts (e.g., relations, rules), actions and action-complexes, and stories, which is usually made explicit as images. (c) Knowledge about the linkages between (b) and (a) items, as reflected in individual experience. (d) Knowledge resulting from the categorizations of items in (a), (b), and (c).

Specific structures and operations of somatic markers

The VM prefrontal cortex is a repository of dispositionally recorded linkages between factual knowledge and bioregulatory states

Structures in VM prefrontal cortex provide the substrate for learning the association between certain classes of complex situation, on the one hand, and the type of bioregulatory state (including emotional state) usually associated with that class of situation in prior individual experience. The ventromedial sector holds linkages between the facts that compose a given situation, and the emotion previously paired with it in an individual's experience. The linkages are 'dispositional' in the sense that they do not hold explicitly the representation of the facts or of the emotional state, but rather the potential to reactivate an emotion by acting on the appropriate cortical or subcortical structures. In other words, the experience of a complex situation and its components (e.g., a certain configuration of actors and actions requiring a response; a set of response options; a set of immediate and long-term outcomes for each response option) is processed in sensory and motor terms. Then, it is recorded in dispositional and categorized form (the records are maintained in distributed form in large-scale systems which involve many cortices including those in prefrontal sectors other than the ventromedial). But the experience of some of those components, individually or in sets, has been associated with emotional responses, which were triggered from cortical and subcortical limbic sites that were dispositionally prepared to organize such

responses. The VM prefrontal cortex establishes a linkage between the disposition for a certain aspect of a situation (for instance, the long-term outcome for a certain response option), and the disposition for the type of emotion that in past experience has been associated with the situation.

Experimental evidence
Somatic responses to emotionally charged stimuli.
One prediction of the hypothesis that the VM cortex links certain classes of complex situation with the type of bioregulatory state associated with that class of situation is that VM patients would fail to generate somatic states to emotionally charged stimuli. We tested this prediction in the following study (Damasio et al., 1990). Three groups of subjects were tested. (1) Patients with bilateral VM damage and impairments in social conduct and decision-making ($n = 5$). (2) Brain-damaged controls, i.e., patients with lesions outside the VM sector, and without defects in decision-making and social conduct ($n = 6$). (3) Normal control subjects, with no history of neurological or psychiatric disease ($n = 5$). The three groups were of comparable age and education. All subjects were shown two types of visual stimuli: (1) target stimuli, including pictures of emotionally charged stimuli such as mutilations and nudes; and (2) non-target stimuli, including pictures of emotionally neutral stimuli such as farm scenes and abstract patterns. The dependent measure was the skin conductance response (SCR), a highly sensitive index of autonomic responding, especially with regard to emotionally charged stimuli.

Prior to showing the subjects the target and non-target pictures, it was determined that they were all capable of generating SCRs to basic physical stimuli such as a loud noise or deep breath. This was in order to rule out the possibility that a lack of SCR to the pictures could be attributed to a general, nonspecific lack of autonomic response. When viewing the target and non-target stimuli, both the normal and brain-damaged control groups displayed large-amplitude SCRs to the target pictures, and little or no response to the non-targets (Fig. 2). By contrast, the VM patients generated almost no response to the target pictures, and completely failed to show the standard target and non-target SCR difference, despite the fact that their ability to generate SCRs to basic physical

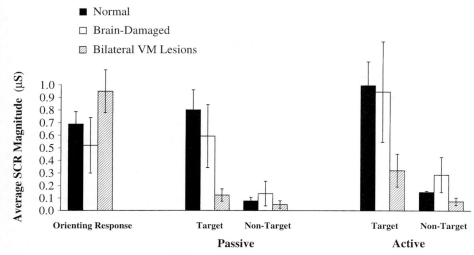

Fig. 2. Skin conductance response (SCR) magnitudes to orienting stimuli, neutral pictures, and emotionally charged pictures in normal controls, brain-damaged controls, and ventromedial (VM) prefrontal patients. In the VM group, the SCR magnitude for the emotionally charged pictures is abnormal. Note that when similar pictures were viewed actively (i.e., patients had to describe the content), the SCR magnitude from VM patients was normal (Damasio et al., 1990).

stimuli was intact. This finding is consistent with the prediction of the somatic marker hypothesis outlined earlier.

Somatic responses from internally generated images of emotional situations. As a further test of the somatic marker hypothesis, we predicted that VM patients would have a reduced ability to experience emotions when they recall specific emotional situations from their personal life. Recently, we have started to address this prediction using a procedure where the subject is asked to think about a situation in their life in which they felt each of the following emotions: happiness, sadness, fear, and anger. After a brief description of each memory is obtained, the subject is then put to a physiological test. The subject is asked to image and re-experience each emotional experience, while their SCR activity is monitored. As a control condition, the subject is asked to recall and image a neutral, non-emotional experience, such as a routine morning of preparing to go to work. At the end of the imagery of the emotional situation, each subject is asked to rate how much they felt the target emotion (on a scale of 0 to 4).

Using this emotional imagery procedure, we tested several patients with bilateral VM lesions. Analysis of the findings is in a preliminary stage,

but so far, it suggests that VM patients were able to retrieve previous emotional experiences, i.e., they were able to recall emotional situations that occurred before their brain lesion (such as weddings, funerals, car accidents, angry situations). However, they had difficulties re-experiencing happy and sad events. The VM patients had difficulties evoking changes in their SCR activity when recalling happy and sad emotions. Also, these patients gave low subjective ratings of the imagery of these emotions. At least with regard to the happy and sad emotions, the VM patients had a memory of the emotion, but they lacked the emotion of that memory.

Although the VM patients had difficulties re-experiencing happy and sad emotions, they gave higher subjective ratings of feeling anger and fear relative to neutral. They also showed higher SCR activity during the imagery of the fearful and angry situations, relative to neutral. Although all the VM patients we tested so far seemed to reliably express the anger emotion, the expression of the fear emotion was less reliable, i.e., some VM patients could not express it, and in those who expressed it, it was less intense. Obviously, these are preliminary findings that require further investigation. However, they suggest that damage to the VM cortex weakens the ability to re-experience an emotion from the recall of an appropriate emotional event, and thus support the prediction of the somatic marker hypothesis that VM patients have a weakened ability to generate somatic states or emotions.

Patients with bilateral VM damage are not emotionless. It is important to emphasize that the VM patients are not emotionless. For example, they can evoke somatic states in response to emotional stimuli if they verbalize the content of the stimuli (Damasio et al., 1990), and they can reliably evoke anger (and to a much lesser extent fear) from recall of personal experiences. These observations are consistent with the original observations of Butter et al. (Butter, Mishkin and Mirsky, 1968; Butter, Mishkin and Rosvold, 1963; Butter and Snyder, 1972) in monkeys with orbital lesions. The orbital lesions produced a clear reduction in the aggressive behavior of these monkeys, but the reduction was situational. In other words, the animals could still demonstrate aggression when brought back to the colony where they had been dominant, suggesting that the capacity to display aggression had not been eliminated in these monkeys, and that the lesions did not produce a permanent or pervasive state of 'bluntness of affect'.

Another important point is that although the VM patients seem unable to couple complex exteroceptive information with interoceptive information (somatic states) derived from complex cognitive scenarios (e.g., emotional pictures, etc.), many of these patients seem able to couple simple exteroceptive with interoceptive information. This finding was revealed in a fear-conditioning study, which consisted of the use of four different colors of monochrome slides as conditioned stimuli (CS), and a startlingly loud sound (100 dB) as the unconditioned stimulus (US) (Bechara, Tranel, Damasio et al., 1995). Electrodermal activity (SCR) served as the dependent measure of autonomic conditioning. The emotional conditioning of each subject included three phases. The first was a habituation phase. The second was a conditioning phase. In this phase, only one of the colors was paired with the US. These CS slides were presented at random among the other colors. The third was an extinction phase.

Using this emotional conditioning procedure, we tested a group of five VM patients and ten matched control subjects. The VM patients did acquire the SCR conditioning with the loud noise (Bechara, Damasio, Damasio and Lee, 1999). This suggests that the VM cortex is not essential for emotional conditioning. However, it is important to note that not all VM patients were able to acquire conditioned SCRs (Tranel, Bechara, Damasio and Damasio, 1996). The reason why some VM patients do not acquire emotional conditioning seems to depend on whether the anterior cingulate and/or the basal forebrain is involved in the lesion. This suggestion is preliminary and the issue is the subject of current investigations. The sparing of emotional conditioning by VM lesions is consistent with conditioning studies in animals showing that the VM cortex is not necessary for acquiring fear conditioning (Morgan and LeDoux, 1995). This is in contrast to the amygdala which appears essential for coupling a stimulus with an emotional state induced by a primary aversive unconditioned stimulus such as an electric shock (Davis, 1992; Kim, Rison and Fanselow, 1993; LeDoux, 1993). Murray and colleagues have elegantly shown that specific neurotoxic lesions of the primate amygdala interfere in learning the association between stimuli and the value of a particular reward (Malkova, Gaffan and Murray, 1997). Human studies have also shown that lesions of the amygdala impair emotional conditioning with an aversive loud sound (Bechara et al., 1995; LaBar, LeDoux, Spencer and Phelps, 1995), and functional neuroimaging studies have shown that the amygdala is activated during such conditioning tasks (LaBar, Gatenby, Gore et al., 1998). The conditioning results suggest that the amygdala and VM cortex play different roles in emotional processing. These distinct roles have implications for understanding the nature of the decision-making deficits observed in both VM and amygdala patients as will be discussed in a later section.

The reactivation of somatic signals related to previous individual contingencies

When a situation arises for which some factual aspect has been previously categorized, related dispositions are activated in higher-order association cortices (including some prefrontal cortices). This leads to the recall of pertinently associated facts, which will be experienced in image form. Simultaneously, or nearly so, the related VM prefrontal linkages are also activated, and as a consequence, the emotional disposition apparatus is activated too. The result of the combined activation is the approximate reconstruction of a previously learned factual–emotional set. In short, when a situation of a given class recurs, factual knowledge pertaining to the situation (possible options of action, outcomes of such actions immediately and at longer term) is evoked in sensory images based on the appropriate sensory cortices. Depending on previous individual contingencies, signals related to some or even many of those images, or even the entire situation, act on the VM prefrontal cortex (which has previously acquired the link between the situation or its components and the class of somatic state). This in turn triggers the re-activation of the somatosensory pattern that describes the appropriate emotion.

Experimental evidence
The gambling task. The key feature of this task (Bechara, Damasio, Damasio and Anderson, 1994) is that it resembles the decisions made in real-life, in the way it factors reward, punishment, and uncertainty. The task involves four decks of cards named A, B, C, and D. The goal in the task is to maximize profit on a loan of play money. Subjects are required to make a series of 100 card selections. However, they are not told ahead of time how many card selections they are going to make. Subjects can select one card at a time from any deck they choose, and they are free to switch from any deck to another at any time, and as often as they wish. However, the subject's decision to select from one deck versus another is largely influenced by various schedules of immediate reward and future punishment. These schedules are pre-programmed and known to the examiner, but not to the subject, and they entail the following principles.

Every time the subject selects a card from deck A or deck B, the subject gets $100. Every time the subject selects deck C or deck D, the subject gets $50. However, in each of the four decks, subjects encounter unpredictable punishments (money loss). The punishment is set to be higher in the high paying decks A and B, and lower in the low paying decks C and D. If ten cards were picked from deck A over the course of trials, one earns $1000. However, in those ten card picks, five unpredictable punishments are

encountered, ranging from $150 to $350, bringing a total cost of $1250. The same goes for deck B. Every ten cards that were picked from deck B earn $1000. However, these ten card picks are encountered by one high punishment for $1250. On the other hand, every ten cards from deck C or D earn only $500, but only cost $250 in punishment. In essence, decks A and B are disadvantageous because they cost more in the long run; i.e., one loses $250 every 10 cards. Decks C and D are advantageous because they result in an overall gain in the end; i.e., one wins $250 every 10 cards. We note that since this original task, we have devised a new computerized version of the task with slightly different schedules of reward and punishment. However, the basic principles of the task remain the same.

We investigated the performance of normal controls ($n = 13$) and VM patients ($n = 6$) on this task. Normal subjects learned to avoid the bad decks A and B and prefer the good decks C and D. In sharp contrast, VM frontal patients did not show this learning, i.e., they preferred the bad decks A and B (Fig. 3). Although this task involves a long series of gains and losses, it is not possible for subjects to perform an exact calculation of the net gains or losses generated from each deck as they play. Thus, from these results we suggested that the patients' performance profile is comparable to their real-life inability to decide advantageously. This is especially true in personal and social matters, a domain for which in life, as in the task, an exact calculation of the future outcomes is not possible and choices must often be based on approximations, hunches, and guesses (Bechara et al., 1994).

Autonomic responses in anticipation of future outcomes. The more pertinent evidence in support of the somatic marker hypothesis and the reactivation of somatic signals related to prior experience is the failure to generate somatic signals when pondering decisions. This evidence comes from a study where we added a physiological measure to the gambling task. The goal was to assess somatic state activation while subjects were making decisions during the gambling task. We studied two groups: normal subjects ($n = 12$) and VM patients ($n = 7$). We had them perform the gambling task while we recorded their SCRs (Bechara, Tranel, Damasio and Damasio, 1996).

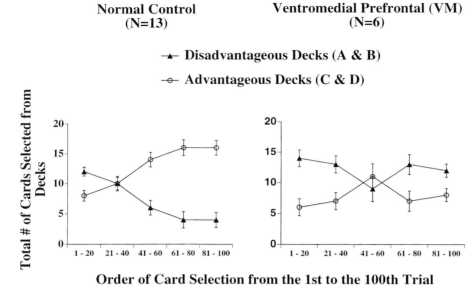

Normal Control **(N=13)**   **Ventromedial Prefrontal (VM)** **(N=6)**

Total # of Cards Selected from Decks

Order of Card Selection from the 1st to the 100th Trial

Fig. 3. Card selection on the gambling task as a function of group (normal control, ventromedial prefrontal (VM)), deck type (disadvantageous versus advantageous), and trial block. Normal control subjects shifted their selection of cards towards the advantageous decks. The VM prefrontal patients did not make a reliable shift and opted for the disadvantageous decks.

Both normal subjects and VM patients generated SCRs after they had picked the card and were told that they won or lost money. The most important difference, however, was that normal subjects, as they became experienced with the task, began to generate SCRs *prior* to the selection of any cards, i.e., during the time when they were pondering from which deck to choose. These anticipatory SCRs were more pronounced before picking a card from the risky decks A and B, when compared to the safe decks C and D. The VM patients entirely failed to generate SCRs before picking a card, and they also failed to avoid the bad decks and choose advantageously (Fig. 4). These results provide strong support for the somatic marker hypothesis notion regarding the reactivation of signals related to previous individual contingencies. They suggest that decision-making is guided by emotional signaling (or somatic states) that are generated in anticipation of future events.

A marker role for signals related to previous emotional state contingencies

It is proposed that somatosensory patterns can be established via one of two chains of events (Fig. 5).

One chain involves the activation of true somatic states in the body proper. When an appropriate emotional (somatic) state is actually re-enacted, signals from its activation are then relayed back to subcortical and cortical somatosensory processing structures, especially in the somatosensory (SI and SII) and insular cortices. This anatomical system is described as the 'body loop'. After emotions have been expressed and experienced at least once, one can form representations of these emotional experiences in the somatosensory/insular cortices. Therefore, after emotions are learned, another possible chain of physiologic events is to by-pass the body altogether, activate the insular/somatosensory cortices directly, and create a fainter image of an emotional body state than if the emotion were actually expressed in the body. This anatomical system is described as the 'as if body loop'.

The establishment of a somatosensory pattern appropriate to the situation, via the 'body loop' or via the 'as if' loop, either overtly or covertly, is co-displayed with factual evocations pertinent to the situation, and qualifies those factual evocations. In doing so, it operates to constrain the process of reasoning over multiple options and multiple future

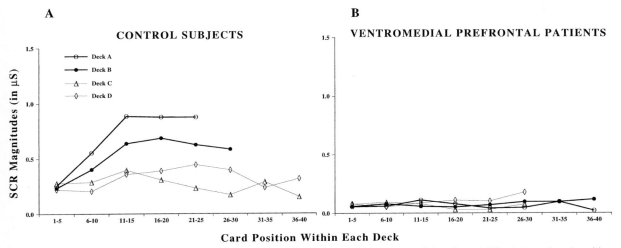

Fig. 4. Magnitudes of anticipatory SCRs as a function of group (control (A) versus ventromedial prefrontal (B)), deck, and card position within each deck. Note that control subjects gradually began to generate high-amplitude SCRs to the disadvantageous decks. The ventromedial prefrontal patients failed to do so.

outcomes. For instance, when the somatosensory image which defines a certain emotional response is juxtaposed to the images which describe a related scenario of future outcome, and which triggered the emotional response via the ventromedial linkage, the somatosensory pattern marks the scenario as good or bad. In other words, the images of the scenario are 'judged' and marked by the juxtaposed images of the somatic state. When this process is overt, the somatic state operates as an alarm signal or an incentive signal. The somatic state alerts one to the goodness or badness of a certain option–outcome pair. The device produces its result at the overtly cognitive level. When the process is covert the somatic state constitutes a biasing signal. Using an indirect and non-conscious influence, e.g., through nonspecific neurotransmitter systems such as dopamine or serotonin, the device influences cognitive processing.

Whether body states are evoked via a 'body loop' or 'as if' loop, the corresponding neural pattern can be made conscious and constitute a feeling. However, although many important choices involve feelings, a number of our daily decisions undoubtedly proceed without feelings. That does not mean that the evaluation that normally leads to a body state has not taken place, or that the body state or its surrogate has not been engaged, or that the dispositional machinery underlying the process has not been

activated. It simply means that the body state or its surrogate have not been attended. Without attention, neither will be part of consciousness, although either can be part of a covert action on the mechanisms that govern, without willful control, our appetitive (approach) or aversive (withdrawal) attitudes toward the world. In short, although the hidden machinery underneath has been activated, we may never know it.

Experimental evidence
Covert biases influence decisions before conscious knowledge does. Evidence suggests that somatic signals generated in anticipation of future outcomes do not need to be perceived consciously. We carried out an experiment similar to the previous one, where we tested normal subjects ($n = 10$) and VM patients ($n = 6$) on the gambling task, while recording their SCRs. However, every time the subject picked ten cards from the decks, we stopped the game briefly, and asked the subject to declare whatever they knew about what was going on in the game (Bechara, Damasio, Tranel and Damasio, 1997a). From the answers to the questions, we were able to distinguish four periods as subjects went from the first to the last trial in the task. The first was a pre-punishment period, when subjects sampled the decks, and before they had yet encountered any punishment. The sec-

125

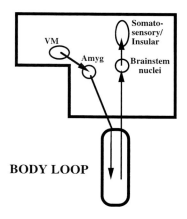

BODY LOOP

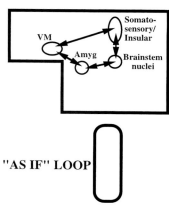

"AS IF" LOOP

Fig. 5. Simple diagrams illustrating the 'body loop' and 'as if loop' chain of physiologic events. In both 'body loop' and 'as if loop' panels, the brain is represented by the top black perimeter and the body by the bottom one.

ond was a pre-hunch period, when subjects began to encounter punishment, but when asked about what was going on in the game, they had no clue. The third was a hunch period, when subjects began to express a hunch about which decks were riskier, but they were not certain. The fourth was a conceptual period, when subjects knew very well the contingencies in the task, and which decks were the good ones, and which decks were the bad ones (Fig. 6).

When examining the anticipatory SCRs from each period, we found that in the normal subjects, there was no significant activity during the pre-punishment period. There was a substantial rise in anticipatory responses during the pre-hunch period, i.e., before

any conscious knowledge developed. This SCR activity was sustained for the remaining periods. When examining the behavior during each period, we found that there was a preference for the high paying decks (A and B) during the pre-punishment period. Furthermore, there was a hint of a shift in the pattern of card selection, away from the bad decks, even in the pre-hunch period. This shift in preference for the good decks, became more pronounced during the hunch and conceptual periods. The frontal patients on the other hand, never reported a hunch about which of the decks were good or bad. Furthermore, they never developed anticipatory SCRs, and they continued to choose more cards from decks A and B relative to C and D. Also, even though 30% of controls did not reach the conceptual period, they still performed advantageously. Although 50% of frontal patients did reach the conceptual period, they still performed disadvantageously (Bechara et al., 1997a).

These results show that VM patients continue to choose disadvantageously in the gambling task, even after realizing the consequences of their action. This suggests that these anticipatory SCRs represent unconscious biases derived from prior experiences with reward and punishment. These biases help deter the normal subject from pursuing a course of action that is disadvantageous in the future. This occurs even before the subject becomes aware of the goodness or badness of the choice s/he is about to make. Without these biases, the knowledge of what is right and what is wrong may still become available. However, by itself, this knowledge may not be sufficient to ensure advantageous behavior. Therefore, although the frontal patient may be fully aware of what is right and what is wrong, s/he fails to act accordingly. Thus, these patients may 'say' the right thing, but they 'do' the wrong thing.

Somatic markers participate in process as well as content

Certain emotion-related somatosensory patterns (somatic states) also act as boosters in the processes of attention and memory. However, it is possible that the mechanism by which somatic states influence decisions is separate from the mechanism by which somatic states boost and improve memory. In addi-

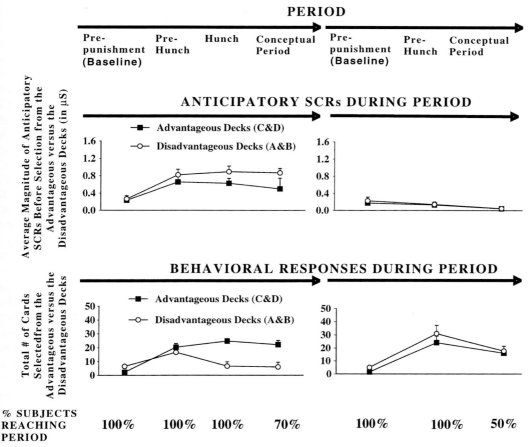

Fig. 6. Anticipatory SCRs and behavioral responses (card selection) as a function of four periods (pre-punishment, pre-hunch, hunch, and conceptual) from normal control subjects and ventromedial prefrontal patients.

tion to assisting with the process of specific experiential contents (e.g., certain combinations of facts and emotions), somatic states may also assist in the process of response inhibition. Somatic states provide the important signals that help decide whether to inhibit, or not, the response under consideration.

Experimental evidence
Somatic states boost memory. It has been known for some time that emotion interacts with and improves memory, although extreme emotion can sometimes weaken memory (Easterbrook, 1959). We asked whether the mechanism by which emotion improves memory is the same as, or different from, the mechanism through which emotion biases decisions. Based on other work (Cahill, Babinsky, Markowitsch and McGaugh, 1995), the amygdala appears to be important for the mechanism by which emotion improves memory. So at least as far as the amygdala is concerned, the mechanisms through which emotion modulates memory and decision-making may be inseparable. We asked whether these mechanisms are separable in the VM cortex. This question is important to address because it relates to whether decision-making is a distinct function facilitated by somatic markers, or simply a memory-related function that can be explained by mechanisms of memory. We tested groups of normal control subjects ($n = 12$)

127

and VM patients (n = 6) with anterior bilateral VM lesions that spared the basal forebrain region. We showed the subjects a series of slides, some of which were neutral (e.g., farm scenes), and some of which were emotional (e.g., scenes of a murder or an accident). We presented each picture for 5 s on a computer screen, followed by a 5-s blank screen. The series of pictures involved sixteen slides. Eight of the slides were neutral and eight were emotional. However, we divided the pictures into four sets, with four pictures in each set. Each set of four pictures contained two neutral and two emotional pictures. The pictures in set 1 were presented once each; in set 2, they were presented twice each; in set 3, they were presented four times each; and in set 4, they were presented eight times each. The order of the pictures was randomized. Then we tested subjects for their recall of the pictures. The recall of picture content was calculated for each subject, as a function of repetition times and emotional content.

Thus, this experiment actually separates the memory curve that is a function of repetition, from the curve that is a function of emotional content. In normal controls, memory improved as a result of repetition. The most important finding, however, was that both groups showed a response to the emotion manipulation, producing a better memory curve for pictures with emotional content (Fig. 7). The results indicate that the VM patients are able to use emotion in order to enhance their memory, suggesting that

the mechanism through which emotion modulates decision-making is different from that through which emotion modulates memory. Most importantly, these results also support the conclusion that emotional memory deficits cannot explain the decision-making impairment of VM patients.

Role of somatic states in impulsiveness and response inhibition. Impulsiveness is a poorly defined term, but it is often linked to dysfunction of the prefrontal cortex (Fuster, 1996; Jones and Mishkin, 1972; Miller, 1992), and it usually means the lack of response inhibition. We distinguish between two types of impulsive behaviors: motor impulsiveness, and cognitive impulsiveness. *Motor impulsiveness* is usually studied in animals under the umbrella of 'response inhibition'. In these paradigms, after establishing a habit of responding to a stimulus that predicts a reward, there is a sudden requirement to inhibit the previously rewarded response. Go/no go tasks, delayed alternation, and reversal learning are prime examples of paradigms that measure this type of behavior (Diamond, 1990; Dias, Robbins and Roberts, 1996; Fuster, 1990; Mishkin, 1964; Stuss, 1992). It has been proposed that in humans, motor impulsiveness can have several forms (Evenden, 1999). This includes (1) impulsive preparation, which involves making a response before all the necessary information has been obtained, and (2) impulsive execution, which involves quick action without thinking. These types of impulsive behaviors can be tested using a variety of procedures including the Matching Familiar Figures Test (MFFT), Proteus Mazes, and the Tower of London (Evenden, 1999). We note that frontal lobe damage, especially damage to the dorsolateral sector of the prefrontal cortex, is associated with impairments on the Wisconsin Card Sorting Task (WCST). These patients persist with the original classification of cards despite being told they are wrong, i.e., they fail to suppress a previously correct response. Analogous deficits associated with damage to the lateral prefrontal cortex (Brodmann area 9) have been reported in monkeys using attentional shift tasks, presumably analogous to the WCST (Dias et al., 1996; Dias, Robbins and Roberts, 1997). The perseveration on the WCST is certainly a deficit in impulse control, and it may be more complex than the motor impulsiveness de-

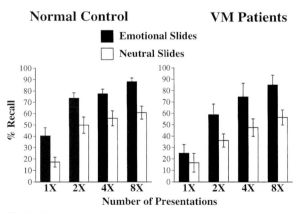

Fig. 7. Recall scores in normal controls and ventromedial prefrontal cortex (VM) patients as a function of repetition and emotional content. These patients showed a strong improvement in recall as a function of the emotional manipulation.

scribed earlier. However, there is some debate as to whether perseveration is actually an indication of impulsivity (Evenden, 1999).

The other type of impulsive behaviors is *cognitive impulsiveness*, which can be seen akin to an inability to delay gratification, and which is more complex than the other forms of impulsive behaviors. The term 'cognitive impulsiveness' has been used previously in human studies (Barratt, 1994), and it may be analogous to the term 'impulsive outcome' referring to a failure to delay gratification and evaluate the outcome of a planned action (Evenden, 1999). Cognitive impulsiveness may be illustrated with the example in which a child sees a candy on the table, the parent says "no, you must wait 30 minutes before you can have the candy". Otherwise, the child will face negative consequences. The child understands the information and may deliberate for a short while, but after 2 minutes, he can no longer resist the temptation, delay the gratification, and inhibit his response to reach for the candy. The child gets up and reaches for the candy.

Based on previous studies, we believe that VM patients with more anterior lesions that spare the basal forebrain do not have motor impulsiveness, although those with lesions involving the basal forebrain may have the defect (Bechara et al., 1998a). The absence of motor impulsiveness is supported by evidence showing, first, that when observing the behavior of VM patients during their performance of the gambling task, one finds that these patients switch decks whenever they receive punishment. They switch decks just like normal controls, a performance that does not indicate lack of inhibition of a previously rewarded response. Second, most VM patients are unimpaired on delay task procedures considered sensitive to deficits in response inhibition.

On the other hand, it is possible that VM patients have cognitive impulsiveness. The behavior of VM patients on the gambling task, and in real-life, can be viewed as similar to the impulsive behavior of the child with the candy. That is, when the patients are presented with a deck of cards with a large immediate reward, but with delayed costs, the patients seek the reward. These VM patients seem unable to delay the gratification of the reward for too long, as indicated by their tendency to return quickly and more

often to the decks that yield high immediate reward, but even a larger future loss. Sahakian and colleagues (Plaisted and Sahakian, 1997; Rahman, Robbins and Sahakian, 1999a; Rahman, Sahakian, Hodges et al., 1999b) have proposed an "inhibition hypothesis" to account for the social cognition deficits observed in patients with frontal lobe dysfunction. The hypothesis is partially based on work in nonhuman primates with the attentional set-shifting paradigm developed by Roberts and colleagues (Dias et al., 1996, 1997). The hypothesis proposes that the inability of the patient to suppress the response evoked by the stimulus present in the immediate environment prevent the patient from selecting an appropriate action plan. Thus, the behavior becomes dominated by the immediate emotional impact of the stimulus at hand. The work of the same group reveals an important finding in that their patients with either orbitofrontal lesions or frontotemporal dementia do not suffer from a complete loss of inhibitory control (Rahman et al., 1999b; Rogers, Everitt, Baldacchino et al., 1999). The patients make disadvantageous choices, but only after taking some time to deliberate. These findings are consistent with the argument made earlier that VM patients may not have motor impulsiveness, but rather they suffer from cognitive impulsiveness.

The inhibition hypothesis and the construct of cognitive impulsiveness is intriguing because it can account for some of the behavioral impairments associated with prefrontal cortex damage. However, there is a need to explain the nature of the mechanism that triggers the inhibition of the response. In other words, what is the nature of the mechanism that decides when to suppress, or not to suppress, a certain response, such as the seeking of a large immediate reward? We argue that the nature of this mechanism is a somatic state, i.e., an emotional signal. Using the impulsive behavior of the child with the candy as an illustrative example, one can see the conflict created by the decision to reach, or not to reach, for the candy. On the one hand, there are positive somatic states generated by the immediate and available reward (the candy). On the other hand, there are negative somatic states generated by the delayed but absent punishment posed by the parent. Based on previous experience, if the punishment was severe enough, then the evoked negative somatic states from the threat of punishment would override the positive somatic states from the reward, and the overall somatic state would signal the mechanisms of response inhibition to inhibit the response to secure the reward. If there was a failure to evoke these negative somatic states, and the child still sought the immediate reward, then the decision to reach for the candy is considered disadvantageous, and the behavior would be reminiscent of that seen in patients with frontal lobe dysfunction. However, there are situations where previous punishment by the parent in similar situations was mild. In this case, the negative somatic state evoked by the threat of punishment is weak in comparison to the positive state from the immediate reward. In other words, the negative somatic states from the punishment may be evoked, but not sufficiently enough to override the positive state from the immediate reward. In this case, if the child reaches for the candy, the behavior may be considered as normal, advantageous, and not impulsive.

These examples illustrate the importance of somatic states in influencing the mechanisms of response inhibition. Specifically, response inhibition and cognitive impulsiveness are constructs that address the inhibition of a behavior that used to predict reward, but now it may lead to punishment. The examples above describe two situations involving an immediate reward and a future punishment, but one situation requires the inhibition of the action to seek the reward, whereas the other situation does not, depending on the weights of reward versus punishment. The constructs of response inhibition and cognitive impulsiveness do not address the issue of when to inhibit, or not inhibit the response under consideration. However, the concept of somatic states does address this issue. The proposal is that somatic states serve as important signals that help decide whether to inhibit, or not to inhibit, a given response.

Thus, although cognitive impulsiveness and response inhibition can account for some of the observed behavior of VM patients (Plaisted and Sahakian, 1997), this explanation does not fully address the underlying cause of the decision-making impairment. There is a need for a mechanism that guides the decision on when to inhibit, and when not to inhibit a particular response, i.e., to explain why the impulsive behavior occurs. We suggest that cognitive impulsiveness result from a failure to evoke somatic states when imagining the scenarios of future punishment. Therefore, we suggest that the anticipatory SCRs observed in the gambling task are indices of these somatic states which serve as important signals that help decide whether to inhibit, or not to inhibit, the selection of cards from a particular deck.

Somatic markers facilitate logical reasoning

It is proposed that somatic markers normally help constrain the decision-making space by making that space manageable for logic-based, cost–benefit analyses. In situations where there is remarkable uncertainty about the future, where many similarly valued response options exist, and in which the decision should be influenced by previous individual experience, such constraints permit the organism to decide efficiently within short time intervals. In the absence of a somatic marker, options and outcomes may become virtually equalized, and the process of choosing will depend entirely on logic operations over many option–outcome pairs. The strategy is necessarily slower and may fail to take into account previous experience. This is the pattern of slow and error-prone decision-making behavior we often see in VM patients. Choosing according to the immediate consequences of an action, irrespective of the long-term consequences, is a related pattern.

Experimental evidence
VM patients take longer to decide. One prediction of the above hypothesis is that VM patients would take longer time to deliberate in situations where there is remarkable uncertainty about the future, and in which the decision should be influenced by previous individual experience. We are currently in the process of testing this prediction with gambling task. However, studies from other laboratories have already provided support for this notion (Rahman et al., 1999b; Rogers et al., 1999). Using a decision-making task similar to the gambling task, it was found that patients with frontal lobe dysfunction would deliberate significantly longer before making their choices than matched normal control subjects (Rahman et al., 1999b; Rogers et al., 1999). It is important to note that such patients often take inordinate amounts of time to decide in real-life situations, such as going to the supermarket and choosing, for

instance, between one brand of canned food and another (Damasio, 1994).

VM patients choose according to the immediate consequences, irrespective of the future consequences of their actions. We asked whether VM patients fail to account for the future, and simply choose according to the immediate consequences of an action. We considered three possibilities that may account for the behavior of VM patients. The first was hypersensitivity to reward, where the prospect of a large immediate gain outweighs any prospect of future loss. The second was insensitivity to punishment, where the prospect of a large loss cannot override any prospect of gain. The third was insensitivity to future consequences, positive or negative, so that the subject is oblivious to the future and is guided by immediate prospects (Bechara, Tranel and Damasio, 2001). In trying to distinguish among these different possibilities, we first designed a variant of the original gambling task with decks EFGH. In the variant task, the advantageous decks (E and G) are those with high immediate punishment (pay $100 each time you pick a card), but higher future reward (win more in the long run). The disadvantageous decks (F and H) are those with low immediate punishment (pay only $50 immediately), but they also yield lower future reward (lose in the long run). Second, when subjects played both the original task (with decks ABCD) and the variant task (with decks EFGH), we recorded and measured SCRs that occurred in response to reward or punishment. The rationale was that hypersensitivity to reward should be associated with abnormally higher SCRs when receiving a reward. On the other hand, insensitivity to punishment should be associated with abnormally lower SCRs when receiving a punishment.

Thus, we used a combined behavioral and psychophysiological approach to distinguish among the three possibilities. We tested normal and VM patients on the variant task. The results revealed that normal controls selected more from the advantageous decks E and G, and less from the disadvantageous decks F and H. By contrast, the VM patients tended to go in the opposite direction, and they selected more from the disadvantageous decks as compared to the advantageous decks. When looking at the SCRs from both the original (ABCD) and vari-

ant (EFGH) tasks, the SCRs of normal controls and VMF subjects after receiving a reward or punishment were not significantly different, as a function of group.

These results are inconsistent with a hypersensitivity to reward explanation for two reasons. First, the large sums of reward are in decks E and G, but most frontal patients did not go for them, i.e., they were not lured by the high reward. Second, the SCRs of VM patients after receiving reward were not significantly different from those of normal subjects. These results are also inconsistent with insensitivity to punishment explanation for two reasons. First, the high punishments are in decks E and G. The VM patients ended up avoiding these decks because they had to pay large immediate penalties. Second, the SCRs of VM patients after receiving punishment were not significantly different from normal subjects.

The combined results from the original and variant versions of the gambling task are inconsistent with the hypersensitivity to reward and insensitivity to punishment explanations. A parsimonious explanation is that in most VM patients, the decision-making impairment is due to insensitivity to future consequences, whatever they may be. The subject appears oblivious to the future and guided predominantly by immediate prospects (Bechara et al., 2001).

A neural system for the activation of somatic states

Based on neuroanatomy from nonhuman primates, the VM prefrontal cortex receives projections from all sensory modalities, directly or indirectly (Chavis and Pandya, 1976; Jones and Powell, 1970; Pandya and Kuypers, 1969; Potter and Nauta, 1979). In turn, the VM prefrontal cortex sends projections to central autonomic control structures (Nauta, 1971), and such projections have a physiological influence on visceral control (Hall, Livingston and Bloor, 1977). The VM cortices have extensive bi-directional connections with the hippocampus and amygdala (Amaral and Price, 1984; Amaral, Price, Pitkanen and Carmichael, 1992; Van Hoesen, 1985). In other words, VM prefrontal cortices are possible and natural recipients of signals concerning the multiple-site representation of scenarios of complex situations, all over early sensory cortices. Moreover, VM cortices are also recipients of signals from somatosensory structures and bioregulatory structures. Finally, VM cortices can originate signals to bioregulatory structures such as the amygdala, hypothalamus, and autonomic centers in the brainstem. The somatic marker hypothesis posits that these VM cortices contain convergence zones which hold a record of temporal conjunctions of activity in varied neural units (i.e., sensory cortices and limbic structures) hailing from both external and internal stimuli. Thus, when parts of certain exteroceptive–interoceptive conjunctions are re-processed, consciously or unconsciously, their activation is signaled to VM cortices, which in turn activate somatic effectors in amygdala, hypothalamus, and brainstem nuclei. This latter activity is an attempt to reconstitute the kind of somatic state that belonged to the conjunction (Fig. 5).

Thus, the neural network mediating the activation of somatic states involves numerous neural regions. The VM prefrontal cortex is one critical region. However, there are other critical components in this neural network, namely, the amygdala, the somatosensory/insular cortices, and the peripheral nervous system.

The VM prefrontal cortex

The prefrontal cortex is a relatively large cortical structure. It has been linked to cognitive functions such as working memory (Fuster, 1991; Goldman-Rakic, 1992; Milner, Petrides and Smith, 1985; Petrides, 1996), planning, and cognitive estimation (Shallice, 1982, 1993; Shallice and Burgess, 1991b). In addition, developmental changes involving neural pruning have been observed to occur several years after birth (Rapoport, Giedd, Blumenthal et al., 1999; Sowell, Thompson, Holmes et al., 1999). Therefore, it is important to ask which sector of the prefrontal cortex is related to decision-making, as opposed to other cognitive functions. Furthermore, given the developmental changes that occur in the prefrontal cortex, it is important to ask whether there is recovery of function if the damage was acquired early in life.

Decision-making and working memory are mediated by separate anatomical sectors
The rationale for the notion that working memory and decision-making are distinct functions comes

from the observations that VM patients suffer from impairments in decision-making, while preserving a normal level of memory and intellect. On the other hand, although some patients with lesions in the dorsolateral (DL) sector of the prefrontal cortex complain of memory impairments, they do not appear to suffer from impairments in decision-making, as judged from their behavior in real-life. Using modified delay-task procedures (delayed response and delayed non-matching to sample) to measure working memory (Fuster, 1991; Goldman-Rakic, 1987), and the gambling task to measure decision-making, the following experiment was performed. A group of normal control subjects ($n = 21$), patients with bilateral VM lesions ($n = 9$), and patients with right or left lesions of the DL sector of the prefrontal cortex ($n = 10$) were tested on the delay and gambling tasks (Bechara et al., 1998a). The gambling task was the same task mentioned previously, and the delay task procedures were modifications of classical delay tasks.

Delay tasks that are used in nonhuman primates are too simple for use with humans. Therefore, a distracter was introduced during the delay between the cue and the response. The purpose of the distracter was to interfere with the ability of the subject to rehearse during the delay, and thus to increase the demands of the tasks on working memory. In the *delayed response* experiment, four cards appeared for 2 s on a computer screen, with two of the cards face down, and the other two face up showing red or black colors. The cards disappeared for 1, 10, 30, or 60 s and then reappeared, but this time all the cards were faces down. The correct response was to select the two cards that were first face up. During the delay, the subject had to read aloud a series of semantically meaningless sentences. The scores were calculated as the percent of correct choices that were made by the subject at the 10, 30, and 60-s delays. Impaired performance on the delayed response task was defined as achieving a percent correct score of 80 or less at the 60-s delay, a cutoff score below which no normal control ever performed (Bechara et al., 1998a). In the *delayed non-matching to sample* experiment, the task was similar to the delayed response task except that only one card appeared initially on the computer screen for 2 s. The card was face up and was either red or black. After the card disappeared for 1, 10, 30, or 60 s, four cards appeared on the screen, all were face up, two of them were red, and two were black. The correct response was to select the two cards of opposite color (non-matching) to the initial sample card.

In this experiment we used two types of delay tasks because studies in nonhuman primates show that different areas of the DL frontal cortex are associated with different domains of working memory. The inferior areas of the DL sector have been associated with object memory, whereas the superior areas have been associated with spatial memory (Goldman-Rakic, 1987, 1992; Wilson, Scalaidhe and Goldman-Rakic, 1993). A similar dissociation was found in humans, using functional neuroimaging techniques (Courtney, Ungerleider, Keil and Haxby, 1996). The delayed response tasks have been designed to tax the spatial (*where*) domain of working memory, whereas the delayed non-matching to sample tasks are supposed to tax the object (*what*) domain of working memory (Fuster, 1990; Wilson et al., 1993). Since the lesions in the patients we studied were not restricted to the inferior or superior regions, and the lesions spanned a wide area of DL frontal cortices, we used both types of delayed tasks because we anticipated that both domains of working memory (spatial and object) may be affected. In other words, our attempt was not to sort out differences between different types of working memory, but rather, to cover a range of working memory with one set of tasks. Therefore, the results we report here are an average of the results obtained from both delay tasks. In the next section, we use the term 'delay tasks' to refer to both procedures (see Bechara et al. (1998a) for detailed results for each individual delay task).

This experiment revealed two intriguing findings. First, working memory is not dependent on the intactness of decision-making, i.e., subjects can have normal working memory in the presence or absence of deficits in decision-making. Some VM frontal patients who were severely impaired in decision-making (i.e., abnormal in the gambling task) had superior working memory (i.e., normal in the delay tasks). On the other hand, decision-making seems to be influenced by the intactness or impairment of working memory, i.e., decision-making is worse in the presence of abnormal working memory. Patients

with right DL frontal lesions and severe working memory impairments showed low normal results in the gambling task (Fig. 8). In summary, working memory and decision-making were asymmetrically dependent. Second, although all VM patients tested in this experiment were impaired on the gambling task, they were split in their performance in the delay tasks. Five patients were abnormal in the delay tasks (Abnormal Gambling/Abnormal Delay), and four patients were normal in the delay tasks (Abnormal Gambling/Normal Delay). The most important finding is that all patients in the Abnormal Gambling/Abnormal Delay group had lesions that extended posteriorly, possibly involving the basal forebrain region. However, the other group (Abnormal Gambling/Normal Delay), had lesions that were more anterior and did not involve the basal forebrain.

It is important to note that in this experiment, only the patients with right DL lesions were impaired on these working memory tasks. All patients with left DL frontal lesions had normal working memory. The absence of a working memory impairment in left DL patients is not surprising because, during the delay, the verbal memorization of cues was probably avoided by the interference procedure, thus rendering the task primarily non-verbal. This is consistent with several functional neuroimaging studies in humans. These studies showed higher activation in the right DL frontal cortex, relative to the left, during the performance of similar delay tasks (D'Esposito, Detre, Alsop et al., 1995a; D'Esposito, Shin, Detre et al., 1995b; Jonides, Smith, Koeppe et al., 1993; McCarthy, Blamire, Puce et al., 1994; Petrides, Alivisatos, Meyer and Evans, 1993; Smith, Jonides, Koeppe et al., 1995; Swartz, Halgren, Fuster et al., 1995). These findings reveal a double dissociation (cognitive and anatomic) between deficits in decision-making (anterior VM) and working memory (right DL). They reinforce the special importance of the VM prefrontal region in decision-making, independently of a direct role in working memory.

Decision-making versus other frontal lobe functions
In a series of studies with nonhuman primates and human subjects, Petrides (Petrides, 1985, 1990) has established a link between frontal lobe damage and learning *conditional associations*, i.e., the ability to associate responses with specific stimuli on the ba-

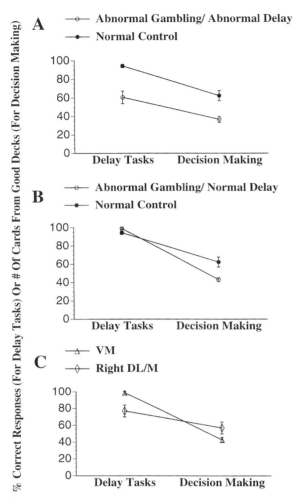

Fig. 8. Means ± s.e.m. of the average of percent correct responses from the two delay tasks, or the total number of cards selected from the good decks. (A) The behavioral results on the gambling task and the delay tasks from the groups of VM patients with lesions that extended more posteriorly and probably affected the basal forebrain. Note their abnormal performance on both the delay tasks and gambling task relative to normal controls. (B) The behavioral results on the gambling task and the delay tasks from the groups of VM patients with lesions that were more anterior and spared the basal forebrain. Note their normal performance on the delay tasks but abnormal performance on the gambling task relative to normal controls. (C) The behavioral results on the gambling task and the delay tasks from the groups of VM patients with more anterior lesions, and from patients with right DL lesions. Note that the VM patients were severely impaired on the gambling task (low number of choices from the good decks), but normal on the delay tasks (high % correct response). On the other hand, the right DL patients were impaired on the delay tasks, and although their gambling task performance is considered advantageous, it falls in the low normal range.

sis of repeated feedback. However, closer investigations revealed that the posterior DL sector may be the critical region for this function. Shallice and colleagues have attempted to investigate deficits in *planning* ability associated with frontal lobe damage using tasks that require the planning and execution of sequences of responses. They used the Tower of Hanoi (and a variant, the Tower of London) (Shallice, 1982), and other tasks that resemble more closely real-world activities (Shallice and Burgess, 1991b). Patients with frontal lobe lesions were found to have deficits in planning as measured by these laboratory tasks. However, the VM sector does not seem critical for mediating this type of planning function since many VM patients perform well on the Tower of Hanoi, and are only mildly impaired on other complex planning tasks (Anderson, Hathaway-Nepple, Tranel and Damasio, 2000). Thus, it seems that this type of planning deficit is more severe when the basal forebrain is damaged, or when the DL sector is damaged. Finally, a link between frontal lobe damage and the ability to make *cognitive estimation*s has been established in a few studies (Shallice and Burgess, 1991a; Smith and Milner, 1984). However, this function also does not seem to depend on the VM sector of the prefrontal cortex, since many VM patients perform normally on this type of task (Saver and Damasio, 1991).

One cognitive function which may relate to the VM sector and decision-making, and which we began to investigate, is *prospective memory*, the capacity of "remembering in the future" (Wilkins and Baddeley, 1978). An example is remembering to call your spouse around 3:00 PM to arrange plans for picking up the children from school. Patients with VM lesions may be defective in this function. The defect may be linked to a failure to appose a somatic marker to this stimulus configuration, so that when 3:00 PM arrives, one is 'reminded' by the brain via a somatic signal (a feeling) that 'something needs to be done', i.e., to call your spouse.

Right versus left VM damage

Given the functional asymmetry of the cerebral hemispheres, it is important to determine whether the decision-making deficit associated with VM damage is produced by right or left lesions. Unfortunately, this question is difficult to address with lesion stud-

BEHAVIOR

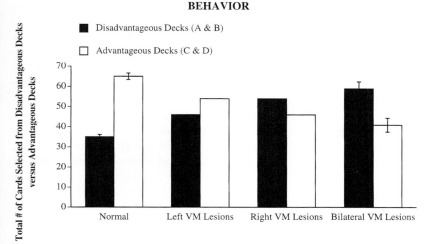

ANTICIPATORY SCRs

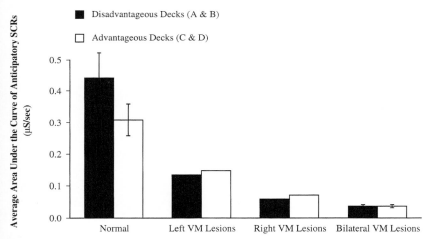

Fig. 9. Responses to the gambling task and anticipatory SCRs generated by groups of normal subjects and patients with bilateral VM lesions compared to patients with either left or right VM lesions.

ies, because of the rarity of patients who have a unilateral damage on the medial and orbital side of the prefrontal cortex. However, from these rare cases, it appears that the decision-making deficit observed in real-life and as measured by the gambling task is most pronounced after right, as opposed to left, frontal lesions (Fig. 9). One left VM patient that we have studied extensively has been able to hold gainful employment, and her performance on the gambling task appears in the advantageous direction, albeit not as advantageous as normal controls. By contrast, a right VM patient that we studied

has been unable to hold employment, and his performance on the gambling task is similar to that of patients with bilateral lesions. The anticipatory SCRs in these patients are also consistent with the behavior. The left VM patient generated higher-magnitude anticipatory SCRs relative to right or bilateral VM patients, but these anticipatory SCRs were abnormal when compared to normal subjects. The anticipatory SCRs from the right VM patient were similar to those of patients with bilateral lesions (Fig. 9). These findings are consistent with those from a case study of a patient with right orbitofrontal lesion (Angrilli,

Palomba, Cantagallo et al., 1999). An intriguing question is whether extensive damage to the VM side of the left prefrontal cortex would be selectively associated with decision-making impairments in the verbal domain (e.g., poor choice of words and things to say).

Early versus adult onset of VM damage

Patients who acquired VM prefrontal lobe damage during childhood are relatively rare. However, recent evidence from two young adults who acquired focal damage to the prefrontal cortex in early childhood, prior to 16 months of age, revealed very important facts (Anderson et al., 1999). The patients with early-onset VM lesions superficially resemble adult-onset patients in terms of disrupted social behavior, which contrasts with normal basic cognitive abilities. These patients show insensitivity to future consequences, and their behavior is guided by immediate consequences, both in the social world and on the gambling task (Fig. 10). The ability to generate somatic signals in anticipation of future outcomes (anticipa-

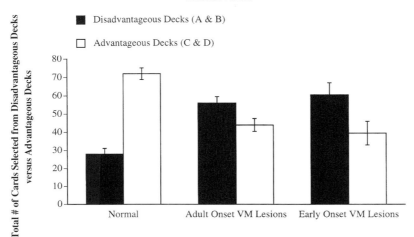

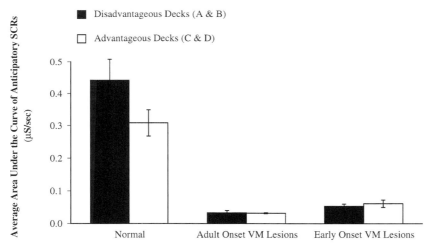

Fig. 10. Responses to the gambling task and anticipatory SCRs generated by groups of normal subjects and patients with adult onset VM lesions compared to patients with early onset VM lesions.

tory SCRs) was also defective. However, a closer analysis revealed several distinctive features. First, the inadequate social behaviors are present throughout development and into adulthood, i.e., there was no recovery of function such as happens with language when the left hemisphere is damaged at an early age. Second, these behavioral defects are more severe in early-onset patients relative to adult-onset. Third, the inadequate emotional responses are also more severe. Finally, the early-onset patients cannot retrieve socially relevant knowledge at factual level as adult-onset patients do (Anderson et al., 1999).

The amygdala

Central autonomic structures, such as the *amygdala*, can activate somatic responses in the viscera, vascular bed, endocrine system, and nonspecific neurotransmitter systems. Furthermore, the amygdala plays an important role in emotion as demonstrated repeatedly in various lesion and functional neuroimaging studies (Davidson and Irwin, 1999). Therefore, we tested the hypothesis that the amygdala plays a role in decision-making.

Amygdala damage impairs decision-making
Using the gambling task as a tool for measuring decision-making, we studied a group of four patients with bilateral amygdala lesions, and thirteen demographically and educationally matched normal subjects. We monitored the SCR activity of these subjects during their performance of the gambling task (Bechara et al., 1999). The results showed that normal controls selected more cards from the advantageous decks C and D, and fewer cards from the disadvantageous decks A and B (high immediate gain, but larger future loss). Similar to the patients with bilateral lesions of the VM cortex, the amygdala patients did the opposite and selected more cards from the disadvantageous decks as compared to the advantageous decks. Since patients with selective bilateral amygdala lesions are rare, and most of our amygdala patients had damage involving the hippocampus, further studies showed that the impaired performance on the gambling task was specific to lesions involving the amygdala, but not the hippocampus (Bechara, Tranel, Damasio and Damasio, 1998b). Indeed, patients with amnesia due to

anoxic encephalopathy (known to damage CA1 cells of the hippocampus; Rempel-Clower, Zola, Squire and Amaral, 1996; Zola-Morgan, Squire and Amaral, 1986) performed normally on the gambling task, albeit at a low normal level of performance (Bechara et al., 1998b).

When examining the anticipatory SCRs of amygdala patients, the amygdala patients were similar to the VM frontal patients in that they also failed to generate anticipatory SCRs before the selection of a card (Bechara et al., 1999). The results support the hypothesis that the amygdala is a critical component of a system involved in decision-making.

The amygdala and VM cortex participate differently in decision-making
Although bilateral damage to the amygdala disrupts the ability to make advantageous decisions both in real-life, and as revealed in the gambling task, there seems to be a difference between the roles played by the amygdala versus the VM cortex in decision-making. Studies have revealed two main differences between amygdala and VM patients (Bechara et al., 1999). The first is that amygdala damage precludes the development of conditioned emotional responses to aversive stimuli such as loud sounds, whereas VM damage does not block such conditioning. Secondly, when patients are given feedback such as 'you won or lost a certain amount of money', the VM patients can generate somatic states in response to this feedback, whereas the amygdala patients fail to do so. We note that in the VM patients, this emotional reaction to feedback is somewhat weaker in comparison to controls (Bechara et al., 1999), but not abolished, thus consistent with the finding of compromised somatic reactions to emotional target pictures (Damasio et al., 1990). VM and amygdala patients are similar in that they fail to generate anticipatory SCRs in anticipation of future outcomes, and they choose disadvantageously in the gambling task.

Studies have shown consistently that monkeys with lesions of the amygdala have an increased tendency to approach objects such as snakes (Aggleton, 1992), as if the object of fear can no longer evoke a state of fear. Using a parallel line of reasoning, it is conceivable and even likely that humans learn throughout development to associate the idea of winning and losing money with items that represent actual, innate, and immediate (i.e., *primary*) reward and punishment, such as food, shelter, sex, pain, and so forth. When the amygdala is damaged, such items can no longer evoke the appropriate somatic states of reward and punishment, just like an object of fear can no longer elicit the state of fear in an amygdalatomized monkey. Consequently, the concepts of winning and losing money can no longer elicit the appropriate somatic states. On the other hand, when the VM cortex is damaged, we believe that the concepts of winning and losing money can still evoke appropriate somatic states. However, if new exteroceptive information was associated with a punishment such as losing money, which is abstract, learned, and remote (i.e., *secondary*), then VM patients begin to show impairments. In other words, the VM patients fail to couple exteroceptive information (e.g., the bad decks) with the somatic state of a secondary punishment (e.g., the feeling that occurs after money loss).

Thus, the amygdala and VM cortex are both essential for the coupling of exteroceptive information with interoceptive information concerning somatic states. These somatic states serve to bias the decision-making process towards an advantageous course of action. The difference, however, between the amygdala and the VM cortex is the following. In the case of the amygdala, the defect is in coupling the exteroceptive information with somatic states generated by a *primary* punishment. In the case of the VMF cortex, the defect is in coupling exteroceptive information with somatic states evoked by a *secondary* punishment. As such, the amygdala can be viewed as an important structure for mediating some forms of first-order conditioning, especially aversive conditioning. The disruption of first-order conditioning in amygdala patients would indirectly preclude the development of anticipatory SCRs, which are necessary to guide decisions in the advantageous direction. In contrast, damage to the VM cortex spares first-order aversive conditioning, such as learning an association between external information and a somatic state evoked by a primary punishment. Considering the nature of the punishment used in the gambling task (money loss), we can state that it entails the learning of an association between a certain stimulus configuration (i.e., certain decks) and the

somatic state evoked by the money loss (secondary punishment). Given these conditions, it is reasonable to assume that this learned association involves high-order conditioning. We use the term 'high-order' instead of 'second, third, or *n*th-order' because we do not really know how many steps this money loss is away from a primary punishment.

Significance of the different roles played by the amygdala and VM cortex in decision-making

There may be an evolutionary significance for the role of the VM cortex in high-order conditioning. It is easy to achieve first-order conditioning, mediated by subcortical structures like the amygdala, in lower animals. Although second-order and perhaps third-order conditioning can be achieved in animals, higher forms of conditioning have not been reported. Perhaps the reason is that higher-order (higher than third-order) conditioning is difficult to achieve in laboratory animals, and when it is achieved, it is never as powerful as when seen in humans. The suggestion that the prefrontal cortex is important for higher-order conditioning might explain why humans could rely on concepts that are far removed from primary emotions (i.e., secondary emotions) to guide their decisions in every-day life.

The proposal that the VM cortex couples exteroceptive information with the somatic states of abstract, learned, and remote punishment, while the amygdala couples exteroceptive information with the somatic states of actual, innate, and immediate punishment, is significant. This distinction may reflect two types of decision-making deficit observable in the real-life activities of these patients. The decision-making impairments of patients with VM cortex lesions have remote consequences, and usually do not cause bodily harm. For instance, VM patients make choices that lead to long-term financial losses, or to the loss of friend and family relationships down the line, but rarely engage in actions that may lead to physical harm for themselves, or to others. In other words, the type of decision-making impairment that the VM patients exhibit in real-life is related to secondary as opposed to primary reward and punishment. On the other hand, although patients with bilateral amygdala lesions do exhibit decision-making impairments in the social realm similar to those of the VMF patients, they can actually pursue actions that eventually lead to physical harm to themselves and to others. Indeed, with one exception (Adolphs, Tranel, Damasio and Damasio, 1995), amygdala patients who participated in this study live under supervised care, and are unable to function alone in society. In two of the cases, the patients have pursued actions that endangered themselves and others.

The somatosensory/insular cortices

The visceral and somatosensory projection systems and cortices, especially the *insula, SII, SI, and cingulate* cortices, receive signals from the soma. Furthermore, based on studies from patients with lesions in the right somatosensory/insular cortices, it has been proposed that these areas may hold representations of the body states such as those occurring during the experience of an emotion (Damasio, 1994). Therefore, we hypothesized that these structures are critical for decision-making and somatic state activation.

Right somatosensory/insular cortex damage impairs decision-making

Using the gambling task as a tool for measuring decision-making, we studied a group of twelve patients with lesions in the right somatosensory/insular cortices, eleven patients with homologous left-sided lesions, and a demographically and educationally matched group of twelve normal subjects. We monitored the SCR activity of these subjects during their performance of the gambling task (Bechara, Tranel, Damasio and Damasio, 1997b). The results are still preliminary, but so far, they show that normal subjects selected more cards from the advantageous decks C and D, and fewer cards from the disadvantageous decks A and B. Patients with lesions in the left SS/I cortices performed like normal subjects on the gambling task. By contrast, patients with right SS/I lesions did the opposite, and selected more cards from the disadvantageous decks as compared to the advantageous decks. As in the case of the VM and amygdala patients, the right somatosensory/insular (but not left) patients failed to generate anticipatory SCRs before the selection of a card. This supports the hypothesis that the right somatosensory/insular cortices are critical components of a system involved in decision-making.

Different lesions interfere with different aspects of the emotional process involved in decision-making

Emotion is not a unitary process, and various studies have shown that it consists of several dissectable components (Davidson and Irwin, 1999; LeDoux, 1996). There are many different aspects of emotion. These aspects include the ability to perceive or recognize emotions in facial expressions (Adolphs et al., 1995), the ability to attach emotional significance to events that are otherwise neutral (i.e., conditioning), and the ability to experience emotion (i.e., express emotions and have feelings). Investigators sometimes assume that studying one aspect of the emotional process automatically provides understanding of its other aspects. However, it is now recognized that the neural mechanisms involved in the perception or recognition of emotions, for example, are not necessarily the same as those involved in the experience of emotions (Davidson and Irwin, 1999). Therefore, we believe that each of these structures provides a different contribution to the global process of decision-making. These different contributions reflect the different roles that each of these structures plays in different aspects of the emotional process. Current studies are aimed at elucidating the specific roles played by each structure in emotional processing, and the differences in the roles of the somatosensory/insular cortex, amygdala, and VM cortex in decision-making.

Conclusion

The somatic marker hypothesis proposes that individuals make judgements not only by assessing the potential benefits of outcomes and their probability of occurrence, but also and perhaps even primarily in terms of their emotional quality. Lesions of the VM prefrontal cortex interfere with the normal processing of somatic or emotional signals, but leave other cognitive functions minimally affected. This damage leads to pathological impairments in the decision-making process, which seriously compromise the efficiency of every-day-life decisions. The somatic marker proposal supports the views of others who invoke a primary role for mood, affect, and emotion in decision-making, e.g., LeDoux (1996), Schwartz and Clore (1983) and Zajonc (1984). However, it differs from the views of Rolls (1999) and Shallice (1993) which are more consistent with the 'as if loop' component of the somatic marker hypothesis. The fundamental notion of the somatic marker hypothesis is that bioregulatory signals, including those that constitute feeling and emotion, provide the principal guide for decisions, and the basis for the development of the 'as if loop' mode of operation.

The somatic marker hypothesis and the experimental strategies used to study decision-making in neurological patients provide parallels and direct implications for understanding the nature of several psychiatric disorders. Substance abuse is one example. Substance abusers are similar to VM patients in that when faced with a choice that brings some immediate reward (i.e., taking a drug), at the risk of incurring the loss of job, home, and family, they choose the immediate reward and ignore the future consequences. Using the gambling task (Bechara, Dolan, Denburg et al., 2000; Grant, Contoreggi and London, 2000; Petry, Bickel and Arnett, 1998) or related decision-making tasks (Rogers et al., 1999), recent studies have indicated that an impairment in decision-making may stand at the core of the problem of substance abuse. Similarly, the personality profile of VM patients bears some striking similarities to psychopathic (or sociopathic) personality, so much so that we have used the term "acquired sociopathy" to describe the condition of patients with VM damage (Damasio et al., 1990). The qualifier 'acquired' signifies that the condition in VM patients follows the onset of brain injury, and occurs in persons whose personalities and social conduct were previously normal. The patients are usually not destructive and harmful to others, a feature that tends to distinguish the 'acquired' form of the disorder from the standard 'developmental' form. As we have indicated earlier, when the VM damage is acquired earlier in life, the antisocial behavior is more severe, which suggests that early dysfunction in prefrontal cortex may, by itself, cause abnormal development of social and moral behavior (Anderson et al., 1999). Recent studies have began to look at the possibility that the psychopathic behavior seen in cases in which no neurological history has been identified may be linked to abnormal operation of the neural system involving the VM prefrontal cortex (Schmitt, Brinkley and Newman, 1999). Finally, in addition to the disorders mentioned above, applications of the

somatic marker hypothesis may extend to psychiatric disorders that include schizophrenia (Wilder, Weinberger and Goldberg, 1998), pathological gambling, depression, and attention deficit and hyperactivity disorders (ADHD).

Acknowledgements

This work was supported in part by NIH Program Project Grant PO1 NS19632, and the Mathers Foundation. Requests for reprints should be addressed to Dr. Antonio R. Damasio, Department of Neurology, University of Iowa Hospitals and Clinics, Iowa City, IA 52242, USA.

References

Ackerly SS, Benton AL: Report of a case of bilateral frontal lobe defect. Proceedings of the Association for Research in Nervous and Mental Disease (Baltimore): 27; 479–504, 1948.

Adolphs R, Tranel D, Damasio H, Damasio AR: Fear and the human amygdala. Journal of Neuroscience: 15; 5879–5892, 1995.

Aggleton JP: The functional effects of amygdala lesions in humans: a comparison with findings from monkeys. In Aggleton JP (Ed), The Amygdala: Neurobiological Aspects of Emotion, Memory, and Mental Dysfunction. New York: Wiley–Liss, pp. 485–504, 1992.

Amaral DG, Price JL: Amygdalo-cortical connections in the monkey (*Macaca fascicularis*). Journal of Comparative Neurology: 230; 465–496, 1984.

Amaral DG, Price JL, Pitkanen A, Carmichael ST: Anatomical organization of the primate amygdaloid complex. In Aggleton JP (Ed), The Amygdala: Neurobiological Aspects of Emotion, Memory, and Mental Dysfunction. New York: Wiley–Liss, pp. 1–66, 1992.

Anderson SW, Bechara A, Damasio H, Tranel D, Damasio AR: Impairment of social and moral behavior related to early damage in the human prefrontal cortex. Nature Neuroscience: 2; 1032–1037, 1999.

Anderson SW, Damasio H, Jones RD, Tranel D: Wisconsin card sorting test performance as a measure of frontal lobe damage. Journal of Clinical and Experimental Neuropsychology: 3; 909–922, 1991.

Anderson SW, Hathaway-Nepple J, Tranel D, Damasio H: Impairments of strategy application following focal prefrontal lesions. Journal of the International Neuropsychological Society: 6; 205, 2000.

Angrilli A, Palomba D, Cantagallo A, Maietti A, Stegagno L: Emotional impairment after right orbitofrontal lesion in a patient without cognitive deficits. NeuroReport: 10; 1741–1746, 1999.

Barratt ES: Impulsiveness and aggression. In Monahan J, Steadman HJ (Eds), Violence and Mental Disorder: Developments in Risk Assessment. Chicago, IL: University of Chicago Press, pp. 61–79, 1994.

Bechara A, Damasio AR, Damasio H, Anderson SW: Insensitivity to future consequences following damage to human prefrontal cortex. Cognition: 50; 7–15, 1994.

Bechara A, Damasio H, Damasio AR, Lee GP: Different contributions of the human amygdala and ventromedial prefrontal cortex to decision-making. Journal of Neuroscience: 19; 5473–5481, 1999.

Bechara A, Damasio H, Tranel D, Anderson SW: Dissociation of working memory from decision making within the human prefrontal cortex. Journal of Neuroscience: 18; 428–437, 1998a.

Bechara A, Damasio H, Tranel D, Damasio AR: Deciding advantageously before knowing the advantageous strategy. Science: 275; 1293–1295, 1997a.

Bechara A, Dolan S, Denburg N, Hindes A, Anderson SW, Nathan PE: Decision-making deficits, linked to a dysfunctional ventromedial prefrontal cortex, revealed in alcohol and stimulant abusers. Neuropsychologia: 39; 376–389, 2001.

Bechara A, Tranel D, Damasio H: Characterization of the decision-making impairment of patients with bilateral lesions of the ventromedial prefrontal cortex. Brain: 123; 2189–2202, 2000.

Bechara A, Tranel D, Damasio H, Adolphs R, Rockland C, Damasio AR: Double dissociation of conditioning and declarative knowledge relative to the amygdala and hippocampus in humans. Science: 269; 1115–1118, 1995.

Bechara A, Tranel D, Damasio H, Damasio AR: Failure to respond autonomically to anticipated future outcomes following damage to prefrontal cortex. Cerebral Cortex: 6; 215–225, 1996.

Bechara A, Tranel D, Damasio H, Damasio AR: An anatomical system subserving decision-making. Society for Neuroscience Abstracts: 23; 495, 1997b.

Bechara A, Tranel D, Damasio H, Damasio AR: The use of psychophysiology to predict structural damage in anoxic encephalopathy. International Journal of Psychophysiology: 30; 147, 1998b.

Brickner RM: An interpretation of frontal lobe function based upon the study of a case of partial bilateral frontal lobectomy. Localization of function in the cerebral cortex. Proceedings of the Association for Research in Nervous and Mental Disease (Baltimore): 13; 259–351, 1932.

Butter CM, Mishkin M, Mirsky AF: Emotional responses toward humans in monkeys with selective frontal lesions. Physiology and Behavior 3; 213–215, 1968.

Butter CM, Mishkin M, Rosvold HE: Conditioning and extinction of a food-rewarded response after selective ablations of frontal cortex in rhesus monkeys. Experimental Neurology: 7; 65–75, 1963.

Butter CM, Snyder DR: Alternations in aversive and aggressive behaviors following orbital frontal lesions in rhesus monkeys. Acta Neurobiologiae Experimentalis: 32; 525–565, 1972.

Cahill L, Babinsky R, Markowitsch HJ, McGaugh JL: The amygdala and emotional memory. Nature: 377; 295–296, 1995.

Chavis DA, Pandya DN: Further observations on corticofrontal connections in the rhesus monkey. Brain Research: 117; 369–386, 1976.

Courtney SM, Ungerleider LG, Keil K, Haxby JV: Object and spatial visual working memory activate separate neural systems in human cortex. Cerebral Cortex: 6; 39–49, 1996.

Damasio AR: The brain binds entities and events by multiregional activation from convergence zones. Neural Computation: 1; 123–132, 1989a.

Damasio AR: Time-locked multiregional retroactivation: a systems-level proposal for the neural substrates of recall and recognition. Cognition: 33; 25–62, 1989b.

Damasio AR: Descartes' Error: Emotion, Reason, and the Human Brain. New York: Grosset/Putnam, 1994.

Damasio AR: The Feeling of What Happens: Body and Emotion in the Making of Consciousness. New York: Harcourt Brace and Company, 1999.

Damasio AR, Damasio H: Cortical systems for retrieval of concrete knowledge: the convergence zone framework. In Koch C (Ed), Large Scale Neuronal Theories of the Brain. Cambridge, MA: MIT Press, pp. 61–74, 1994.

Damasio AR, Tranel D, Damasio H: Individuals with sociopathic behavior caused by frontal damage fail to respond autonomically to social stimuli. Behavioural Brain Research: 41; 81–94, 1990.

Damasio AR, Tranel D, Damasio H: Somatic markers and the guidance of behavior: Theory and preliminary testing. In Levin HS, Eisenberg HM, Benton AL (Eds), Frontal Lobe Function and Dysfunction. New York: Oxford University Press, pp. 217–229, 1991.

Damasio H: Human Brain Anatomy in Computerized Images. New York: Oxford University Press, 1995.

Damasio H, Grabowski T, Frank R, Galburda AM, Damasio AR: The return of Phineas Gage: Clues about the brain from the skull of a famous patient. Science: 264; 1102–1104, 1994.

Davidson RJ, Irwin W: The functional neuroanatomy of emotion and affective style. Trends in Cognitive Sciences: 3; 11–21, 1999.

Davis M: The role of the amygdala in conditioned fear. In Aggleton JP (Ed), The Amygdala: Neurobiological Aspects of Emotion, Memory, and Mental Dysfunction. New York: Wiley–Liss, 353–375, 1992.

D'Esposito M, Detre JA, Alsop DC, Shin RK, Atlas S, Grossman M: The neural basis of central execution systems of working memory. Nature: 378; 279–281, 1995a.

D'Esposito M, Shin RK, Detre JA, Incledon S, Annis D, Aguirre GK, Grossman M, Alsop DC: Object and spatial working memory activates dorsolateral prefrontal cortex: a functional MRI study. Society for Neuroscience Abstracts: 21; 1498, 1995b.

Diamond A (Ed): The Development and Neural Bases of Higher Cognitive Functions. Annals of the New York Academy of Sciences: 608, 1990.

Dias R, Robbins TW, Roberts AC: Dissociation in prefrontal cortex of affective and attentional shifts. Nature: 380; 69–72, 1996.

Dias R, Robbins TW, Roberts AC: Dissociable forms of inhibitory control within prefrontal cortex with an analog of the Wisconsin Card Sort Test: restrictions to novel situations and independence from 'on-line' processing. Journal of Neuroscience: 17; 9285–9297, 1997.

Easterbrook JA: The effect of emotion on cue utilization and the organization of behavior. Psychological Review: 66; 183–201, 1959.

Eslinger PJ, Damasio AR: Severe disturbance of higher cognition after bilateral frontal lobe ablation: patient EVR. Neurology: 35; 1731–1741, 1985.

Evenden J: Impulsivity: a discussion of clinical and experimental findings. Journal of Psychopharmacology: 13; 180–192, 1999.

Fuster JM: Prefrontal cortex and the bridging of temporal gaps in the perception–action cycle. In Diamond A (Ed), The Development and Neural Bases of Higher Cognitive Functions. Annals of the New York Academy of Science: 608; 318–336, 1990.

Fuster JM: The prefrontal cortex and its relation to behavior. In Holstege G (Ed), Progress in Brain Research. New York: Elsevier, pp. 201–211, 1991.

Fuster JM: The Prefrontal Cortex. Anatomy, Physiology, and Neuropsychology of the Frontal Lobe (3rd ed.). New York: Raven Press, 1996.

Goldman-Rakic PS: Circuitry of primate prefrontal cortex and regulation of behavior by representational memory. In Plum F (Ed), Handbook of Physiology; The Nervous System. Bethesda, MD: American Physiological Society, pp. 373–401, 1987.

Goldman-Rakic PS: Working memory and the mind. Scientific American: 267; 111–117, 1992.

Grant S, Contoreggi C, London ED: Drug abusers show impaired performance in a laboratory test of decision-making. Neuropsychologia: 38; 1180–1187, 2000.

Hall RE, Livingston RB, Bloor CM: Orbital cortical influences on cardiovascular dynamics and myocardial structure in conscious monkeys. Journal of Neurosurgery: 46; 638–647, 1977.

Harlow JM: Passage of an iron bar through the head. Boston Medical and Surgical Journal: 39; 389–393, 1848.

Harlow JM: Recovery from the passage of an iron bar through the head. Publications of the Massachusetts Medical Society: 2; 327–347, 1868.

Janis IL, Mann L: Decision-Making: A Psychological Analysis of Conflict, Choice, and Commitment. New York: Free Press, 1977.

Jones B, Mishkin M: Limbic lesions and the problem of stimulus — reinforcement associations. Experimental Neurology: 36; 362–377, 1972.

Jones EG, Powell TPS: An anatomical study of converging sensory pathways within the cerebral cortex of the monkey. Brain: 93; 793–820, 1970.

Jonides J, Smith EE, Koeppe RA, Awh E, Minoshima S, Mintun MA: Spatial working memory in humans as revealed by PET. Nature: 363; 623–625, 1993.

Kim JJ, Rison RA, Fanselow MS: Effects of amygdala, hippocampus, and periaqueductal gray lesions on short and long term contextual fear. Behavioral Neuroscience: 107; 1093–1098, 1993.

LaBar KS, Gatenby JC, Gore JC, LeDoux JE, Phelps EA: Human amygdala activation during conditioned fear acquisition and extinction: a mixed-trial fMRI study. Neuron: 20; 937–945, 1998.

LaBar KS, LeDoux JE, Spencer DD, Phelps EA: Impaired fear conditioning following unilateral temporal lobectomy in humans. Journal of Neuroscience: 15; 6846–6855, 1995.

LeDoux J: The Emotional Brain: The Mysterious Underpinnings of Emotional Life. New York: Simon and Schuster, 1996.

LeDoux JE: Emotional memory systems in the brain. Behavioural Brain Research: 58; 69–79, 1993.

Malkova L, Gaffan D, Murray EA: Excitotoxic lesions of the amygdala fail to produce impairment in visual learning for auditory secondary reinforcement but interfere with reinforcer devaluation effects in rhesus monkeys. Journal of Neuroscience: 17; 6011–6020, 1997.

Mann L: Stress, affect, and risk taking. In Frank YJ (Ed), Risk-Taking Behavior. Chichester: Wiley, pp. 202–230, 1992.

McCarthy G, Blamire AM, Puce A, Nobre AC, Boch G, Hyder F, Goldman-Rakic P, Shulman RG: Functional magnetic resonance imaging of human prefrontal cortex activation during a spatial working memory task. Proceedings of the National Academy of Sciences, USA: 91; 8690–8694, 1994.

Miller LA: Impulsivity, risk-taking, and the ability to synthesize fragmented information after frontal lobectomy. Neuropsychologia: 30; 69–79, 1992.

Milner B, Petrides M, Smith ML: Frontal lobes and the temporal organization of memory. Human Neurobiology: 4; 137–142, 1985.

Mishkin M: Perseveration of central sets after frontal lesions in monkeys. In Warren JM, Akert K (Eds), The Frontal Granular Cortex and Behavior. New York: McGraw-Hill, pp. 219–241, 1964.

Morgan MA, LeDoux JE: Differential contribution of dorsal and ventral medial prefrontal cortex to the acquisition and extinction of conditioned fear in rats. Behavioral Neuroscience: 109; 681–688, 1995.

Nauta WJH: The problem of the frontal lobes: a reinterpretation. Journal of Psychiatric Research: 8; 167–187, 1971.

Pandya DN, Kuypers HGJM: Cortico-cortical connections in the rhesus monkey. Brain Research: 13; 13–36, 1969.

Petrides M: Deficits on conditional associative learning tasks after frontal and temporal lobe lesions in man. Neuropsychologia: 23; 601–614, 1985.

Petrides M: Nonspatial conditional learning impaired in patients with unilateral frontal but not unilateral temporal lobe excisions. Neuropsychologia: 28; 137–149, 1990.

Petrides M: Specialized systems for the processing of mnemonic information within the primate frontal cortex. Philosophical Transactions of the Royal Society of London: 351; 1445–1457, 1996.

Petrides M, Alivisatos B, Meyer E, Evans AC: Functional activation of the human frontal cortex during the performance of verbal working memory tasks. Proceedings of the National Academy of Sciences of the United States of America: 90; 878–882, 1993.

Petry NM, Bickel WK, Arnett M: Shortened time horizons

and insensitivity to future consequences in heroin addicts. Addiction: 93; 729–738, 1998.

Plaisted KC, Sahakian BJ: Dementia of frontal lobe type-living in the here and now. Aging and Mental Health: 1; 293–295, 1997.

Potter H, Nauta WJH: A note on the problem of olfactory associations of the orbitofrontal cortex in the monkey. Neuroscience: 4; 316–367, 1979.

Rahman S, Robbins TW, Sahakian BJ: Comparative cognitive neuropsychological studies of frontal lobe function: implications for therapeutic strategies in frontal variant frontotemporal dementia. Dementia and Geriatric Cognitive Disorders: 10; 15–28, 1999a.

Rahman S, Sahakian BJ, Hodges JR, Rogers RD, Robbins TW: Specific cognitive deficits in mild frontal variant frontotemporal dementia. Brain: 122; 1469–1493, 1999b.

Rapoport JL, Giedd JN, Blumenthal J, Hamburger S, Jeffries N, Fernandez T, Nicolson R, Bedwell J, Lenane M, Zijdenbos A, Paus T, Evans A: Progressive cortical change during adolescence in childhood-onset schizophrenia: A longitudinal magnetic resonance imaging study. Archives of General Psychiatry: 56; 649–654, 1999.

Rempel-Clower NL, Zola SM, Squire LR, Amaral DG: Three cases of enduring memory impairment after bilateral damage limited to the hippocampal formation. Journal of Neuroscience: 16; 5233–5255, 1996.

Rogers RD, Everitt BJ, Baldacchino A, Blackshaw AJ, Swainson R, Wynne K, Baker NB, Hunter J, Carthy T, Booker E, London M, Deakin JFW, Sahakian BJ, Robbins TW: Dissociable deficits in the decision-making cognition of chronic amphetamine abusers, opiate abusers, patients with focal damage to prefrontal cortex, and tryptophan-depleted normal volunteers: evidence for monoaminergic mechanisms. Neuropsychopharmacology: 20; 322–339, 1999.

Rolls ET: The Brain and Emotion. Oxford: Oxford University Press, 1999.

Saver JL, Damasio AR: Preserved access and processing of social knowledge in a patient with acquired sociopathy due to ventromedial frontal damage. Neuropsychologia: 29; 1241–1249, 1991.

Schmitt WA, Brinkley CA, Newman JP: Testing Damasio's somatic marker hypothesis with psychopathic individuals: risk-takers or risk-averse? Journal of Abnormal Psychology: 108; 538–543, 1999.

Schwartz N, Clore GL: Mood, misattribution, and judgements of well-being: information and directive functions of affective states. Journal of Personality and Social Psychology: 45; 513–523, 1983.

Shallice T: Specific impairments of planning. Philosophical Transactions of the Royal Society, London: 298; 199–209, 1982.

Shallice T: From Neuropsychology to Mental Structure. Cambridge: Cambridge University Press, 1993.

Shallice T, Burgess P: Higher-order cognitive impairments and frontal lobe lesions in man. In Levin HS, Eisenberg HM, Benton AL (Eds), Frontal Lobe Function and Dysfunction. New York: Oxford University Press, pp. 125–138, 1991a.

Shallice T, Burgess PW: Deficits in strategy application following frontal lobe damage in man. Brain: 114; 727–741, 1991b.

Smith EE, Jonides J, Koeppe RA, Awh E, Schumacher EH, Minoshima S: Spatial versus object working memory: PET investigations. Journal of Cognitive Neuroscience: 7; 337–356, 1995.

Smith ML, Milner B: Differential effects of frontal lobe lesions on cognitive estimation and spatial memory. Neuropsychologia: 22; 697–705, 1984.

Sowell ER, Thompson PM, Holmes CJ, Batth R, Jernigan TL, Toga AW: Localizing age-related changes in brain structure between childhood and adolescence using statistical parametric mapping. Neuroimage: 9; 587–597, 1999.

Stuss DT: Biological and psychological development of executive functions. Brain and Cognition: 20; 8–23, 1992.

Swartz BE, Halgren E, Fuster JM, Simpkins F, Gee M, Mandelkern M: Cortical metabolic activation in humans during a visual memory task. Cerebral Cortex: 3; 205–214, 1995.

Tranel D, Bechara A, Damasio H, Damasio AR: Fear conditioning after ventromedial frontal lobe damage in humans. Society for Neuroscience Abstracts: 22; 1108, 1996.

Van Hoesen GW: Neural systems of the non-human primate forebrain implicated in memory. Annals of the New York Academy of Sciences: 444; 97–112, 1985.

Welt L: Uber Charakterveränderungen des Menschen infolge von Läsionen der Stirnhirns. Deutsches Archiv fuer Klinische Medizin: 42; 339–390, 1888.

Wilder KE, Weinberger DR, Goldberg TE: Operant conditioning and the orbitofrontal cortex in schizophrenic patients: unexpected evidence for intact functioning. Schizophrenia Research: 30; 169–174, 1998.

Wilkins A, Baddeley A: Remembering to recall in everyday life: an approach to absentmindedness. In Gruneberg MM, Morris PE, Sykes RN (Eds), Practical Aspects of Memory. New York: Academic Press, pp. 27–34, 1978.

Wilson FAW, Scalaidhe SPO, Goldman-Rakic PS: Dissociation of object and spatial processing domains in primate prefrontal cortex. Science: 260; 1955–1958, 1993.

Zajonc RB: On the primacy of affect. American Psychologist: 39; 117–123, 1984.

Zola-Morgan S, Squire LR, Amaral DG: Human amnesia and the medial temporal region: enduring memory impairment following a bilateral lesion limited to field CA1 of the hippocampus. Journal of Neuroscience: 6; 2950–2967, 1986.

Handbook of Neuropsychology, 2nd Edition, Vol. 7
J. Grafman (Ed)

CHAPTER 7

Neuropsychological consequences of dysfunction in human dorsolateral prefrontal cortex

Steven W. Anderson * and Daniel Tranel

Department of Neurology, Division of Behavioral Neurology and Cognitive Neuroscience, The University of Iowa College of Medicine, Iowa City, IA 52242, USA

Introduction

The dorsolateral prefrontal (DLP) area includes the expanse of cortex on the lateral convexity anterior to the premotor region, and corresponds primarily to Brodmann's areas 45, 46, and the lateral aspects of 9 and 10. Damage to DLP cortex can result in a multitude of neuropsychological impairments, including dysfunction in diverse aspects of cognition such as attention, language and executive functions. The nature and scope of abilities impacted by DLP damage point toward a role of this polymodal association cortex in the coordination and integration of human mental abilities. With rich bidirectional connections to limbic areas, major subcortical nuclei, the ventromedial prefrontal region, and posterior unimodal and multimodal sensory cortices, the DLP cortex is well-suited for such a role (see Barbas, 1995; Damasio, 1996; Nauta, 1972).

The anatomical boundaries for this review are admittedly somewhat arbitrary, and there is good reason to believe that frontal areas not covered in this discussion, namely premotor, orbital and mesial prefrontal regions, contribute substantially to many of the cognitive operations to be considered. Further, the DLP cortex, as defined above, is likely to have important functional subdivisions. The anatomical boundaries also are blurred because of the limita-

tions of available human lesion data. Focal lesions restricted to the DLP region are rare. Most surgical resections in this region involve concomitant resection of mesial prefrontal cortex and often the anterior cingulate, and stroke in the anterior distribution of the middle cerebral artery most often involves additional damage to parietal, temporal, or subcortical structures Fig. 1).

Interpretation of neuropsychological findings regarding DLP damage is further complicated by issues of neuronal plasticity and functional recovery. Marked post-lesion reorganization of functionally relevant ipsilateral and contralateral motor and premotor cortex has been demonstrated (e.g., Chollet, DiPiero, Wise et al., 1991; Weiller, Chollet, Friston et al., 1992), and it would be surprising if DLP cortex did not have similar plasticity. Both age at the time of lesion onset and the time from onset to testing are likely to be important factors shaping performances on clinical and experimental behavioral measures. Most of the research discussed below was based on subjects evaluated in some chronic phase of recovery (i.e., several months or years following onset) from prefrontal damage acquired in adulthood. However, even under these conditions, the behavioral profiles are not necessarily stable; for at least some impairments (e.g., Broca's aphasia), substantial recovery can continue for years after onset, particularly in young, intelligent, and motivated patients.

Despite these complications, there is increasing evidence that damage to the DLP cortex has unique

* Corresponding author. E-mail: steven-anderson@uiowa.edu

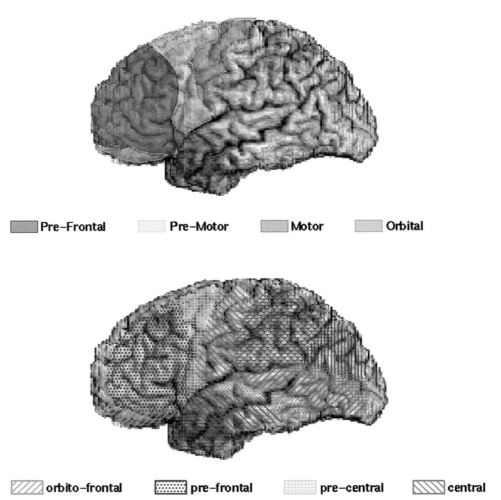

Pre-Frontal Pre-Motor Motor Orbital

orbito-frontal pre-frontal pre-central central

Fig. 1. Primary divisions and vascular territories of the dorsolateral frontal lobe. Upper panel: lateral view of the left hemisphere, with the primary divisions of the lateral frontal lobe indicated. Lower panel: lateral view of the left hemisphere with the vascular territories of the middle cerebral artery indicated.

functional implications. Although focal damage in the DLP region is not a common occurrence, specification of the neuropsychological consequences of lesions restricted to DLP cortex can help guide models of cognitive-neural function, as well as clinical evaluation of patients with known or suspected DLP damage in the context of more widespread brain dysfunction. We consider below the consequences of DLP damage for human cognition, with the findings from human lesion studies supplemented by selected findings from animal research and functional imaging studies of normal humans. The discussion is

organized around the general cognitive domains of attention and memory, language, visuospatial abilities, executive functions, and emotion and social behavior, with recognition that there is overlap and interaction among these functions.

Attention and memory

Damage to DLP cortex does not cause amnesia in the conventional sense, but it can disrupt aspects of learning and memory at several stages of processing, from basic allocation of attention to metamemory skills. Patients with DLP lesions typically can perform adequately on standardized tests of learning and memory, in which the memory challenges are presented with considerable external structure. For example, Swick and Knight (1996) found normal performances on measures of cued recall and repetition priming by patients with right DLP lesions, and only mild impairments in patients with left DLP lesions (see Petrides, 2000, for a review). However, in some instances free recall of recently presented or remote information may be defective. At least in part, this appears to be due to reduced executive control of active learning and recall strategies (Eslinger and Grattan, 1994; Gershberg and Shimamura, 1995; Mangels, Gershberg, Shimamura and Knight, 1996). Schacter, Curran, Galluccio et al. (1996) demonstrated a high rate of false recognition in a patient with a right DLP lesion, suggestive of a defect in monitoring mnemonic operations. There is evidence of neural pathways which could provide a substrate for prefrontal influence on memory. Neurons in prefrontal cortex project to the hippocampal region via at least two distinct routes (Goldman-Rakic, 1984), and recent neurophysiological studies in monkeys have supported the concept of prefrontal 'top down' control of mesial temporal lobe memory functions (Tomita, Ohbayashi, Nakahara et al., 1999).

Patients with DLP damage (or their relatives and caretakers) often complain of 'forgetfulness' in the execution of daily activities, despite relatively normal scores on standard tests of anterograde memory, especially tests that utilize recognition formats. One factor contributing to this discrepancy is impairment of attentional processes, particularly increased susceptibility to distraction, which tends to be greater in the real world than in the exam room. The acquisition of new, functionally relevant information requires the filtering or gating of irrelevant stimuli, and DLP cortex appears to contribute to this process by exerting inhibitory effects on posterior brain regions involved in perception. Holmes (1938) argued that an important role of the frontal lobes was suppression of reflexive ocular behavior, and recently, damage to inferior DLP cortex has been associated with impairments on the 'anti-saccade' paradigm, which requires inhibition of reflexive glances to peripheral stimuli (Walker, Husain, Hodgson et al., 1998).

Patients with DLP lesions show heightened vulnerability to distracting stimuli in other modalities, as well as electrophysiological evidence of disinhibition in sensory regions. For example, Chao and Knight (1998) found that patients with DLP lesions generated enhanced event-related potentials in primary auditory cortex in response to distracting noises. It has been proposed that the dependency on immediately present environmental cues shown by some patients with DLP damage may be due to release of parietal lobe activity resulting from loss of frontal lobe inhibition (Lhermitte, Pillon and Serdaru, 1986). Dias, Robbins and Roberts (1997) have provided evidence suggesting that DLP damage in primates may have a greater impact on inhibitory control of attention selection, while orbitofrontal damage may have a greater effect on affectively related inhibition. Impairment of inhibitory control following DLP damage appears to be a common mechanism affecting not just allocation of attention, but several aspects of cognition and behavior.

Many of the reasoning abilities considered uniquely characteristic of humans, such as long-term planning, hypothetical reasoning, and reorganization of complex concepts, require *working memory*, or the transient maintenance of representations in an activated or accessible state while the reasoning is taking place. Working memory is not a unitary process and likely is an essential function of many nonfrontal brain regions. However, the DLP cortex appears to be important for working memory tasks which involve bridging of temporally separate elements and the comparison or manipulation of several pieces of information (e.g., Fuster, 1995; Goldman-Rakic, 1984; Kim and Shadlen, 1999; Petrides, 1995). Functional imaging studies have linked different subregions of the DLP cortex to working memory for different types of stimulus material, such as stimulus identity, spatial location, and aspects of verbal processing (e.g., Cohen, Forman, Braver et al., 1994; Crossen, Rao, Woodley et al., 1999; Smith, Jonides and Koeppe, 1996). Despite evidence of prefrontal involvement in working memory, patients with DLP damage generally do not have impairments on standard measures of working memory, such as verbal or spatial span, although delayed response performance can be impaired under certain conditions (D'Esposito and Postle, 1999). There is growing reason to think that much of the contribution of DLP to working memory task performance may be in executive control over mnemonic processing, rather than working memory per se (Postle, Berger and D'Esposito, 1999).

We examined performance on delayed response and delayed non-matching to sample tasks in patients with lesions in ventromedial (VM) or dorsolateral/high mesial (DL/M) prefrontal areas (Bechara, Damasio, Tranel and Anderson, 1998). Only patients with right DL/M or posterior VM lesions (with likely involvement of the basal forebrain) were impaired on these spatial working memory tasks. The posterior VM patients, but not the DL/M patients, were impaired on an experimental decision-making task, and patients with VM lesions were able to perform normally on the working memory tasks despite impairments of decision-making. These findings suggest a dissociation of working memory and decision-making in the human prefrontal cortex. Within the prefrontal region, damage to DLP cortex appears to have the greatest impact on working memory, possibly through disrupted regulation of posterior regions more directly involved in memory encoding and storage.

Several lines of research suggest that additional mnemonic processes might be impacted by DLP damage, but the data to this point do not allow conclusions regarding the contributions of specific regions of dysfunction within the prefrontal cortex to these memory impairments. For example, there is considerable evidence that frontal lobe damage results in impairment in aspects of memory for *contextual information*, such as memory for temporal order (Jurado, Junque, Vendrell et al., 1998; McAndrews and Milner, 1991; Milner, Corsi and Leonard, 1991; Shimamura, Janowsky and Squire, 1990) and memory for the source of newly acquired information (Janowsky, Shimamura and Squire, 1989; Schacter, Harbluck and McLaughlin, 1984). Also, learning of conditional associations and visuomotor skills may be impacted by prefrontal lobe lesions (e.g., Beldarrain, Grafman, Pascual-Leone and Garcia-Monco, 1999; Petrides, 1985). The precise role of DLP damage in these various impairments of memory, as well as possible implications for clinical evaluation or patient management, remain to be determined.

Language

Damage to the posterior portion of the left DLP cortex can result in a variety of speech and language defects (discussed in greater detail in Volume 3 of the *Handbook*). The precise roles of DLP cortex in linguistic functioning remain to be defined, and the task of disentangling the effects of DLP damage from the effects of damage to neighboring premotor and subcortical areas is particularly difficult in the realm of language. Lack of appropriate subjects is one factor, as surgeons generally avoid language areas, and vascular lesions do not respect the anatomical boundaries in question. Also, it is clear that DLP and neighboring areas work in concert to process aspects of language, as can be illustrated by consideration of frontal lobe aphasic syndromes.

Broca's area, as traditionally defined, straddles the premotor–prefrontal boundary (areas 44 and 45), and there is evidence suggestive of shared functions of the DLP and premotor sectors of Broca's area. Verbal working memory tasks activate Broca's area, including premotor regions (Crossen et al., 1999). Further, left DLP and premotor areas are activated during lexical retrieval, with the demands of mental search and different lexical categories contributing to activation of various subregions (Grabowski, Damasio and Damasio, 1998). Circumscribed damage to Broca's area results in limited disturbance of speech. For the full syndrome of Broca's aphasia to appear, with nonfluent agrammatical speech and impairments in several aspects of language, it is necessary for the damage to extend to additional cortical areas, underlying white matter, and subcortical structures (Mohr, 1976). In another permutation, focal damage in the vicinity of the left DLP-premotor junction superior to Broca's area can result in transcortical motor aphasia, which generally resembles a mild Broca's aphasia except for relatively preserved verbatim repetition (Damasio, 1998; Rubens, 1976; Tranel and Anderson, 1999). When the lesion associated with Broca's aphasia is accompanied by additional damage to posterior language areas, global aphasia results. The basic principles illustrated here, i.e., that the linguistic functions of DLP cortex are tightly integrated with those of other brain regions, and that the pattern of damage to nearby cortical, white matter, and subcortical structures will have a major impact on the consequences of DLP lesions, are likely to apply in similar fashion to other aspects of cognition.

Several functional imaging studies have reported left inferior frontal gyrus activation when normal subjects are shown concrete nouns and required to generate semantically appropriate verbs (the so-called 'verb generate' task), but the exact role of this brain region in the task has not been clear (e.g., Petersen, Fox, Posner et al., 1989; Raichle, Fiez, Videen et al., 1994). Thompson-Schill, Swick, Farah et al. (1998) found that patients with lesions in the left inferior frontal gyrus were impaired on the verb generation task when there were high demands for selection among competing responses, but not when there were low selection demands (i.e., in response to nouns with only one commonly associated verb, such as 'scissors'). This suggests that the role of the inferior frontal gyrus in verb generation tasks may not be in semantic retrieval per se, but rather in the process of selection among multiple competing sources of information, a function which may apply across a wider range of tasks.

Additional aspects of language beyond the scope of this chapter may be affected by DLP damage. For example, the juncture in the left frontal lobe of language-related cortices and the region controlling skilled movement of the dominant hand provides a critical part of the neural substrate for handwriting, and DLP lesions often result in agraphia (Anderson, Saver, Damasio and Tranel, 1993). Damage to left DLP cortex affects not only language expression, but also auditory comprehension and reading (Alexander, Benson and Stuss, 1989). Further, right DLP damage may result in restriction of speech prosody.

Visuospatial abilities

Damage to large sectors of the lateral surface of the right hemisphere, such as the fronto-parieto-temporal lesions caused by infarction of the middle cerebral artery, typically causes impairment of visuospatial and visuoconstructional abilities, but there is limited information regarding specific contributions of DLP cortex damage to this profile. DLP and posterior parietal cortices are anatomically interconnected and appear to function in a coordinated fashion in the performance of many visuospatial and visuomo-

TABLE 1

Visuospatial and visuoconstructional abilities following DLP damage (see also Fig. 2)

	JLO	Blocks	CFT	PIQ
Right DLP				
AH1331	23	12	33	95
SB2046	26	12	31	104
KS2224	24	7	32	90
Left DLP				
JP1172	29	12	32	98
MV1852	25	11	34	98
NS2078	26	16	32	128

JLO = Judgment of Line Orientation, raw score. Blocks = WAIS-R Block Design age-corrected scale score. CFT = Rey–Osterreith Complex Figure, copy raw score out of 36 possible. PIQ = WAIS-R Performance IQ.

tor tasks (e.g., Quintana and Fuster, 1999), but DLP damage alone generally does not impair performance on such tasks in the absence of significant integrative or decision-making demands. There are exceptions — for example, isolated damage to the right DLP can result in transient neglect (Heilman and Valenstein, 1972), although parietal or combined parieto-frontal damage is more common. Although anatomical details have been sparse, frontal lobe damage appears to be less important than more posterior damage in the development of basic visuospatial deficits (Damasio and Anderson, 1993; Stuss and Benson, 1986).

In Table 1, we present findings from six patients with DLP lesions (3 right and 3 left) on standardized measures of visuospatial and visuoconstructional abilities (see also Fig. 2). The scores were uniformly within normal limits on the following measures: (1) Judgment of Line Orientation (Benton, Hamsher, Varney and Spreen, 1983), which requires matching lines of various spatial orientations without a significant motor component; (2) the Block Design subtest from the Wechsler Adult Intelligence Scale-Revised (WAIS-R; Wechsler, 1981), which requires construction of block patterns to match models; (3) the Rey–Osterreith Complex Figure Test (Lezak, 1995), which requires drawing a complex geometric figure to match a model; and (4) the composite Performance IQ measure from the WAIS-R, which reflects the average performance across a number of primarily nonverbal

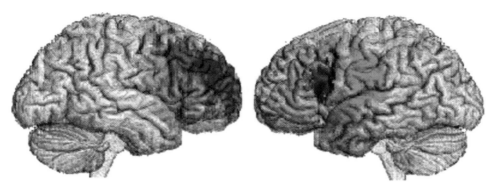

Fig. 2. MAP-3 image of lateral views of the left and right hemispheres indicating locations of lesions of the subjects included in Table 1. Darker colors indicate overlap of lesions.

reasoning tasks (e.g., identifying missing components of pictures, constructing puzzles, arranging picture sequences).

These findings illustrate that unilateral damage to either left or right DLP cortex does not substantially interfere with performance on basic visuospatial tasks, and are consistent with the notion that the DLP activation seen some spatial tasks, e.g., working memory, may primarily reflect executive demands of the tasks rather than representation of spatial information (Postle et al., 1999).

Executive functions

Damage to the human DLP cortex has long been linked to impairments of *executive functions*. This is an umbrella term encompassing a number of cognitive abilities which generally have been conceptualized as controlling or guiding behavior in a top-down fashion, such as decision-making, planning, self-monitoring, and behavior initiation, organization, and inhibition. Each of these executive abilities may involve multiple component cognitive operations, working in concert to achieve combinatorial or relational consideration of environmental information together with stored representations in the guidance of behavior (see Luria, 1966; Shallice and Burgess, 1991). There is clear overlap of executive functions with cognitive processes described earlier, such as working memory and response inhibition, and this linkage figures substantially in the relationship between DLP damage and executive dysfunction (e.g., Konishi, Kawazu, Uchida et al., 1999;

Petrides, 1995; Sarazin, Pillon, Giannakopoulos et al., 1998).

One of the most widely used measures of executive function is the Wisconsin Card Sorting Test (WCST). Early lesion studies in rhesus monkeys at the University of Wisconsin Primate Laboratory showed that the ability to alter responses to stimuli on the basis of changing patterns of reinforcement was impaired by frontal lobe ablations (Settlage, Zable and Harlow, 1948; Zable and Harlow, 1946), and Weigl (1941) demonstrated that the ability to sort stimuli according to dimensions such as color or form was impaired in some brain-damaged subjects. The WCST was developed to provide a measure of the ability to identify abstract categories and shift cognitive set, and was conceptually linked to frontal lobe functions from its inception (Berg, 1948; Grant and Berg, 1948).

The WCST requires patients to sort a deck of cards according to various stimulus dimensions. The patient is not informed of the sorting principles, but rather must infer these from information given by the examiner after each response. In addition to the ability to form cognitive sets, the test requires the patient to shift sets in response to changing contingencies, as the sorting principles change throughout the test without any clue other than a changing pattern of feedback (see Heaton, Chelune, Talley et al., 1993, for a complete description of the test). The WCST is neither particularly sensitive nor specific to prefrontal damage (Anderson, Damasio, Jones and Tranel, 1991; Grafman, Jonas and Salazar, 1990), but many patients with DLP damage have difficulty on this task (Stuss, Levine, Alexander et al., 2000). Lesion and functional imaging data, as well as consideration of the task demands, converge to suggest that performance on the WCST involves DLP activity in the context of more widespread brain activity.

The performance measure from the WCST which has proven most sensitive to prefrontal damage has been the number of perseverative errors committed, highlighting the contribution of this region to controlling inhibition of prepotent responses. Decreased ability to inhibit repetition of ineffective responses appears to be one of the basic defects arising from DLP damage, affecting performance on a variety of tasks. It is possible that variations in inhibitory control may underlie important cognitive differences between humans and other species, as well as differences which arise during the course of human development (Hauser, 1999).

Lesion and functional imaging studies also converge to suggest that DLP cortex is important for the mental act of *planning*, which is the prospective construction and mental testing of a series of intermediate steps required to achieve a goal. The mental act of planning should be distinguished from the execution of the plan, although this distinction often is blurred in behavioral measures. The relationship of planning defects to DLP damage is due in part to the major working memory and temporal sequencing demands of planning (Owen, 1997; Sirigu, Zalla, Pillon et al., 1995). Recent neurophysiological evidence from monkeys also indicates that neurons in prefrontal cortex are involved in prospective coding of anticipated reward and expected objects (Rainer, Rao and Miller, 1999; Watanabe, 1996), and functional imaging in normal humans has suggested that anterior DLP cortex may have a unique role in maintaining goals in working memory while proceeding with subgoals (Koechlin, Basso, Peitrini et al., 1999).

Much of the evidence for an impairment of planning following prefrontal damage has been based on the 'Tower' tasks (e.g., Tower of Hanoi, Tower of London), which require movement of a set of disks to a goal position, following certain rules which require planning of a series of steps (Owen, Downes, Sahakian et al., 1990; Shallice, 1982). The linkage between performance on such tasks and DLP activity has been supported by functional imaging studies (e.g., Dagher, Owen, Boecker and Brookds, 1999).

Goel and Grafman (1995) have pointed out that the Tower of Hanoi task may have different cognitive demands than the Tower of London, with the critical step in TOH being the ability to see and resolve a goal–subgoal conflict (i.e., to perform a counterintuitive backward move), rather than planning.

Despite the association between DLP damage and impairments on laboratory tests involving planning, Sarazin et al. (1998) have provided evidence that orbital prefrontal dysfunction, rather than DLP damage, is most highly correlated with planning defects in real world activities. Planning clearly is not a single cognitive operation. Rather, planning involves conflation of multiple component processes and occurs over a broad range of circumstances involving widely divergent time frames, levels of complexity and intentionality, and methods of execution. It appears that broad sectors of the frontal lobes, including DLP and orbital prefrontal regions, in addition to premotor cortex, are involved in aspects of planning. Further specification of the planning impairments resulting from DLP damage is likely to be a fruitful area for future research.

One of the puzzles in the neuropsychology of human prefrontal cortex is the observation that some patients with prefrontal damage seem able to perform normally on a broad range of standardized neuropsychological tests, including those directed at probing executive functions, despite substantial disruption of their daily behavior (e.g., Eslinger and Damasio, 1985). This discrepancy presents a major challenge in the clinical evaluation of patients with prefrontal damage, and it is not uncommon for children or adults with prefrontal dysfunction to fail to qualify for needed services because no deficit is evident on their standardized test scores. The discrepancy also provides guidance in trying to conceptualize and study the impairments which arise from damage to prefrontal regions. Problem-solving and decision-making in the real world involve integration of information from diverse sources over extended time frames, a seemingly unlimited choice of response options, and often an absence of specific criteria for adequate task performance. The complexity of real-world problem-solving and decision-making is not approximated by standard clinical or laboratory tasks, which typically are characterized by having a single explicit problem, brief trials, and clear cues for task initiation and completion. It appears likely that much of the dissociation between standardized test performance and real world dysfunction arises because prefrontal cortex damage leaves relatively preserved an ability to respond appropriately when sufficient constraints and structure are provided by immediately present environmental stimuli. There is a fundamental conflict between the needs of standardized testing and the nature of the impairments which arise from prefrontal damage.

Shallice and Burgess (1991) developed a quantifiable analog of the type of relatively unstructured, open-ended, multiple subgoal tasks which seem to cause so much difficulty in the daily life of patients with frontal lobe damage. Subjects were required to complete a set of everyday tasks (e.g., buy a certain item, meet someone at a given time, obtain basic meteorological and consumer information), which were designed to place limited demands on non-executive cognitive functions. They found that three patients with traumatic head injuries involving prefrontal damage approached the tasks inefficiently and tended to violate rules (task imposed and societal norms). We have replicated this experiment in a group of 34 patients with focal brain lesions caused by stroke or surgery (17 prefrontal, 17 nonfrontal), and also found that the behavior of patients with prefrontal lesions was marked by inefficiencies, failure to complete tasks, and rule infractions (Anderson, Hathaway-Nepple, Tranel and Damasio, 2000). Although DLP damage was associated with defective performances on these 'real world' strategy application tasks, the impairments tended to be most severe in patients with damage to orbital prefrontal regions. The findings from this and the other lesion studies described above suggest that DLP damage can result in impairments in several aspects of executive functions, but the implications of these deficits for daily activities may be less severe than when orbital and mesial prefrontal areas are damaged.

Emotion and social behavior

It is well established that impairments of emotion and social behavior follow prefrontal cortex damage, but the specific contributions of DLP cortex damage to such defects remain to be defined. Alterations of emotional experience and social functioning following prefrontal damage are not independent of cognitive impairments of the type described above. For example, disruption of the neural representation of emotions likely plays a key role in impairments of judgement and decision-making following prefrontal damage (Damasio, 1994), and adaptive social behavior clearly draws upon many of the cognitive abilities described above. The most dramatic impairments of social behavior appear to require damage to the ventromedial prefrontal region, and occur in the context of altered emotion and impairments of judgement and decision-making (e.g., Anderson, Bechara, Damasio et al., 1999; Barrash, Tranel and Anderson, 2000; Damasio, 1994; Sarazin et al., 1998). However, isolated damage to either the left or right DLP cortex also can affect aspects of emotion and social behavior.

Although depression is not unique to DLP damage, various aspects of depressive symptomatology occur following damage to this region at a frequency which suggests more than a reaction to non-specific brain damage. Blumer and Benson (1975) referred to a pseudodepressed syndrome with decreased self-initiation following DLP damage, and Cummings (1985) described apathy, indifference, and psychomotor retardation as frequent features. Depressive symptomatology and social unease were found to be common in patients three months after DLP damage from stroke or trauma (Paradiso, Chemerinski, Yazici et al., 1999). Lesions in the left DLP appear to create the strongest predisposition to develop depression (Robinson, Kubos, Starr et al., 1984; Robinson, Starr, Lipsey et al., 1985). Damage in this area, for example that resulting from stroke in the anterior distribution of the middle cerebral artery, often has devastating consequences (paresis of the dominant side, aphasia), and unlike comparable lesions in the right DLP, generally does not impair awareness of acquired impairments. However, during the acute phase following stroke, depression cannot be fully explained by these factors, suggesting a neurophysiological basis for the relationship between left DLP damage and depression. Reaction to functional impairments likely plays a more important role in depression with onset six months or more following stroke (Robinson, Bolduc and Price, 1987).

The emotional consequences of focal damage to the right DLP have not been well characterized, but may include both restriction of affect and emotional dyscontrol. Damage to the right frontal lobe also may contribute to anosognosia for hemiparesis and unawareness of acquired cognitive impairments. Stuss and Benson (1986) described self-awareness as a frontally mediated ability at the top of a hierarchical arrangement of human cognitive abilities, and as described earlier, self-monitoring may be disrupted by DLP damage.

The perception or comprehension of emotional information also appears to be altered in some patients with DLP damage. The families of patients with frontal lobe injuries often complain that the patient has impaired empathy, but little is known regarding anatomical–functional correlations for this deficit. Damage to right somatosensory cortex appears to be the most important factor contributing to impairments in the recognition of emotional facial expressions (Adolphs, Damasio, Tranel et al., 2000), but damage to various sectors of prefrontal cortex also may disrupt empathic processing. Eslinger (1998) has raised the possibility that damage to DLP areas may impair certain cognitive aspects of empathy (e.g., perspective taking), while orbital damage may impact more on emotional aspects of empathy. In another example of disrupted processing of social–emotional information, damage to the right frontal lobe has been associated with impaired appreciation of humor, possibly through disruption of component processes such as emotional reactivity and abstract thought (Shammi and Stuss, 1999). Involvement of superior and polar right prefrontal regions appeared to be important, but it is not yet possible to describe specific contributions of DLP damage to this deficit.

Conclusions

The consequences of damage to human DLP cortex are diverse and not yet well defined. Isolated damage to this area is not common in most clinical situations, but rather the effects of DLP damage are seen in interaction with damage to other brain regions. Impairments in aspects of working memory and inhibitory control appear to be basic-order defects which result from DLP damage and have many manifestations, particularly in situations in which representations of multiple items or events must be

activated for purposes of comparison or integration. Dysfunction in this region often impacts on the so-called executive abilities which seem to be uniquely characteristic of human cognition, in large part due to disruption of essential component cognitive processes. Much work remains to be done to delineate these impairments and their relationships to particular patterns of DLP damage, but the rewards may be substantial in terms of better understanding of a number of neurological and psychiatric conditions which involve dysfunction in this area.

Acknowledgements

We thank Dr. Hanna Damasio for the neuroanatomical images. Supported by NINDS PO1 NS19632 and the Mathers Foundation.

References

Adolphs R, Damasio H, Tranel D, Cooper G, Damasio AR: A role for somatosensory cortices in the visual recognition of emotion as revealed by three-dimensional lesion mapping. The Journal of Neuroscience: 20; 2683–2690, 2000.

Alexander MP, Benson DF, Stuss DT: Frontal lobes and language. Brain and Language: 37; 1989.

Anderson SW, Bechara A, Damasio H, Tranel D, Damasio AR: Impairment of social and moral behavior related to early damage in the human prefrontal cortex. Nature Neuroscience: 2; 1032–1037, 1999.

Anderson SW, Damasio H, Jones RD, Tranel D: Wisconsin Card Sorting Test performance as a measure of frontal lobe damage. Journal of Clinical and Experimental Neuropsychology: 13; 909–922, 1991.

Anderson SW, Hathaway-Nepple J, Tranel D, Damasio H: Impairments of strategy application following focal prefrontal lesions. Journal of the International Neuropsychological Society: 6; 205, 2000.

Anderson SW, Saver J, Damasio H, Tranel D: Acquired agraphia caused by focal brain lesions. Acta Psychologica: 82; 193–210, 1993.

Barbas H: Anatomic basis of cognitive–emotional interactions in the primate prefrontal cortex. Neuroscience and Biobehavioral Reviews: 19; 499–510, 1995.

Barrash J, Tranel D, Anderson SW: Acquired personality disturbances associated with bilateral damage to the ventromedial prefrontal region. Developmental Neuropsychology: 18; 355–381, 2000.

Bechara A, Damasio H, Tranel D, Anderson SW: Dissociation of working memory from decision-making within the human prefrontal cortex. The Journal of Neuroscience: 18; 428–437, 1998.

Beldarrain G, Grafman J, Pascual-Leone A, Garcia-Monco JC:

Procedural learning is impaired in patients with prefrontal lesions. Neurology: 52; 1853–1860, 1999.

Benton AL, Hamsher K, Varney NR, Spreen O: Contributions to Neuropsychological Assessment. New York: Oxford University Press, 1983.

Berg EA: A simple objective technique for measuring flexibility in thinking. The Journal of General Psychology: 39; 15–22, 1948.

Blumer D, Benson DF: Personality changes with frontal and temporal lobe lesions. In Blumer D, Benson DF (Eds), Psychiatric Aspects of Neurologic Disease. New York: Grune and Stratton, 1975.

Chao LL, Knight RT: Contribution of human prefrontal cortex to delay performance. Journal of Cognitive Neuroscience: 10; 167–177, 1998.

Chollet F, DiPiero V, Wise RJS, Brooks DJ, Dolan RJ, Frackowiak RSJ: The functional anatomy of motor recovery after stroke in humans: a study with positron emission tomography. Annals of Neurology: 29; 63–71, 1991.

Cohen JD, Forman SD, Braver TS, Casey BJ, Servan-Schreiber D, Noll DC: Activation of the prefrontal cortex in a nonspatial working memory task with functional MRI. Human Brain Mapping: 1; 293–304, 1994.

Crossen B, Rao SM, Woodley SJ, Rosen AC, Bobholz JA, Mayer A, Cunningham JM, Hammeke TA, Fuller SA, Binder JR, Cox RW, Stein EA: Mapping of semantic phonological and orthographic verbal working memory in normal adults with functional magnetic resonance imaging. Neuropsychology: 13; 171–187, 1999.

Cummings JL: Clinical Neuropsychiatry. New York: Grune and Stratton, 1985.

Dagher A, Owen AM, Boecker H, Brookds DJ: Mapping the network for planning: a correlational PET activation study with the Tower of London task. Brain: 122; 1973–1987, 1999.

Damasio AR: Descartes' Error: Emotion, Reason, and the Human Brain. New York: Grosset/Putnam, 1994.

Damasio AR, Anderson SW: The frontal lobes. In Heilman KM, Valenstein E (Eds), Clinical Neuropsychology, 3rd ed. New York: Oxford, 1993

Damasio H: Human neuroanatomy relevant to decision-making. In Damasio AR, Damsio H, Christen Y (Eds), Neurobiology of Decision-Making. Berlin: Springer, 1996.

Damasio H: Neuroanatomical correlates of the aphasias. In Sarno MT (Ed), Acquired Aphasia, 3rd ed. San Diego, CA: Academic Press.

D'Esposito M, Postle BR: The dependence of span and delayed-response performance on prefrontal cortex. Neuropsychologia: 37; 1303–1315, 1999.

Dias R, Robbins TW, Roberts AC: Dissociable forms of inhibitory control within prefrontal cortex with an analog of the Wisconsin card sort test: restriction to novel situations and independence from 'on-line' processing. The Journal of Neuroscience: 17, 9285–9297, 1997.

Eslinger PJ: Neurological and neuropsychological bases of empathy. European Neurology: 39; 193–199, 1998.

Eslinger PJ, Damasio AR: Severe disturbances of higher cognition after bilateral frontal lobe ablation: Patient EVR. Neurology: 34; 1731–1741, 1985.

Eslinger PJ, Grattan LM: Altered serial position learning after frontal lobe lesion. Neuropsychologia: 32; 729–739, 1994.

Fuster JM: Temporal processing. In Grafman J, Holyoak KJ, Boller F (Eds), Structure and Functions of the Human Prefrontal Cortex. Annals of the New York Academy of Science: 769; 173–181, 1995.

Gershberg FB, Shimamura AP: Impaired use of organizational strategies in free recall following frontal lobe damage. Neuropsychologia: 13; 1305–1333, 1995.

Goel V, Grafman J: Are the frontal lobes implicated in 'planning' functions? Interpreting data from the Tower of Hanoi. Neuropsychologia: 33; 623–642, 1995.

Goldman-Rakic PS: Modular organization of the prefrontal cortex. Trends in Neuroscience: 7; 419–424, 1984.

Grabowski TJ, Damasio H, Damasio AR: Premotor and prefrontal correlates of category-related lexical retrieval. Neuroimage: 7; 232–243, 1998.

Grafman J, Jonas B, Salazar A: Wisconsin Card Sorting Test performance based on location and size of neuroanatomical lesion in Vietnam veterans with penetrating head injuries. Perceptual and Motor Skills: 71; 1120–1122, 1990.

Grant DA, Berg EA: A behavioral analysis of degree of reinforcement and ease of shifting to new resonses in a Weigl-type card sorting problem. Journal of Experimental Psychology: 38; 404–411, 1948.

Hauser MD: Perseveration, inhibition and the prefrontal cortex: a new look. Current Opinion in Neurobiology: 9; 214–222, 1999.

Heaton RK, Chelune GJ, Talley JL, Kay GG, Curtiss G: Wisconsin Card Sorting Test Manual. Odessa, FL: Psychological Assessment Resources, 1993.

Heilman KM, Valenstein E: Frontal lobe neglect in man. Neurology: 22; 660–664, 1972.

Holmes G: The cerebral integration of ocular movements. British Medical Journal: 2; 107–112, 1938.

Janowsky JS, Shimamura AP, Squire LP: Source memory impairment in patients with frontal lobe lesions. Neuropsychologia: 27; 1043–1056, 1989.

Jurado MA, Junque C, Vendrell P, Treserras P, Grafman J: Overestimation and unreliability in 'feeling of doing' judgments about temporal ordering performance: Impaired self-awareness following frontal lobe damage. Journal of Clinical and Experimental Neuropsychology: 20; 353–364, 1998.

Kim J-N, Shadlen MN: Neural correlates of a decision in the dorsolateral prefrontal cortex of the macaque. Nature Neuroscience: 2; 176–185, 1999.

Koechlin E, Basso G, Peitrini P, Panzer S, Grafman J: The role of the anterior prefrontal cortex in human cognition. Nature: 399; 148–151, 1999.

Konishi S, Kawazu M, Uchida I, Kikyo H, Asakura I, Miyashita Y: Contribution of working memory to transient activation in human inferior prefrontal cortex during performance of the Wisconsin Card Sorting Task. Cerebral Cortex: 9; 745–753, 1999.

Lezak MD: Neuropsychological Assessment, 3rd ed. New York: Oxford University Press, 1995.

Lhermitte F, Pillon B, Serdaru M: Human autonomy and the frontal lobes, Part 1. Imitation and utilization behavior. Annals of Neurology: 19; 326–334, 1986.

Luria AR: Higher Cortical Functions in Man. New York: Plenum Press, 1966.

Mangels JA, Gershberg FB, Shimamura AP, Knight RT: Impaired retrieval from remote memory in patients with frontal lobe damage. Neuropsychology: 10; 32–41, 1996.

McAndrews MP, Milner B: The frontal cortex and memory for temporal order. Neuropsychologia: 29; 849–859, 1991.

Milner B, Corsi P, Leonard G: Frontal lobe contribution to recency judgments. Neuropsychologia: 29; 601–618, 1991.

Mohr JP: Broca's area and Broca's aphasia. In Whitaker H, Whitaker HA (Eds), Studies in Neurolinguistics, Vol. 1. New York: Academic Press, 1976.

Nauta WJH: Neural associations of the frontal cortex. Acta Neurobiologia: 32; 125–140, 1972.

Owen AM: Cognitive planning in humans: neuropsychological, neuroanatomical and neuropharmacological perspectives. Progress in Neurobiology: 53; 431–450, 1997.

Owen AM, Downes JD, Sahakian BJ, Polkey CE, Robbins TW: Planning and spatial working memory following frontal lobe lesions in man. Neuropsychologia: 28; 1021–1034, 1990.

Paradiso S, Chemerinski E, Yazici KM, Tartaro A, Robinson RG: Frontal lobe syndrome reassessed: comparison of patients with lateral or medial frontal brain damage. Journal of Neurology, Neurosurgery, and Psychiatry: 67; 664–667, 1999.

Petersen SE, Fox PT, Posner MI, Mintun MA, Raichle MA: Positron emission tomographic studies of the processing of single words. Journal of Cognitive Neuroscience: 1; 153–170, 1989.

Petrides M: Deficits on conditional associative learning tasks after frontal and temporal lobe lesions in man. Neuropsychologia: 23; 601–614, 1985.

Petrides M: Functional organization of the human frontal cortex for mnemonic processing. In Grafman J, Holyoak KJ, Boller F (Eds), Structure and Functions of the Human Prefrontal Cortex. Annals of the New York Academy of Science: 769; 85–96, 1995.

Postle BR, Berger JS, D'Esposito M: Functional neuroanatomical double dissociation of mnemonic and executive control processes contributing to working memory performance. Proceedings of the National Academy of Science: 96; 12959–12964, 1999.

Quintana J, Fuster JM: From perception to action: temporal integrative functions of prefrontal and parietal neurons. Cerebral Cortex: 9; 213–221, 1999.

Raichle ME, Fiez JA, Videen TO, MacLeod AM, Pardo JV, Fox PT, Petersen SE: Practice-related changes in human brain functional anatomy during nonmotor learning. Cerebral Cortex: 4; 8–26, 1994.

Rainer G, Rao SC, Miller EK: Prospective coding for object in primate prefrontal cortex. The Journal of Neuroscience: 19; 5493–5505, 1999.

Robinson RG, Bolduc PL, Price TR: Two-year longitudinal study

of poststroke mood disorders: diagnosis and outcome at one and two years. Stroke: 18; 837–843, 1987.

Robinson RG, Kubos KL, Starr LB, Rao K, Price TR: Mood disorders in stroke patients. Importance of location of lesion. Brain: 107; 81–93, 1984.

Robinson RG, Starr LB, Lipsey JR, Rao K, Price TR: A 2-year longitudinal study of poststroke mood disorders: in-hospital prognostic factors associated with 6-month outcome. Journal of Nervous and Mental Disease: 173; 221–226, 1985.

Rubens AB: Transcortical motor aphasia. In Whitaker H, Whitaker HA (Eds), Studies in Neurolinguistics. New York: Academic Press, 1976.

Sarazin M, Pillon B, Giannakopoulos P, Rancurel G, Samson Y, Dubois B: Clinicometabolic dissociation of cognitive functions and social behavior in frontal lobe lesions. Neurology: 51; 142–148, 1998.

Schacter DL, Curran T, Galluccio L, Milberg WP, Bates JF: False recognition and the right frontal lobe: a case study. Neuropsychologia: 34; 793–808, 1996.

Schacter DL, Harbluck J, McLaughlin D: Retrieval without recollection: an experimental analysis of source amnesia. Journal of Verbal Learning and Verbal Behavior: 23; 593–611, 1984.

Settlage P, Zable M, Harlow HF: Problem solution by monkeys following bilateral removal of the frontal areas, VI. Performance on tests requiring contradictory reactions to similar and identical stimuli. Journal of Experimental Psychology: 38; 50–65, 1948.

Shallice T: Specific impairments of planning. Philosophical Transactions of the Royal Society of London (Biology): 298; 199–209, 1982.

Shallice T, Burgess P: Higher-order cognitive impairments and frontal lobe lesions. In Levin HS, Eisenberg HM, Benton AL (Eds), Frontal Lobe Function and Dysfunction. New York: Oxford University Press, 1991.

Shammi P, Stuss DT: Humour appreciation: a role of the right frontal lobe. Brain: 122; 657–666, 1999.

Shimamura AP, Janowsky JS, Squire LR: Memory for the temporal order of events in patients with frontal lobe lesions and amnesic patients. Neuropsychologia: 28; 803–814, 1990.

Sirigu A, Zalla T, Pillon B, Grafman J, Dubois B, Agid Y: Planning and script analysis following prefrontal lobe lesions. In Grafman J, Holyoak K, Boller F (Eds), Structure and Functions of the Human Prefrontal Cortex. Annals of the New York Academy of Sciences: 769; 277–288, 1995.

Smith EE, Jonides J, Koeppe A Jr: Dissociating verbal and spatial working memory using PET. Cerebral Cortex: 6; 11–20, 1996.

Stuss DT, Benson DF: The Frontal Lobes. New York: Raven Press, 1986.

Stuss DT, Levine B, Alexander MP, Hong J, Palumbo C, Hamer L, Murphy KJ, Izukawa D: Wisconsin Card Sorting Test performance in patients with focal frontal and posterior brain damage: effects of lesion location and test structure on separable cognitive processes. Neuropsychologia: 38; 388–402, 2000.

Swick D, Knight RT: Is prefrontal cortex involved in cued recall?

A neuropsychological test of PET findings. Neuropsychologia: 34; 1019–1028, 1996.

Thompson-Schill SL, Swick D, Farah M, D'Esposito M, Kan IP, Knight RT: Verb generation in patients with focal frontal lesions: a neuropsychological test of neuroimaging findings. Proceedings of the National Academy of Sciences: 95; 15855–15860, 1998.

Tomita H, Ohbayashi M, Nakahara K, Hasegawa I, Miyashita Y: Top-down signal from prefrontal cortex in executive control of memory retrieval. Nature: 401; 699–703, 1999.

Tranel D, Anderson SW: Syndromes of aphasia. In Fabbro F (Ed), Concise Encyclopedia of Language Pathology. New York: Elsevier, 305–319, 1999.

Walker R, Husain M, Hodgson TL, Harrison J, Kennard C: Saccadic eye movement and working memory deficits following damage to human prefrontal cortex. Neuropsychologia: 26; 1141–1159, 1998.

Watanabe M: Reward expectancy in primate prefrontal neurons. Nature: 382; 629–632, 1996.

Wechsler D: WAIS-R Manual. New York: The Psychological Corporation, 1981.

Weigl E: On the psychology of the so-called process of abstraction. Journal of Abnormal and Social Psychology: 36; 3–33, 1941.

Weiller C, Chollet F, Friston KJ, Wise RJS, Frackowiak RSJ: Functional reorganization of the brain in recovery from striatocapsular infarction in man. Annals of Neurology: 31; 463–472, 1992.

Zable M, Harlow HF: The performance of rhesus monkeys on a series of object quality and positional discriminations and discrimination reversals. Journal of Comparative Psychology: 39; 13–23, 1946.

Handbook of Neuropsychology, 2nd Edition, Vol. 7
J. Grafman (Ed)

CHAPTER 8

The human prefrontal cortex has evolved to represent components of structured event complexes

Jordan Grafman *

Cognitive Neuroscience Section, National Institute of Neurological Disorders and Stroke, National Institutes of Health, Building 10, Room 5C205, 10 Center Drive, MSC 1440, Bethesda, MD 20892-1440, USA

Purpose and plan of this chapter

Perhaps there is no region of human cerebral cortex whose functional assignments are as puzzling to us as the human prefrontal cortex (HPFC). Over one hundred years of observation and experimentation has led to several general conclusions about its overall functions. The consensus of investigators is that the prefrontal cortex is important for modulating social behavior, reasoning, planning, working memory, thought, concept formation, inhibition, attention, and abstraction. Yet unlike the research conducted in other cognitive domains such as object recognition or lexical/semantic storage, there has been little effort to propose and investigate in detail the underlying cognitive architecture(s) that would capture the essential features and computational properties of the higher cognitive processes presumably supported by the HPFC. Since the processes that are attributed to the HPFC appear to constitute the most complex and abstract of cognitive functions, many of which are responsible for the internal guidance of human behavior, a critical step in understanding the functions of the human brain requires us to adequately describe the cognitive topography of the HPFC. The purpose of this chapter is to argue for the validity of one specific research framework to understand HPFC functioning in humans. This framework suggests that by postulating the form of the various units of rep-

resentation (in essence, the elements of memory) stored in prefrontal cortex, it will be much easier to derive clear and testable hypotheses that will enable rejection or validation of this and other frameworks. My colleagues and I have labeled the set of HPFC representational units alluded to above as a structured event complex (SEC). Below I will detail my hypotheses about the SECs representational structure and features and I will attempt to distinguish the SEC framework from other cognitive models of HPFC function. Before doing so, I will briefly summarize the key elements of the biology and structure of the HPFC, the evidence of its general role in cognition based on convergent evidence from lesion and neuroimaging studies, some key models postulating the functions of the HPFC, then briefly argue for a rationale that specifies why there must be representational knowledge stored in the prefrontal cortex, present a short primer on the SEC framework I have adapted, and finally offer some suggestions about future directions for research of HPFC functions using the SEC framework.

Introduction

What we know about the anatomy and physiology of the HPFC is inferred almost entirely from work in the primate and lower species. It is likely that additional complexity has been added in the HPFC — to reveal it will require decades of further study of human brain anatomy. Nevertheless, it is likely that the connectivity already described in other species also exists in the HPFC (Petrides and Pandya, 1994;

* Tel.: +1 (301) 496-0220; Fax: +1 (301) 480-2909;
E-mail: grafmanj@ninds.nih.gov

Barbas, Gashghaei, Rempel-Clower and Xiao, 2002, this volume). The HPFC is composed of Brodmann's areas 8–14 and 24–47. Grossly, it can be subdivided into lateral, medial, and orbital regions with Brodmann's areas providing morphological subdivisions within (and occasionally across) each of the gross regions (Barbas, 2000). Some regions of the prefrontal cortex have a total of 6 layers, other regions are agranular meaning that the granule cell layer is absent. The HPFC has a columnar design like other cortical regions. All regions of the HPFC are interconnected. The HPFC is also richly interconnected with other areas of brain. For example, the HPFC has at least five distinct regions each of which is independently involved in separate cortico-striatal loops (Alexander, Crutcher and DeLong, 1990). The general functional role of each relatively segregated circuit has been popularized by Cummings and his colleagues (Masterman and Cummings, 1997). The HPFC also has strong limbic system connections via its medial and orbital efferent connections that terminate in the amygdala, thalamus, and parahippocampal regions (Groenewegen and Uylings, 2000; Price, 1999). Finally, the HPFC has long pathway connections to association cortex in the temporal, parietal, and occipital lobes. Almost all of these pathways are reciprocal.

When compared with the prefrontal cortex of other species, most investigators have claimed that the HPFC is proportionally (compared to the remainder of the cerebral cortex) much larger (Rilling and Insel, 1999; Semendeferi, Armstrong, Schleicher et al., 2001). There is other recent research indicating that the size of the HPFC is not proportionally larger than that of other primates, but that its internal neural architecture must be more sophisticated, or at least differentially organized in order to support superior human functions (Chiavaras, LeGoualher, Evans and Petrides, 2001; Petrides and Pandya, 1999). The functional argument being that in order to subserve such higher-order cognitive functions as extended reactive planning and complex reasoning that are not obviously apparent in other primates or lower species, the HPFC must have a uniquely evolved neural architecture (Elston, 2000).

Developmentally, the HPFC is not considered mature until the early teenage years. Maturation means adult-like size and neural architecture complexity based on pruning and elaborated connectivity. This maturation process occurs later in development in the HPFC than almost all other cortical association areas. The fact that the HPFC does not mature until this late date suggests that those cognitive processes mediated by the prefrontal cortex are not fully operational until that time (Diamond, 2000).

The HPFC is innervated by a number of different neurotransmitter and peptide systems, most prominent among them being the dopaminergic, serotonergic, and cholinergic transmitters and their varied receptor subtypes (Robbins, 2000). The functional role of each of these neurotransmitters in the HPFC is not entirely clear. Mood disorders which involve alterations in serotonergic functions lead to reduced blood flow in HPFC. Several degenerative neurologic disorders are at least partially due to disruption in the production and transfer of dopamine from basal ganglia structures to the HPFC. This loss of dopamine may cause deficits in cognitive flexibility. Serotonergic receptors are distributed throughout the HPFC and have a role in motivation and intention. Finally, the basal forebrain in ventral and posterior HPFC is richly innervated by cholinergic terminals whose loss can cause impaired memory and attention.

A unique and key property of neurons in the prefrontal cortex of monkeys (and presumably humans) is their ability to fire during an interval between a stimulus and a delayed probe (Levy and Goldman-Rakic, 2000). Neurons in other brain areas are either directly linked to the presentation of a single stimulus or the probe itself and if they demonstrate continuous firing, the explanation has been that they are driven by neurons in the prefrontal cortex. If the firing of neurons in the prefrontal cortex is linked to activity that 'moves' the subject towards a goal rather than the appearance of a single stimulus, then potentially, those neurons could continuously fire across many stimuli or events until the goal was achieved or the behavior of the subject disrupted. This observation of sustained firing of prefrontal cortex neurons across time and events has led many investigators to suggest that the HPFC must be involved in the maintenance of a stimulus across time, i.e., working memory (Fuster, Bodner and Kroger, 2000).

Besides the property of sustained firing, Elston (2000) has recently demonstrated a unique structural feature of neurons in the prefrontal cortex. He found that pyramidal cells in the prefrontal cortex of macaque monkeys are significantly more spinous compared to pyramidal cells in other cortical areas suggesting that they are capable of handling a larger amount of excitatory inputs compared to pyramidal cells elsewhere. This could be one of several structural explanations for the HPFC's ability to integrate input from many sources in order to implement more abstract behaviors.

Functional studies

The traditional approach to understanding the functions of the HPFC is to perform cognitive studies testing the ability of normal and impaired humans on tasks designed to induce the activation of processes or representational knowledge presumably stored in the HPFC (Grafman, 1999). Both animals and humans with brain lesions can be studied to determine the effects of a prefrontal cortex lesion on task performance. Lesions in humans, of course, are due to an act of nature whereas lesions in animals can be more precisely and purposefully made by investigators. Likewise, 'intact' animals can be studied using precise electrophysiological recordings of single neurons or neural assemblies. In humans, the powerful new neuroimaging techniques such as functional Magnetic Resonance Imaging (fMRI) have been used to demonstrate frontal lobe activation during the performance of a range of tasks in normal subjects and patients. A potential advantage in studying humans (instead of animals) comes from the presumption that since the HPFC represents the kind of higher-order cognitive knowledge that distinguishes humans from other primates, an understanding of its underlying cognitive and neural architecture can only come from the study of humans.

Patients with frontal lobe lesions are generally able to understand conversation and commands, recognize and use objects, express themselves adequately enough to navigate through many social situations in the world, learn and remember routes, and even make decisions. They have documented deficits in sustaining their attention and anticipating what will happen next, in dividing their resources, in inhibiting pre-potent behavior, in adjusting to some situations requiring social cognition, in processing the theme or the moral of a story, in forming con-

cepts, abstracting, reasoning, and planning (Arnett, Rao, Bernardin et al., 1994; Carlin, Bonerba, Phipps et al., 2000; Dimitrov, Grafman, Soares and Clark, 1999a; Dimitrov, Granetz, Peterson et al., 1999b; Goel and Grafman, 1995; Goel, Grafman, Tajik et al., 1997; Grafman, 1999; Jurado, Junque, Vendrell et al., 1998; Vendrell, Junque, Pujol et al., 1995; Zahn, Grafman and Tranel, 1999). These deficits have been observed and confirmed across many studies over the last 40 years of clinical and experimental research.

Neuroimaging investigators have published studies that show prefrontal cortex activation during encoding, retrieval, decision-making and response conflict, task-switching, reasoning, planning, forming concepts, understanding the moral or theme of a story, inferring the motives or intentions of others, and similar high-level cognitive processing (Goel, Grafman, Sadato and Hallett, 1995; Koechlin, Basso, Pietrini et al., 1999; Koechlin, Corrado, Pietrini and Grafman, 2000; Nichelli, Grafman, Pietrini et al., 1994, 1995b; Wharton, Grafman, Flitman et al., 2000). The major advantage of these studies is that they provide convergent evidence for the involvement of the HPFC in controlling endogenous and exogenous cognitive processes especially those that are engaged by the abstract characteristics of a task.

Neuropsychological frameworks that try to account for HPFC functions

Working memory

Working memory has been described as the cognitive process that allows for the temporary activation of information in memory for rapid retrieval or manipulation (Ruchkin, Berndt, Johnson et al., 1997). It was first proposed some thirty years ago to account for a variety of human memory data that was not addressed by contemporary models of short-term memory (Baddeley, 1998b). Of note is that subsequent researchers have been unusually successful in describing the circumstances under which the so-called 'slave-systems' employed by working memory would be used. These slave-systems allowed for the maintenance of the stimuli in a number of different forms that could be manipulated by the central executive component of the working memory system (Baddeley, 1998a). Neuroscience support for the model followed quickly. Joaquin Fuster was among the first neuroscientists to recognize that neurons in the prefrontal cortex appeared to have a special capacity to discharge over time intervals when the stimulus was not being shown prior to a memory-driven response by the animal (Fuster et al., 2000). He interpreted this neuronal activity as being concerned with the cross-temporal linkage of information processed at different points in an ongoing temporal sequence. Goldman-Rakic and her colleagues later elaborated on this notion and suggested that these same PFC neurons were fulfilling the neuronal responsibility for working memory (Levy and Goldman-Rakic, 2000). In her view, PFC neurons *temporarily* hold in active memory modality-specific information *until* a response is made. This implies a restriction on the kind of memory that may be stored in prefrontal cortex. That is, this point of view suggests that there are no long-term representations in the prefrontal cortex until an explicit intention to act is required and then a temporary representation is created. Miller has challenged some of Goldman-Rakic's views about the role of neurons in the prefrontal cortex and argues that many neurons in the monkey prefrontal cortex are modality nonspecific and may serve a broader integrative function rather than a simple maintenance function (Miller, 2000). The research programs of Fuster, Goldman-Rakic, and Baddeley have had a major influence on the functional neuroimaging research programs of Courtney (Courtney, Petit, Haxby and Ungerleider, 1998), Smith and Jonides (1999), and Cohen (Nystrom, Braver, Sabb et al., 2000), all of whom have studied normal subjects in order to remap the HPFC in the context of working memory theory.

Executive function and attentional/control processes

Although rather poorly described in the scientific literature, it is premature to simply dismiss the general notion of a central executive (Baddeley, 1998a; Grafman and Litvan, 1999b). Several investigators have described the prefrontal cortex as the seat of attentional and inhibitory processes that govern the focus of our behaviors and therefore why not ascribe the notion of a central executive operating within the confines of the HPFC. Norman and Shallice (1986) proposed a dichotomous function of the central ex-

ecutive in HPFC. They argued that the HPFC was primarily specialized for the supervision of attention towards unexpected occurrences. Besides this supervisory attention system, they also hypothesized the existence of a contention scheduling system that was specialized for the initiation and efficient running of automatized behaviors such as repetitive routines, procedures and skills. Shallice, Burgess, Stuss and others have attempted to expand this idea of the prefrontal cortex as a voluntary control device and have further fractionated the supervisory attention system into a set of parallel attention processes that work together to manage complex multi-task behaviors (Burgess, 2000; Burgess, Veitch, de Lacy Costello and Shallice, 2000; Shallice and Burgess, 1996; Stuss, Toth, Franchi et al., 1999).

Social cognition

The role of the HPFC in working memory and executive processes has been extensively examined, but there is also substantial evidence that the prefrontal cortex is involved in controlling certain aspects of social and emotional behavior (Dimitrov, Grafman and Hollnagel, 1996; Dimitrov, Phipps, Zahn and Grafman, 1999c). Although the classic story of the 19th century patient Gage who suffered a penetrating prefrontal cortex lesion has been used to exemplify the problems that patients with ventromedial prefrontal cortex lesions have in obeying social rules, recognizing social cues, and making appropriate social decisions, the details of this social cognitive impairment have occasionally been inferred or even embellished to suit the enthusiasm of the story teller, at least regarding Gage (Macmillan, 2000). Nevertheless, Damasio and his colleagues have repeatedly confirmed the association of ventromedial prefrontal cortex lesions and social behavior and decision-making abnormalities (Anderson, Bechara, Damasio et al., 1999; Bechara, Damasio, Damasio and Lee, 1999; Bechara, Damasio and Damasio, 2000; Damasio, 1996; Eslinger, 1998; Kawasaki, Kaufman, Damasio et al., 2001). The exact functional assignment of that area of HPFC is still subject to dispute but convincing evidence has been presented that indicates that it serves to associate somatic markers (autonomic nervous system modulators that bias activation and decision making) with social knowledge enabling rapid social decision-making, particularly for over-learned associative knowledge. The somatic markers themselves are distributed across a large system of brain regions including limbic system structures such as the amygdala (Damasio, 1996).

Action models

The HPFC is sometimes thought of as an extension of the motor areas of the frontal lobes (Gomez Beldarrain, Grafman, Pascual-Leone and Garcia-Monco, 1999) leading to the idea that it must play an essential role in determining action sequences in the real world. In keeping with that view, a number of investigators have focused their investigations on concrete action series that have proved difficult for patients with HPFC lesions to adequately perform. By analyzing the pattern of errors committed by these patients, it is possible to construct models of action execution and the role of the HPFC in such performance. In some patients, while the total number of errors they commit is greater than that seen in controls, the pattern of errors committed by patients is similar to that seen in controls (Schwartz, Buxbaum, Montgomery et al., 1999). Reduced arousal or effort can also contribute to a breakdown in action production in patients (Schwartz et al., 1999). However, other studies indicate that action production impairment can be due to a breakdown in access to a semantic network that represents aspects of action schema and prepotent responses (Forde and Humphreys, 2000). Action production must rely upon an association between the target object or abstract goal and specific motoric actions (Humphreys and Riddoch, 2000). In addition, the magnitude of inhibition of inappropriate actions appears related to the strength in associative memory of object/goal associations (Humphreys and Riddoch, 2000). Retrieving or recognizing appropriate actions may even help a subject detect a target (Humphreys and Riddoch, 2001). It should be noted that action disorganization syndromes in patients are usually elicited with tasks that traditionally have been part of the examination of ideomotor or ideational praxis such as brushing your teeth or preparing a cup of coffee and it is not clear whether findings in patients performing such tasks apply to a breakdown in action organization at a higher level such as planning a vacation.

Computational frameworks

A number of computational models of potential HPFC processes as well as of the general architecture of the HPFC have been developed in recent years. Some models have offered a single explanation for performance on a wide range of tasks. For example, Kimberg and Farah (1993) showed that the weakening of associations within a working memory component of their model led to impaired simulated performance on a range of tasks such as the Wisconsin Card Sorting test and the Stroop test that patients with HPFC lesions are known to perform poorly on. In contrast, other investigators have argued for a hierarchical approach to modeling HPFC functions that incorporates a number of layers, with the lowest levels regulated by the environment and the highest levels regulated by internalized rules and plans (Changeux and Dehaene, 1998). In addition to the cognitive levels of their model, Changeux and Dehaene, relying on simulations, suggest that control for transient 'pre-representations' that are modulated by reward and punishment signals improved their model's ability to predict patient performance data on the Tower of London test (Changeux and Dehaene, 1998). Norman and Shallice (1986) first ascribed two major control systems to the HPFC. As noted earlier in this chapter, one system was concerned with rigid, procedurally based, and over-learned behaviors whereas the other system was concerned with the supervisory control over novel situations. Both systems could be simultaneously active although one system's activation usually predominated performance. The Norman and Shallice model has been incorporated into a hybrid computational model that blends their control system idea with a detailed description of selected action sequences and their errors (Cooper and Shallice, 2000). The Cooper and Shallice model can account for sequences of responses unlike some recurrent network models and like the Changeux and Dehaene model is hierarchical in nature and based on interactive activation principles. It also was uncanny in predicting the kinds of errors of action disorganization described by Schwartz and Humphreys in their papers. Other authors have implemented interactive control models that use production rules with scheduling strategies for activation and execution to simulate executive control (Meyer and Kieras, 1997). Tackling the issue of how the HPFC mediates schema processing, Botvinick and Plaut have recently argued that schemas are emergent system properties rather than explicit representations (Botvinick and Plaut, 2000). They developed a multi-layered recurrent connectionist network model to simulate action sequences that is somewhat similar to the Cooper and Shallice model described above. In their simulation, action errors occurred when noise in the system caused an internal representation for one scenario to resemble a pattern usually associated with another scenario. Their model also indicated that noise introduced in the middle of a sequence of actions was more disabling than noise presented closer to the end of the task.

The biological plausibility of all these models has not been formally compared yet but it is just as important to determine whether these models can simulate the behaviors and deficits of interest. The fact that models such as the ones described above are now being implemented is a major advance in the study of the functions of the HPFC.

Commonalities and weaknesses of the frameworks used to describe HPFC functions

The cognitive and computational models briefly described above (and further explicated in other chapters in this volume) have commonalities that point to the general role of the prefrontal cortex in maintaining information across time intervals and intervening tasks, in modulating social behavior, in the integration of information across time, and in the control of behavior via temporary memory representations and thought rather than allowing behavior to depend upon environmental contingencies alone. None of the major models have adequately articulated the features of a representational knowledge base that would support such HPFC functions making these models difficult to test and reject using an analysis of patient data or functional neuroimaging.

Say I was to describe cognitive processing in the brain in the following way. The role of cortex was to rapidly process information and encode its features, and bind these features together. This role was rather dependent on bottom-up environmental input but represented the elements of this processed information in memory. Perhaps this is not too controversial

a way to describe the role of the occipital, parietal or temporal cortex in processing objects or words. For the cognitive neuropsychologist, however, it would be critical to define the features of the word or object, the characteristics of the memory representation that led to easier encoding or retrieval of the object or word, and the psychological structure of the representational neighborhood (how different words or objects were related to each other in psychological and potentially neural space). Although there are important philosophical, psychological, and biological arguments about how best to describe a stored unit of memory (be it an orthographic representation of a word, a visual scene, or a conceptualization), there is general agreement that memories = representations. There is less agreement as to the difference between a representation and a cognitive process. It could be argued that processes are simply the sustained activation of a set of representations.

My view is that the description of the functional roles of the HPFC summarized in most of the models and frameworks described in this chapter is inadequate to obtain a clear understanding of its role in behavior. To obtain a clear understanding of the HPFC, a theory or model must describe the *nature of the representational networks* that are stored in prefrontal cortex, the principles by which the representations are stored, the levels and forms of the representations, hemispheric differences in the representational component stored based on the underlying computational constraints imposed by the right and left prefrontal cortex, and it must lead to predictions about the ease of retrieving representations stored in the prefrontal cortex under normal conditions, when additional tasks are required, and after various forms of brain injury. None of the models noted above provide answers to any of these questions except in the most descriptive and general manner.

Process versus representation — how to think about memory in the HPFC

My framework for understanding the nature of the knowledge stored in HPFC depends upon the idea that different forms of knowledge are stored in the HPFC as representations. In this sense, a representation is a unique element of knowledge that, when activated, corresponds to a unique static or dynamic brain state signified by the strength and pattern of neural activity in a local brain sector. This representational element is a 'permanent' unit of memory that can be strengthened by repeated exposure to the same or similar knowledge element and is a member of a psychological and neural network composed of multiple similar representations. Defining the specific forms of the representations I claim are stored in HPFC is crucial since an inappropriate representational depiction can compromise a model or theory as a description of a targeted phenomena. It is likely that these HPFC representations are parsed at multiple grain sizes (that are shaped by behavioral, environmental, and neural constraints).

What should a representational theory claim? It should claim that a process is a representation (or set of representations) in action, essentially a representation that when activated, stays activated over a limited or extended time domain. In order to be activated a representation has to be primed by input from a representation located outside its region or by associated representations within its region. This can occur via bottom-up or top-down information transfer. A representation, when activated, may or may not fit within the typical time window described as working memory. When it does, we are conscious of the representation. When it does not, we can still process that representation but we may not have direct conscious access to all of its contents.

The idea that representations are embedded in computations performed by local neural networks and are permanently stored within those networks so that they can be easily resurrected in a similar form whenever that network is stimulated by stimuli that are the external world's example of that representation or via associated knowledge is not novel nor free of controversy. But similar ideas of representation have dominated the scientific understanding of face, word and object recognition and have been recognized as an acceptable way to describe how the surface and lexical features of information could be encoded and stored in the human brain. Despite the adaptation of this notion of representation to the development of cognitive architectures for various stimuli based on 'lower-level' stimulus features, the application of similar representational theory to better understand the functions of the HPFC has moved much more slowly and in a more limited way.

Evolution of cognitive abilities

There is both folk wisdom about, and research support for, the idea that certain cognitive abilities are uniquely captured in the human brain with little evidence for these same sophisticated cognitive abilities found in other primates. Some examples of these cognitive processes include complex language abilities, social inferential abilities, or reasoning. It is not that these and other complex abilities are not present in other species, but probably, they exist only in a more rudimentary form.

The HPFC, as generally viewed, is most developed in humans. Therefore, it is likely that it has supported the transition of certain cognitive abilities from a rudimentary level to a more sophisticated one. I have already touched upon what kinds of abilities are governed by the HPFC. It is likely, however, that such abilities depend upon a set of fundamental computational processes unique to humans that support distinctive representational forms in the prefrontal cortex (Grafman, 1995). My goal in the remainder of this chapter is to argue for the principles by which such unique representations would be distinctively stored in the HPFC.

The structured event complex (SEC)

The archetype SEC

There must be a few fundamental principles governing evolutionary cognitive advances from other primates to humans. A key principle is the ability of neurons to sustain their firing and code the temporal and sequential properties of ongoing events in the environment or in mind over longer and longer periods of time. This sustained firing has enabled the human brain to code, store, and retrieve the more abstract features of behaviors whose goal or end-stage would not occur until well after the period of time that exceeds the limits of consciousness in 'the present'. Gradually in evolution, this period of time must have extended itself to encompass and encode all sorts of complex behaviors (Nichelli, Clark, Hollnagel and Grafman, 1995a; Rueckert and Grafman, 1996, 1998). Many aspects of such complex behaviors must require compressed (and multiple modes of) representation (such as a verbal listing of a series of things to do and the same set of actions in visual memory) while others may have real-time representational unpacking (unpacking means the amount of time and resources required to activate an entire representation and sustain it for behavioral purposes — for example, an activity composed of several linked events that take 10 minutes to perform would activate some component representations of that activity that would be active for the entire 10 minutes).

The event sequence

Neurons firing over extended periods of time in the HPFC process *sets of input* that can be defined as *events*. Single physical events rarely take long to complete (e.g., a gesture) although another physical event may take many hours (thawing of a turkey). Thus along with extended firing of neurons that allow the processing of behaviors *across time*, there must have also developed special neural parsers that enabled the *editing* of these behaviors into linked sequential but individual events (much like speech can be parsed into phonological units or sentences into grammatical constituents) (Sirigu, Zalla, Pillon et al., 1996; Sirigu, Cohen, Zalla et al., 1998). The event sequences, in order to be goal-oriented and cohere, must obey a logical structure within the constraints of the physical world, the culture that the individual belongs to, and/or the individual's personal preferences. These event sequences can be conceptualized as units of memory within domains of knowledge (e.g., a social attitude, a script that describes cooking a dinner, or a story that has a logical plot). We purposely labeled the archetype event sequence the structured event complex (SEC) in order to emphasize that we believed it to be the general form of representation within the HPFC and to avoid being too closely tied to a particular description of higher-level cognitive processes contained in story, narrative processing, script or schema frameworks.

Goal-oriented

Structured event complexes (SEC) are not random chains of behavior performed by normally functioning adults. They tend to have boundaries that signal their onset and offset. These boundaries can be determined by temporal cues, cognitive cues, or environmental/perceptual cues. Each SEC, however, has some kind of goal whose achievement precedes the offset of the SEC. The nature of the goal can be as different as putting a bookshelf together or determining a present to impress your wife on your wedding anniversary. Some events must be more central or important to an SEC than others. Subjects can have some agreement on which ones they are when explicitly asked. Some SECs are well structured with all the cognitive and behavioral rules available for the sequence of events to occur and there is a clear definable goal. Other SECs are ill-structured requiring the subject to adapt to unpredictable events using analogical reasoning or similarity judgment to determine the sequence of actions on-line (by retrieving a similar SEC from memory) as well as developing a quickly fashioned goal. SEC goals are not only central to their execution, but the process of reaching the goal can be rewarding. Goal achievement itself is probably routinely accompanied by a reward that is mediated by the brain's neurochemical systems. Depending on the salience of this reward cue, it can become essential to the subject's subsequent execution of that same or similar SEC. Goal attainment is usually transparent and subjects can consciously move onto another SEC in its aftermath.

Representational format

I hypothesize that the SEC is actually composed of a set of differentiated representational forms that would be stored in different regions of the HPFC but are activated in parallel to reproduce all the elements of a typical episode. These distinctive memories would represent thematic knowledge, morals, abstractions, concepts, social rules, features of specific events, and grammars for the variety of SECs embodied in actions, stories and narratives, scripts and schemas.

Memory characteristics

As just described, SECs are essentially distributed memory units with different components of the SEC stored in various regions within the prefrontal cortex. The easiest assumption to make then, is that they obey the same principles as other memory units in the brain. These principles revolve around frequency of activation based on use or exposure, association to other memory units, category specificity of the memory unit, plasticity of the representation, priming mechanisms, and binding of the memory unit and its neighborhood memory units to memory units in more distant representational networks both in and remote from the territory of the prefrontal cortex.

Frequency of use and exposure

As a characteristic that predicts a subject's ability to retrieve a memory, frequency is a powerful variable. For the SEC, the higher the frequency of the memory units composing the SEC, the more resilient they should be in the face of prefrontal cortex damage. That is, it is predicted that patients with frontal lobe damage would be most preserved performing or recognizing those SECs that they usually do as a daily routine and most impaired when asked to produce or recognize novel or rarely executed SECs. This retrieval deficit would be affected by the frequency of the specific kind of SEC memory units stored in the damaged prefrontal cortex region.

Associative properties within an HPFC functional region

In order to hypothesize the associative properties of an SEC, it is necessary to adapt some general information processing constraints imposed by each of the hemispheres (Beeman, 1998; Nichelli et al., 1995b; Partiot, Grafman, Sadato et al., 1996). A number of theorists have suggested that hemispheric asymmetry of information coding revolves around two distinct notions of information coding. In this view, the left hemisphere is thought to be specialized for finely tuned rapid encoding that is best at processing within-event information and coding for the boundaries between events. For example, the left prefrontal cortex might be able to best process the primary meaning of an event. The right hemisphere is thought to be specialized for coarse slower coding allowing for the processing of information that is more distantly related (to the information currently being processed) and could be adept at integrating information across events in time. For example, the right prefrontal cortex might be best able to process and integrate information across events in order to

obtain the theme or moral of a story that is being processed for the first time. When left hemisphere fine coding mechanisms are relied upon, a local memory element would be rapidly activated along with a few related neighbors with a relatively rapid deactivation. When right hemisphere coarse coding mechanisms are relied upon, there should be weaker activation of local memory elements but a greater spread of activation across a larger neighborhood of representations and for a sustained period of time corresponding to the length of the SEC currently being processed. This dual form of coding probably occurs in parallel with subjects shifting between the two depending on environmental and strategic demands. Furthermore, a parsimonious explanation of the organization of a population of SEC components within a functionally defined region should be based on the same principles argued for other forms of associative representation with both inhibition of unrelated memory units and facilitation of neighboring (and presumably related) memory units.

Order of events

The HPFC is specialized for the processing of events over time. One aspect of the SEC that is key to its representation is event order. Order can be coded by the sequence of events. The sequence itself must be parsed as each event begins and ends in order to explicitly recognize the nature, duration and number of events that compose the sequence (Hanson and Hanson, 1996; Zacks and Tversky, 2001). I hypothesize that in childhood because of the neural constraints of an immature HPFC, *individual* events are initially represented as independent memory units and only later in development are they linked together to form an SEC. Thus, in adulthood, there should be some redundancy of representation of the independent event and the membership of that same event within the SEC. Adult patients with HPFC lesions would be expected to commit errors of order in developing or executing SECs but could wind up defaulting to retrieving the independently stored but associated events in an attempt to slavishly carry out fragments of an activity. Subjects are aware of the sequence of events that make up an SEC and can even judge their relative importance or centrality to the overall SEC theme or goal. Each event has a typical duration and an expected onset and offset time within the time frame of the entire SEC that is coded. The order of the independent events that make up a particular SEC must be routinely adhered to by the performing subject in order to develop a more deeply stored SEC representation and to improve the subject's ability to predict the sequence of events. The repeated performance of an SEC leads to the systematic and rigidly ordered execution of events — an observation compatible with the AI notion of total order planning. In contrast, new SECs are constantly being encoded given the variable and occasionally unpredictable nature of strategic thought or environmental demands. This kind of adaptive planning in AI is known as partial order planning since event sequences are composed on-line, with the SEC consisting of previously experienced events inter-digitating with novel events. Since there must be multiple SECs that are activated in a typical day, it is likely that they too (like the events within an SEC) can be activated in sequence, or additionally in a cascading or parallel manner.

Category specificity

There is compelling evidence that the HPFC can be divided into regions that have predominant connectivity with specific cortical and subcortical brain sectors. This has led to the hypothesis that SECs may be stored in the HPFC on a category-specific basis. For example, it appears that patients with ventral or medial prefrontal cortex lesions are especially impaired in performing social and reward-related behaviors whereas patients with lesions to the dorsolateral prefrontal cortex appear most impaired on mechanistic planning tasks (Dimitrov et al., 1999c; Grafman, Schwab, Warden et al., 1996; Partiot, Grafman, Sadato et al., 1995; Pietrini, Guazzelli, Basso et al., 2000; Zalla, Koechlin, Pietrini et al., 2000). Further delineation of category specificity within the HPFC awaits more precise testing using various SEC categories as stimuli (Crozier, Sirigu, Lehericy et al., 1999; Sirigu et al., 1998).

Neuroplasticity of HPFC

We know relatively little about the neurobiological rules governing plasticity of the HPFC. It is probable that the same plasticity mechanisms that accompany learning and recovery of function in other cortical areas operate in the frontal lobes too (Grafman and Litvan, 1999a). For example, a change in prefrontal cortex regional functional map size with learning has been noted. Shrinkage of map size is usually associated with learning whereas an increase in map size over time may reflect the general category of representational form being activated but not a specific element of memory within the category. In addition, after left brain damage, right homologous HPFC assumption of at least some of the functions previously associated with Broca's area can occur. How the unique characteristics of prefrontal cortex neurons (e.g., sustained reentrant firing patterns or idiosyncratic neural architectures) interact with the general principles of cortical plasticity has been little explored to date. In terms of the flexibility of representations in the prefrontal cortex, it appears that this area of cortex can rapidly reorganize itself to respond to new environmental contingencies or rules. Thus, although the general underlying principles of how information is represented may be similar within and across species, individual experience manifested by species or individuals within a species will be influential in what is stored in prefrontal cortex and important to control for when interpreting the results of experiments trying to infer HPFC functional organization.

Priming

At least two kinds of priming (Schacter and Buckner, 1998) should occur when an SEC is activated. First of all, within an SEC, there should be priming of forthcoming adjacent and distant events by previously occurring events. Thus, in the case of the event that indicates you are going into a restaurant, subsequent events such as paying the bill or ordering from the menu will be primed at that moment. This priming would activate those event representations even though they had not occurred yet. The activation might be too far below threshold for conscious recognition that the event has been activated but there is probably a relationship between the intensity of the primed activation of a subsequent event and the temporal and cognitive distance the current event is from the primed event. The closer the primed event is in sequence and time to the priming event, the more activated it should be. The second kind of priming induced by SEC activation would involve SECs in the immediate neighborhood of the one currently activated. Closely related SECs (or components of SECs) in the immediate neighborhood should be activated to a lesser degree than the targeted SEC regardless of hemisphere. More distantly related SECs (or components of SECs) would be inhibited in the dominant hemisphere. More distantly related SECs (or components of SECs) would be weakly activated, rather than inhibited, in the nondominant hemisphere.

Binding

Another form of priming, based on the principle of binding (Engel and Singer, 2001) of distinct representational forms *across* cortical regions, should occur with the activation of an SEC. The sort of representational forms I hypothesize are stored in the human prefrontal cortex, such as thematic knowledge, are linked to more primitive representational forms such as objects, faces, words, stereotyped phrases, scenes and emotions. This linkage or binding enables humans to form a distributed episode for later retrieval. The binding also enables priming across representational forms to occur. For example, by activating an event within an SEC that is concerned with working in the office, activation thresholds should be decreased for recognizing and thinking about objects normally found in an office such as a telephone. In addition, the priming of forthcoming events within an SEC referred to above would also result in the priming of the objects associated with the subsequent event. Each representational form linked to the SEC should improve the salience of the bound configuration of representations. Absence of highly SEC-salient environmental stimuli or thought processes would tend to diminish the overall activation of the SEC-bound configuration of representations and bias which specific subset of prefrontal cortex representational forms are activated.

Hierarchical representation of SECs

I have previously argued for a hierarchy of SEC representation (Grafman, 1995). That is, I predicted that

SECs, within a domain, would range from specific to generalized episodes. For example, you could have an SEC representing the actions and themes of a single evening at a specific restaurant, an SEC representing the actions and themes of how to behave at restaurants in general, and an SEC representing actions and themes related to 'eating' that are context-independent — all of these SECs are predicted to be stored within the same HPFC regions. In this view, SEC episodes are formed first during development of the HPFC, followed by more general SECs, and then the context-free and abstract SECs. As the HPFC matures, it is the more general, context-free, and abstract SECs that allow for adaptive and flexible planning. Since these SECs do not represent specific episodes, they can be retrieved and applied to novel situations for which a specific SEC does not exist.

Relationship to other forms of representation

Basal ganglia functions

The basal ganglia receive direct connections from different regions of the HPFC and some of these connections may carry cognitive 'commands'. The basal ganglia, in turn, send back to the prefrontal cortex, via the thalamus, signals that reflect their own processing. Even if the basal ganglia work in concert with the prefrontal cortex, their exact role in cognitive processing is still debatable. They appear to play a role in the storage of visuomotor sequences (Pascual-Leone, Grafman, Clark et al., 1993; Pascual-Leone, Grafman and Hallett, 1995), in reward-related behavior (Zalla et al., 2000), and in automatic cognitive processing such as over-learned word retrieval. It is likely that the SECs in the prefrontal cortex bind with the visuomotor representations stored in the basal ganglia to produce an integrated set of cognitive and visuomotor actions (Koechlin et al., 2000, 2001; Pascual-Leone, Wassermann, Grafman and Hallett, 1996).

Hippocampus and amygdala functions

Both the amygdala and the hippocampus have reciprocal connections with the prefrontal cortex. The amygdala, in particular, has extensive connections with ventromedial prefrontal cortex (Price, 1999; Zalla et al., 2000). The amygdala's signals may provide a somatic marker or cue to the stored information ensemble in the ventromedial prefrontal cortex representing social attitudes, rules, and knowledge. The more salient the input provided by the somatic cue, the more important the somatic marker becomes for biasing the activation of social knowledge and actions. The connections between prefrontal cortex and the hippocampus serve to enlist the SEC as a contextual cue that forms part of an episodic ensemble of information (Thierry, Gioanni, Degenetais and Glowinski, 2000). The more salient the context, the more important it becomes for enhancing the retrieval or recognition of episodic memories. Thus, the hippocampus also serves to help bind the activation of objects, words, faces, scenes, procedures and other information stored in posterior cortices and basal structures to SEC-based contextual information such as themes or plans. Furthermore, the hippocampus may be involved in the linkage of sequentially occurring events. The ability to explicitly predict a subsequent event requires conscious recollection of forthcoming events and that should require the participation of a normally functioning hippocampus. Since the hippocampus is not needed for certain aspects of lexical or object priming, for example, it is likely that components of the SEC that can also be primed (see above) do not require the participation of the hippocampus. Thus subjects with amnesia might gain confidence and comfort in interactions in a context if they were re-exposed to the same context (sic SEC) that they had experienced before. In that case, the representation of that SEC would be strengthened even without conscious recollection of experiencing it (SEC representational priming in amnesia should be governed by the same restraint that affects word or object priming in amnesia).

Temporal–parietal cortex functions

The computational processes representing the major components of what we recognize as a word, object, face, or scene are stored in posterior cortex. These representations are crucial components of a context and can provide the key cue to initiate the activation of an SEC event or its inhibition. Thus, the linkage between anterior and posterior cortices is very

important for providing evidence that contributes to identifying the temporal and physical boundaries delimiting the 'events' that make up an SEC.

Evidence for and against the SEC framework

The advantage of this SEC formulation of the representations and processes stored in the HPFC is that it resembles other cognitive architecture models that are constructed so as to provide testable hypotheses regarding their validity. When hypotheses are supported, they lend confidence to the structure of the model as predicated by its architects. When hypotheses are rejected, they occasionally lead to the rejection of the entire model, but may also lead to a revised view of a component of the model. The SEC model lends itself to this kind of investigation.

The other major driving forces in conceptualizing the role of the prefrontal cortex have, in general, avoided the level of detail required of a cognitive or computational model and instead have opted for functional attributions which can hardly be disproved. This is not entirely the fault of the investigator as the forms of knowledge or processes stored in prefrontal cortex have perplexed and eluded investigators. What I have tried to do by formulating the SEC framework is take the trends in cognitive capabilities observed across evolution and development, that includes greater temporal and sequential processing and more capacity for abstraction, and assume what representational state(s) those trends would lead to.

The current evidence for an SEC type representational network is supportive but still rather sparse. SECs appear to be selectively processed by anterior prefrontal cortex regions (Koechlin et al., 1999, 2000). Errors in event sequencing can occur with preservation of aspects of event knowledge (Sirigu, Zalla, Pillon et al., 1995a). Thematic knowledge can be impaired even though event knowledge is preserved (Zalla, Phipps and Grafman, 2001). Frequency of the SEC can affect the ease of retrieval of SEC knowledge (Sirigu et al., 1995a; Sirigu, Zalla, Pillon et al., 1995b). There is evidence for category specificity in that ventromedial prefrontal cortex appears specialized for social knowledge processing (Dimitrov et al., 1999c). The HPFC is a member of many extended brain circuits. There is evidence that the hippocampus and the HPFC cooperate when the sequence of events have to be anticipated (Dreher, Koechlin, Ali and Grafman, 2001). The amygdala and the HPFC cooperate when SECs are goal- and reward-oriented or emotionally relevant (Zalla et al., 2000). The basal ganglia, cerebellum, and the HPFC cooperate as well (Grafman, Litvan, Massaquoi et al., 1992; Hallett and Grafman, 1997; Pascual-Leone et al., 1993). When the SEC is novel or multitasking is involved, anterior frontopolar prefrontal cortex is recruited, but when SECs are over-learned, slightly more posterior frontomedial prefrontal cortex is recruited (Koechlin et al., 2000). When subjects rely upon the visuomotor components of a task, the basal ganglia and cerebellum are more involved, but when subjects have to rely upon the cognitive aspects of the task, the HPFC is more involved in performance (Koechlin et al., 2001). Thus, there is positive evidence for an SEC representation within the HPFC. There has been little in the way of negative studies of this framework but many predictions of the SEC framework in the areas of goal orientation, neuroplasticity, priming, associative properties, and binding have not been fully explored to date.

Future directions

The representational model of the structured event complex described above lends itself to the generation of testable predictions or hypotheses. Like the majority of representational formats hypothesized for object, face, action, and word stores, the SEC subcomponents can each be characterized by the following features: frequency of exposure/activation, imagability, association to other items/exemplars in that particular representational store, centrality of the feature to the SEC (i.e., what proportional relevance does the feature have to recognizing or executing the SEC), length of the SEC in terms of number of events and duration of each event and the SEC as a whole, implicit or explicit activation, and association to other representational forms that are stored in other areas of the HPFC or in more posterior cortex/subcortical regions.

All these features can be characterized psychometrically by quantitative values based on subject surveys and other experimental methods that have obtained the same or similar values for words, pho-

tographs, objects, and faces. Unfortunately, there have only been a few attempts to collect some of these data for scripts, plans and similar stimuli. If these values for all of the features of interest of an SEC were obtained, one could then make predictions about changes in SEC performance after HPFC lesions. For example, one hypothesis from the SEC representational model described above is that the frequency of activation of a particular representation will determine its accessibility following HPFC lesions. A patient with an HPFC lesion will have had many different experiences eating dinner including eating food with their hands as a child, later eating more properly at the dining room table, eating at fast food restaurants, eating at favorite regular restaurants, and eventually eating occasionally at special restaurants or a brand new restaurant. Only one study has attempted to directly test this idea with modest success (see above (Sirigu et al., 1995a). After an HPFC lesion of moderate size, a patient should be *limited* in retrieving various subcomponents of the SEC stored in the lesioned sector of the HPFC. Thus, such a patient would be expected to behave more predictably and reliably when eating dinner at home then when eating in a familiar restaurant, and worse of all when eating in a new restaurant with an unusual seating or dining procedure for the first time. The kinds of errors that would characterize the inappropriate behavior would depend on the particular subcomponent(s) of the SEC (and thus regions within or across hemispheres) that was (were) damaged. For example, if the lesion were in the right dorsolateral prefrontal cortex, the patient might have difficulty integrating knowledge across dining events so that he or she would be impaired in determining the (unstated) theme of the dinner or restaurant particularly if the restaurant procedures were unfamiliar enough that the patient could not retrieve an analogous SEC. This is just one example of many predictions that emerge from a model with components that have representational features. The claim that SEC representational knowledge is stored in the HPFC in various cognitive sub-architectures is compatible with claims made for models for other forms of representational knowledge stored in other areas of brain and leads to the same kind of general predictions regarding SEC component accessibility made for these other forms of knowledge following brain damage. Thus, future studies need to test these predictions if the strong claims I have made about the SEC representational format are to be accepted.

Representation versus process revisited

The kind of representational model I have proposed for the SEC, I believe, is important for the cognitive neuroscience investigation of HPFC-related functions since it balances the over-reliance upon so-called process models such as working memory that dominate the field today. Process models rely upon a description of performance (holding or manipulating information) without necessarily being concerned about the details of the form of representation (i.e., memory) activated that allows for the performance to occur. Paradoxically, the majority of neuropsychological models of non-frontal functions are at least partially dependent upon representation-based models of memory although there are various levels of commitment to how the function of interest could be represented as a memory.

Promoting a strong claim that the prefrontal cortex is concerned with processes and not representations is a fundamental shift of thinking away from how we have previously tried to understand the format in which information is stored in memory. It suggests that the prefrontal cortex has little neural commitment to long-term storage of knowledge in contrast to posterior cortex. Such a fundamental shift in brain functions devoted to memory requires a much stronger philosophical, neuropsychological, and neuroanatomical defense for the process approach than previously offered by its proponents. The representational point of view that I offer regarding HPFC knowledge stores is more consistent with previous cognitive neuroscience approaches to understanding how other forms of knowledge such as words or objects are represented in the brain. It also allows for many hypotheses to be derived for further study and therefore can motivate more competing representational models of HPFC functions.

Clinical applications

There is no doubt that the impairments caused by lesions to the HPFC can be very detrimental to a person's ability to maintain their previous level of work,

responsibility to their family, and social commitments (Grafman and Litvan, 1999b). There is some evidence that deficits in executive functions can have a more profound effect on daily activities and routines than injuries causing sensory deficits, aphasia, or agnosia (Schwab, Grafman, Salazar and Kraft, 1993). Rehabilitation specialists are aware of the seriousness of deficits in executive impairments but there are precious few group studies detailing specific or general improvements in executive functions that are maintained in the real world and that lead to a positive functional outcome (Levine, Robertson, Clare et al., 2000; Stablum, Umilta, Mogentale et al., 2000).

The SEC representational framework proposed above gives some advantages and clues to rehabilitation specialists working on problems of executive dysfunction. One advantage that the application of SEC theory to rehabilitation methods has is that an understanding of the schemas SECs represent and the types of breakdowns in performance that may occur in carrying out SECs is easily grasped by the family members. Another advantage is in the design of tasks to use to try and modify patient behavior. In keeping with the frequency characteristic of SEC components, a therapist might want to choose a SEC to work with that has a mid-range frequency of experience by the patient. This gives the patient some familiarity with the activity but the activity is not so simple and the patient must take some care to perform it correctly. In addition to developing an error analysis of patient performance, by manipulating SEC level of frequency, difficulty, and SEC event centrality, it should be much easier to see systematic differences in patient performance as they tackle executing and understanding more difficult SECs. The SEC framework proposes that a set of stored representations that are bound together form the unified SEC. Thus, it should be possible to have several different variables measuring specific aspects of SEC performance including accuracy of over-learned or new visuomotor procedures, acquisition or expression of integrative or event-specific thematic content, as well as structural knowledge of the SEC (i.e., order and timing of events). Thus, the potential richness of the SEC framework allows for fine analysis of patient breakdown and improvement in performance. An additional advantage of this approach is that in the rehabilitation or research setting, SECs can be titrated to specific activities that are unique to a particular patient's experience at work, school, or home so that behaviors can be targeted for their relevance to the patient's daily life.

Conclusions

In this chapter I have argued that the best way to understand the functions of the HPFC is to adapt the representational model that has been the predominant approach to understanding the neuropsychological aspects of, for example, language processing and object recognition. The representational approach I developed is based on the Structured Event Complex framework. This framework claims that there are multiple sub-components of higher-level knowledge that are stored throughout the HPFC as distinctive domains of memory. I also have argued that there are topographical distinctions in where these different aspects of knowledge are stored in the HPFC. Each memory domain component of the SEC can be characterized by psychological features such as frequency of exposure, category specificity, associative properties, sequential dependencies, and goal orientation which governs the ease of retrieving an SEC. In addition, when these memory representations become activated via environmental stimuli or by automatic or reflective thought, they are activated for longer periods of time than knowledge stored in other areas of brain giving rise to the impression that performance dependent upon SEC activation is based on a specific form of memory called 'working memory'. Rather, it is the duration of the information that is processed and stored in the HPFC that determines the length of time information is active. There is no need to posit a separate memory system for this purpose. The 'working memory' terminology, in my view, is misleading as it leads to an investigational emphasis on 'process' that ignores the richer and, potentially, more important aspects of representational knowledge. In contrast, adapting a representational framework such as the SEC framework should lead to a richer corpus of predictions about subject performance that can be rejected or validated via experimental studies. Furthermore, the SEC framework lends itself quite easily to rehabilitation practice. Finally, there is now a substantial set

of research that suggests studying the nature of the SEC is a competitive and fruitful way to understand the role of the HPFC in behavior.

References

Alexander GE, Crutcher MD, DeLong MR: Basal ganglia–thalamocortical circuits: parallel substrates for motor, oculomotor, 'prefrontal' and 'limbic' functions. Progress in Brain Research: 85(1); 119–146, 1990.

Anderson SW, Bechara A, Damasio H, Tranel D, Damasio AR: Impairment of social and moral behavior related to early damage in human prefrontal cortex. Nature Neuroscience: 2(11); 1032–1037, 1999.

Arnett PA, Rao SM, Bernardin L, Grafman J, Yetkin FZ, Lobeck L: Relationship between frontal lobe lesions and Wisconsin Card Sorting Test performance in patients with multiple sclerosis. Neurology: 44(3 Pt 1); 420–425, 1994.

Baddeley A: The central executive: a concept and some misconceptions. Journal of the International Neuropsychological Society: 4(5); 523–526, 1998a.

Baddeley A: Recent developments in working memory. Current Opinion in Neurobiology: 8(2); 234–238, 1998b.

Barbas H: Complementary roles of prefrontal cortical regions in cognition, memory, and emotion in primates. Advances in Neurology: 84(11); 87–110 [Record as supplied by publisher], 2000.

Barbas H, Gashghaei HT, Rempel-Clower NL, Xiao D: Anatomic basis of functional specialization in prefrontal cortices in primates. In Boller F, Grafman J (Eds), Handbook of Neuropsychology, 2nd edn, Vol. 7. Amsterdam: Elsevier, pp. 1–27, 2002.

Bechara A, Damasio H, Damasio AR: Emotion, decision making and the orbitofrontal cortex. Cerebral Cortex: 10(3); 295–307, 2000.

Bechara A, Damasio H, Damasio AR, Lee GP: Different contributions of the human amygdala and ventromedial prefrontal cortex to decision-making. Journal of Neuroscience: 19(13); 5473–5481, 1999.

Beeman M: Coarse semantic coding and discourse comprehension. In Beeman M, Chiarello C (Eds), Right Hemisphere Language Comprehension. Mahwah, NJ: Lawrence Erlbaum, pp. 255–284, 1998.

Botvinick M, Plaut DC: Doing without schema hierarchies: a recurrent connectionist approach to routine sequential action and its pathologies. Paper presented at the Annual Meeting of the Cognitive Neuroscience Society (Poster presentation), San Francisco, CA, April 11th, 2000.

Burgess PW: Strategy application disorder: the role of the frontal lobes in human multitasking. Psychological Research: 63(3–4); 279–288, 2000.

Burgess PW, Veitch E, de Lacy Costello A, Shallice T: The cognitive and neuroanatomical correlates of multitasking. Neuropsychologia: 38(6); 848–863, 2000.

Carlin D, Bonerba J, Phipps M, Alexander G, Shapiro M, Grafman J: Planning impairments in frontal lobe dementia and frontal lobe lesion patients. Neuropsychologia: 38(5); 655–665, 2000.

Changeux JP, Dehaene S: Hierarchical neuronal modeling of cognitive functions: from synaptic transmission to the Tower of London. Comptes Rendus de l'Academie des Sciences Serie III: 321(2–3); 241–247, 1998.

Chiavaras MM, LeGoualher G, Evans A, Petrides M: Three-dimensional probabilistic atlas of the human orbitofrontal sulci in standardized stereotaxic space. Neuroimage: 13(3); 479–496, 2001.

Cooper R, Shallice T: Contention scheduling and the control of routine activities. Cognitive Neuropsychology: 17(4); 297–338, 2000.

Courtney SM, Petit L, Haxby JV, Ungerleider LG: The role of prefrontal cortex in working memory: examining the contents of consciousness. Philosophical Transactions of the Royal Society of London Series B: Biological Sciences: 353(1377); 1819–1828, 1998.

Crozier S, Sirigu A, Lehericy S, van de Moortele PF, Pillon B, Grafman J, Agid Y, Dubois B, LeBihan D: Distinct prefrontal activations in processing sequence at the sentence and script level: an fMRI study. Neuropsychologia: 37(13); 1469–1476, 1999.

Damasio AR: The somatic marker hypothesis and the possible functions of the prefrontal cortex. Philosophical Transactions of the Royal Society of London Series B: Biological Sciences: 351(1346); 1413–1420, 1996.

Diamond A: Close interrelation of motor development and cognitive development and of the cerebellum and prefrontal cortex. Child Development: 71(1); 44–56, 2000.

Dimitrov M, Grafman J, Hollnagel C: The effects of frontal lobe damage on everyday problem solving. Cortex: 32(2); 357–366, 1996.

Dimitrov M, Grafman J, Soares AH, Clark K: Concept formation and concept shifting in frontal lesion and Parkinson's disease patients assessed with the California Card Sorting Test. Neuropsychology: 13(1); 135–143, 1999a.

Dimitrov M, Granetz J, Peterson M, Hollnagel C, Alexander G, Grafman J: Associative learning impairments in patients with frontal lobe damage. Brain and Cognition: 41(2); 213–230, 1999b.

Dimitrov M, Phipps M, Zahn TP, Grafman J: A thoroughly modern Gage. Neurocase: 5(4); 345–354, 1999c.

Dreher JC, Koechlin E, Ali O, Grafman J: Dissociation of task timing expectancy and task order anticipation during task switching. Submitted, 2001.

Elston GN: Pyramidal cells of the frontal lobe: all the more spinous to think with. Journal of Neuroscience: 20(18); RC95 (1–4), 2000.

Engel AK, Singer W: Temporal binding and the neural correlates of sensory awareness. Trends in Cognitive Science: 5(1); 16–25, 2001.

Eslinger PJ: Neurological and neuropsychological bases of empathy. European Neurology: 39(4); 193–199, 1998.

Forde EME, Humphreys GW: The role of semantic knowledge and working memory in everyday tasks. Brain and Cognition: 44; 214–252, 2000.

Fuster JM, Bodner M, Kroger JK: Cross-modal and cross-temporal association in neurons of frontal cortex. Nature: 405(6784); 347–351, 2000.

Goel V, Grafman J: Are the frontal lobes implicated in 'planning' functions? Interpreting data from the Tower of Hanoi. Neuropsychologia: 33(5); 623–642, 1995.

Goel V, Grafman J, Sadato N, Hallett M: Modeling other minds. Neuroreport: 6(13); 1741–1746, 1995.

Goel V, Grafman J, Tajik J, Gana S, Danto D: A study of the performance of patients with frontal lobe lesions in a financial planning task. Brain: 120(Pt 10)(1); 1805–1822, 1997.

Gomez Beldarrain M, Grafman J, Pascual-Leone A, Garcia-Monco JC: Procedural learning is impaired in patients with prefrontal lesions. Neurology: 52(9); 1853–1860, 1999.

Grafman J: Similarities and distinctions among current models of prefrontal cortical functions. Annals of the New York Academy of Sciences: 769(1); 337–368, 1995.

Grafman J: Experimental assessment of adult frontal lobe function. In Miller BL, Cummings J (Eds), The Human Frontal Lobes: Function and Disorder. New York: The Guilford Press, pp. 321–344, 1999.

Grafman J, Litvan I: Evidence for four forms of neuroplasticity. In Grafman J, Christen Y (Eds), Neuronal Plasticity: Building a Bridge from the Laboratory to the Clinic. Berlin: Springer, pp. 131–140, 1999a.

Grafman J, Litvan I: Importance of deficits in executive functions. Lancet: 354(9194); 1921–1923, 1999b.

Grafman J, Litvan I, Massaquoi S, Stewart M, Sirigu A, Hallett M: Cognitive planning deficit in patients with cerebellar atrophy. Neurology: 42(8); 1493–1496, 1992.

Grafman J, Schwab K, Warden D, Pridgen A, Brown HR, Salazar AM: Frontal lobe injuries, violence, and aggression: a report of the Vietnam Head Injury Study. Neurology: 46(5); 1231–1238, 1996.

Groenewegen HJ, Uylings HB: The prefrontal cortex and the integration of sensory, limbic and autonomic information. Progress in Brain Research: 126; 3–28, 2000.

Hallett M, Grafman J: Executive function and motor skill learning. International Review of Neurobiology: 41(2); 297–323, 1997.

Hanson C, Hanson SE: Development of schemata during event parsing: Neisser's perceptual cycle as a recurrent connectionist network. Journal of Cognitive Neuroscience: 8(2); 119–134, 1996.

Humphreys GW, Riddoch MJ: One more cup of coffee for the road: object–action assemblies, response blocking and response capture after frontal lobe damage. Experimental Brain Research: 133(1); 81–93, 2000.

Humphreys GW, Riddoch MJ: Detection by action: neuropsychological evidence for action-defined templates in search. Nature Neuroscience: 4(1); 84–88, 2001.

Jurado MA, Junque C, Vendrell P, Treserras P, Grafman J: Overestimation and unreliability in 'feeling-of-doing' judgements about temporal ordering performance: impaired self-awareness following frontal lobe damage. Journal of Clinical and Experimental Neuropsychology: 20(3); 353–364, 1998.

Kawasaki H, Kaufman O, Damasio H, Damasio AR, Granner M,

Bakken H, Hori T, Howard MA, Adolphs R: Single-neuron responses to emotional visual stimuli recorded in human ventral prefrontal cortex. Nature Neuroscience: 4(1); 15–16, 2001.

Kimberg DY, Farah MJ: A unified account of cognitive impairments following frontal lobe damage: the role of working memory in complex, organized behavior. Journal of Experimental Psychology: General: 122(4); 411–428, 1993.

Koechlin E, Basso G, Pietrini P, Panzer S, Grafman J: The role of the anterior prefrontal cortex in human cognition. Nature: 399(6732); 148–151, 1999.

Koechlin E, Corrado G, Pietrini P, Grafman J: Dissociating the role of the medial and lateral anterior prefrontal cortex in human planning. Proceedings of the National Academy of Sciences of the USA: 97(13); 7651–7656, 2000.

Koechlin E, Danek A, Burnod Y, Grafman J: Brain mechanisms underlying the acquisition of behavioral and cognitive sequences. Submitted, 2001.

Levine B, Robertson IH, Clare L, Carter G, Hong J, Wilson BA, Duncan J, Stuss DT: Rehabilitation of executive functioning: an experimental–clinical validation of goal management training. Journal of the International Neuropsychological Society: 6(3); 299–312, 2000.

Levy R, Goldman-Rakic PS: Segregation of working memory functions within the dorsolateral prefrontal cortex. Experimental Brain Research: 133(1); 23–32, 2000.

Macmillan M: An Odd Kind of Fame: Stories of Phineas Gage. Cambridge: MIT Press, 2000.

Masterman DL, Cummings JL: Frontal–subcortical circuits: the anatomic basis of executive, social and motivated behaviors. Journal of Psychopharmacology: 11(2); 107–114, 1997.

Meyer DE, Kieras DE: A computational theory of executive cognitive processes and multiple-task performance, Part 1. Basic mechanisms. Psychological Review: 104(1); 3–65, 1997.

Miller EK: The prefrontal cortex and cognitive control. Nature Reviews: Neuroscience: 1(1); 59–65, 2000.

Nichelli P, Clark K, Hollnagel C, Grafman J: Duration processing after frontal lobe lesions. Annals of the New York Academy of Sciences: 769(1); 183–190, 1995a.

Nichelli P, Grafman J, Pietrini P, Alway D, Carton JC, Miletich R: Brain activity in chess playing. Nature: 369(6477); 191, 1994.

Nichelli P, Grafman J, Pietrini P, Clark K, Lee KY, Miletich R: Where the brain appreciates the moral of a story. Neuroreport: 6(17); 2309–2313, 1995b.

Norman DA, Shallice T: Attention to action: willed and automatic control of behavior. In Davidson RJ, Schwartz GE, Shapiro D (Eds), Consciousness and Self-Regulation, Vol. 4. New York: Plenum Press, pp. 1–18, 1986.

Nystrom LE, Braver TS, Sabb FW, Delgado MR, Noll DC, Cohen JD: Working memory for letters, shapes, and locations: fMRI evidence against stimulus-based regional organization in human prefrontal cortex. Neuroimage: 11(5 Pt 1); 424–446, 2000.

Partiot A, Grafman J, Sadato N, Flitman S, Wild K: Brain activation during script event processing. Neuroreport: 7(3); 761–766, 1996.

Partiot A, Grafman J, Sadato N, Wachs J, Hallett M: Brain activation during the generation of non-emotional and emotional plans. Neuroreport: 6(10); 1397–1400, 1995.

Pascual-Leone A, Grafman J, Clark K, Stewart M, Massaquoi S, Lou JS, Hallett M: Procedural learning in Parkinson's disease and cerebellar degeneration. Annals of Neurology: 34(4); 594–602, 1993.

Pascual-Leone A, Grafman J, Hallett M: Procedural learning and prefrontal cortex. Annals of the New York Academy of Sciences: 769(1); 61–70, 1995.

Pascual-Leone A, Wassermann EM, Grafman J, Hallett M: The role of the dorsolateral prefrontal cortex in implicit procedural learning. Experimental Brain Research: 107(3); 479–485, 1996.

Petrides M, Pandya DN: Comparative architectonic analysis of the human and macaque frontal cortex. In Boller F, Grafman J (Eds), Handbook of Neuropsychology, 1st ed., Vol. 9. Amsterdam: Elsevier, pp. 17–58, 1994.

Petrides M, Pandya DN: Dorsolateral prefrontal cortex: comparative cytoarchitectonic analysis in the human and the macaque brain and corticocortical connection patterns. European Journal of Neuroscience: 11(3); 1011–1036, 1999.

Pietrini P, Guazzelli M, Basso G, Jaffe K, Grafman J: Neural correlates of imaginal aggressive behavior assessed by positron emission tomography in healthy subjects. American Journal of Psychiatry: 157(11); 1772–1781, 2000.

Price JL: Prefrontal cortical networks related to visceral function and mood. Annals of the New York Academy of Sciences: 877(3); 383–396, 1999.

Rilling JK, Insel TR: The primate neocortex in comparative perspective using magnetic resonance imaging. Journal of Human Evolution: 37(2); 191–223, 1999.

Robbins TW: Chemical neuromodulation of frontal-executive functions in humans and other animals. Experimental Brain Research: 133(1); 130–138, 2000.

Ruchkin DS, Berndt RS, Johnson R, Ritter W, Grafman J, Canoune HL: Modality-specific processing streams in verbal working memory: evidence from spatio-temporal patterns of brain activity. Brain Research. Cognitive Brain Research: 6(2); 95–113, 1997.

Rueckert L, Grafman J: Sustained attention deficits in patients with right frontal lesions. Neuropsychologia: 34(10); 953–963, 1996.

Rueckert L, Grafman J: Sustained attention deficits in patients with lesions of posterior cortex. Neuropsychologia: 36(7); 653–660, 1998.

Schacter DL, Buckner RL: Priming and the brain. Neuron: 20(2); 185–195, 1998.

Schwab K, Grafman J, Salazar AM, Kraft J: Residual impairments and work status 15 years after penetrating head injury: report from the Vietnam Head Injury Study. Neurology: 43(1); 95–103, 1993.

Schwartz MF, Buxbaum LJ, Montgomery MW, Fitzpatrick-DeSalme E, Hart T, Ferraro M, Lee SS, Coslett HB: Naturalistic action production following right hemisphere stroke. Neuropsychologia: 37(1); 51–66, 1999.

Semendeferi K, Armstrong E, Schleicher A, Zilles K, Van Hoesen GW: Prefrontal cortex in humans and apes: a comparative study of area 10. American Journal of Physical Anthropology: 114; 224–241, 2001.

Shallice T, Burgess PW: The domain of supervisory processes and temporal organization of behavior. Philosophical Transactions of the Royal Society of London Series B: Biological Sciences: 351; 1405–1412, 1996.

Sirigu A, Cohen L, Zalla T, Pradat-Diehl P, Van Eeckhout P, Grafman J, Agid Y: Distinct frontal regions for processing sentence syntax and story grammar. Cortex: 34(5); 771–778, 1998.

Sirigu A, Zalla T, Pillon B, Grafman J, Agid Y, Dubois B: Selective impairments in managerial knowledge following prefrontal cortex damage. Cortex: 31(2); 301–316, 1995a.

Sirigu A, Zalla T, Pillon B, Grafman J, Agid Y, Dubois B: Encoding of sequence and boundaries of scripts following prefrontal lesions. Cortex: 32(2); 297–310, 1996.

Sirigu A, Zalla T, Pillon B, Grafman J, Dubois B, Agid Y: Planning and script analysis following prefrontal lobe lesions. Annals of the New York Academy of Sciences: 769(1); 277–288, 1995b.

Smith EE, Jonides J: Storage and executive processes in the frontal lobes. Science: 283(5408); 1657–1661, 1999.

Stablum F, Umilta C, Mogentale C, Carlan M, Guerrini C: Rehabilitation of executive deficits in closed head injury and anterior communicating artery aneurysm patients. Psychological Research: 63(3–4); 265–278, 2000.

Stuss DT, Toth JP, Franchi D, Alexander MP, Tipper S, Craik FI: Dissociation of attentional processes in patients with focal frontal and posterior lesions. Neuropsychologia: 37(9); 1005–1027, 1999.

Thierry AM, Gioanni Y, Degenetais E, Glowinski J: Hippocampo-prefrontal cortex pathway: anatomical and electrophysiological characteristics. Hippocampus: 10(4); 411–419, 2000.

Vendrell P, Junque C, Pujol J, Jurado MA, Molet J, Grafman J: The role of prefrontal regions in the Stroop task. Neuropsychologia: 33(3); 341–352, 1995.

Wharton CM, Grafman J, Flitman SS, Hansen EK, Brauner J, Marks A, Honda M: Toward neuroanatomical models of analogy: a positron emission tomography study of analogical mapping. Cognitive Psychology: 40(3); 173–197, 2000.

Zacks JM, Tversky B: Event structure in perception and conception. Psychological Bulletin: 127(1); 3–21, 2001.

Zahn TP, Grafman J, Tranel D: Frontal lobe lesions and electrodermal activity: effects of significance. Neuropsychologia: 37(11); 1227–1241, 1999.

Zalla T, Koechlin E, Pietrini P, Basso G, Aquino P, Sirigu A, Grafman J: Differential amygdala responses to winning and losing: a functional magnetic resonance imaging study in humans. European Journal of Neuroscience: 12(5); 1764–1770, 2000.

Zalla T, Phipps M, Grafman J: Story processing in patients with damage to the prefrontal cortex. Cortex: in press, 2001.

CHAPTER 9

The processing of temporal information in the frontal lobe

Paolo Nichelli [*]

Dipartimento di Patologia Neuropsicosensoriale, Università di Modena e Reggio Emilia, Via Del Pozzo 71, 41100 Modena, Italy

Introduction

One can hardly imagine anything that is more typically associated with life than time. Indeed, time shapes our behavior in almost all areas of our daily lives. This is true for humans as well as for any animal. For a mobile organism, rapid and fluent movements would not be possible without a precise timing of the sequential order of activation of its different parts. In this example, timing is an emergent propriety of an operation. It might be the outcome of the dynamic features of the system or derive from other movement parameters, such as force and amplitude. Yet, in many other instances, time needs to be explicitly computed so that a living organism can anticipate events, rather than simply react to them.

Events can either be synchronized to a natural periodicity or have variable delays from an arbitrary event. These different aspects of timing are likely to be subserved by distinct mechanisms: periodic timers and interval timers. An animal with a periodic timer may use it to identify the time of the day (or of the year) at which food appears at a particular place. The best studied of the periodic timers is circadian (its period lasts around one day) and it allows to anticipate the time of day when something is going to occur. For instance, it has been long known that bees can record the time of the day at which they obtained food at a source, and use the recorded time to schedule their subsequent visits (Beling, 1929). For humans, it is a common experience to wake up

in the morning at the same time, without the need of an alarm clock. Ultradian (with periods shorter than 20 h) and infradian (with periods longer than 28 h) oscillations have been also hypothesized (Gallistel, 1990) to explain the extraordinary precision of animals to detect the time of occurrence of an event when it recurs every many days or weeks.

Interval timers, like stopwatches, can measure the time between two events. Animals need them to determine which temporal intervals in foraging give the best rate of return.

Everyday life gives many examples that humans do routinely estimate durations ranging from milliseconds up to several seconds. In verbal communication we can hear that a 'beach' is not a 'bitch'. In music we can differentiate the duration of notes and of silence (Macar, 1998). In approaching an intersection, when the traffic light changes from green to yellow, depending on the duration of the yellow light, we can decide whether to immediately stop or to cross it before the light turns from yellow to red.

Yet, human representation of time is far more complex than that provided by these examples. It has been suggested (Campbell, 1996) that animals can only have a cyclic representation of time, while humans represent time as linear. We can, for example, think not only of yesterday having been Sunday, but it also having been Sunday three weeks ago. In other words, we are capable of orientation with respect to particular times, whereas animals seem only capable of orienting with respect to phase. Even having mechanisms for both periodic and interval timing does not seem enough for grasping a linear time. This would at best provide a series of "unconnected islands of time" (Campbell, 1996) unrelated to one

---
[*] Tel.: +39 (059) 422-4298; Fax: +39 (059) 422-4299;
E-mail: nichelli@unimo.it

another. Instead, we can think of what happened on one particular Sunday as casually affecting what happened on a later Sunday. Indeed, it is only because we can build up complex causal relationships between events happening at different times that we can think in terms of plans and projects for the future, even including the will to be opened after our death.

Human temporal experience encompass two different concepts: duration and succession (Fraisse, 1984; Wittmann, 1999). The range of phenomena begins with temporal frames of a few milliseconds (which is relevant in differentiating simultaneity and succession) and ends with decades (as in the retrospective view of an individual life's span).

The complexity of human representation of time projects an important role for human temporal processing to the prefrontal cortex, the part of the brain that is enlarged in humans more than any other area. For this reason, the present chapter will be especially focused on the role of the frontal lobe in processing temporal information. However, for better understanding the role of the frontal lobe in temporal processing, it is useful to briefly review the current cognitive models as well as what is known about the neural correlates of timing, in terms of both neurotransmitters and specific brain structures.

Providing a comprehensive framework to understand how the human brain can regulate internal timing of movements, discriminate intervals, estimate durations, determine the order of two events, and eventually obtain a linear representation of time is probably beyond what can be reasonably achieved based on our current neuropsychological knowledge. In the last two decades, there have been several attempts to build up information processing theories of interval timing. Yet, it is only very recently that studies aimed at understanding its neural substrate have been published. Research on this topic is therefore at its beginning.

In this chapter I will first present some background information that is needed for interpreting data from temporal information processing studies. Then, I will present a cognitive model for duration processing. I will eventually come to discuss the neural substrate of the model, with a particular attention to the role of the frontal lobe in different time-related tasks.

Simultaneity, succession and temporal order

Fusion threshold and temporal-order thresholds separate the experiences of simultaneity, succession, and temporal order. In fusion threshold tests, subjects are required to evaluate whether two stimuli or only a single stimulus is perceived (e.g., visual flicker or auditory flutter fusion). In temporal-order threshold tasks, they have to indicate the correct physical order of two stimuli.

Fusion thresholds differ substantially between sensory modalities, while temporal-order thresholds do not. For instance, simultaneity for auditory stimuli is detected with time intervals below approximately 2 to 3 ms (Hirsh and Sherrick, 1961), while critical flicker fusion for vision is about 80 ms (Piéron, 1952, pp. 396–297).

The detection of the temporal order of events is only possible when they are separated by 20–40 ms (Hirsh, 1959; Pastore and Farrington, 1996) and although it is not totally independent of stimulus parameters (Pastore, Harris and Kaplan, 1982), it remains fairly constant across different sensory modalities.

Pöppel (1978) proposed that a central oscillatory processing system with a period of 30 ms can explain the constancy of the temporal-order threshold. According to his model, data picked up within 30 ms are treated as co-temporal, i.e., their temporal order cannot be established. System states of 30 ms duration could be responsible for binding intra- and intermodal information (Pöppel, Schill and von Steinbüchel, 1990). Interestingly, on the same time scale, neurophysiological observations have revealed rhythmic activity of around 40 Hz (the gamma band) that is considered the basis of perceptual feature binding (for a comprehensive overview, see Basar, 1999; for a shorter review see Basar-Eroglu, Strüber, Schürman et al., 1996).

Factors affecting human timing

Duration estimation is extremely sensitive to the contextual conditions under which the estimation task is performed. A fundamental distinction is that between *prospective* and *retrospective* time estimation. In a prospective paradigm subjects make duration estimates knowing that they will be requested to do that before the beginning of the interval. In a retrospective paradigm they do not know it, until the duration is terminated. Unfortunately, since retrospective estimates can only be obtained if subjects are unaware that they will be requested to tell the elapsed time, reliable experimental testing of remembered duration is limited by the fact that retrospective judgments can be investigated only once. The information *processing load* during the interval has opposite effects on the subject's estimate (Hicks, Miller and Kinsbourne, 1976). Within prospective paradigms, the greater the processing load, the shorter the temporal estimate. On the contrary, retrospective estimates tend to be longer when a greater load has to be processed during the interval.

The focus of subject's *attention* is another important determinant of subjective duration: subject's focusing on the passage of time results in lengthening of subjective duration (which explains the old saying "a watched pot never boils") (Block, George and Reed, 1980; Cahoon and Edmonds, 1980).

Attentional focusing and time estimation paradigms have been manipulated in several experimental settings (e.g., Brown, 1985; Casini and Macar, 1997; Zakay, Nitzan and Glicksohn, 1983). In everyday life, it has been proposed that attention to time increases whenever time becomes more important for adaptation to the environment (Zakay, 1992). This happens, for example, when facing a deadline, when hoping a specific event to continue, or when desire for an event to end arises. In such cases prospective time estimation emerge and the perceptual system becomes particularly sensitive to temporal cues.

Time estimation vs. time perception

According to Fraisse (1984), time perception involves the "psychological present" and concerns events lasting between 100 ms and 5 s. Intervals lasting less than 100 ms seem instantaneous (i.e., without duration), while those lasting more than 5 s require duration estimation processes based on linking two past events or on associating a moment in the past with a moment in the present.

While the validity of the boundaries of Fraisse's categories may be questioned (Allan, 1979) the implications of this distinction should be incorporated in any model of temporal information processing.

Also, it is important to realize, that with intervals longer than 1000 ms, humans might get advantage by using chronometric counting (Wearden and Lejune, 1993).

Methods for measuring subjective time

The four basic paradigms for measuring subjective time are: temporal discrimination, verbal estimation, temporal production, and temporal reproduction.

There are several ways to test *temporal discrimination*. In the method of comparison two durations are presented, one after the other, and the subject is requested to decide whether they are same or different. In the single stimulus method, on each trial, subjects are presented with one of two possible durations and they are requested to decide whether it was the short or the long value. A particular modification of the single stimulus method is the time bisection procedure. Here, subjects are first trained to discriminate a standard short and a standard long stimulus. Then, the two standard durations are presented along with intervals with intermediate duration and subjects are requested to classify each interval as more similar to the short or to the long standard. By plotting the proportion of intervals judged to be long as a function of interval duration (see Fig. 1) it is possible to obtain both measures of perceptual bias (the bisection point) and of discrimination accuracy (the difference limen and the Weber fraction).

In *verbal estimation* tasks subjects are required to judge durations in terms of conventional time units. This method can be used both with prospective and retrospective paradigms.

Temporal production experiments require the production of an interval of a specified duration. In a variant of this task (Fortin and Breton, 1995; Fortin and Rousseau, 1987; Fortin, Rousseau, Bourque and Kirouac, 1993), subjects are first trained to produce an interval of a fixed duration (e.g., 2 s), then they are asked to provide the answer for a concurrent task (e.g., recognizing a probe in a string of digits) when the trained duration has elapsed. In this way, various experimental manipulations of the concurrent task can help in identifying the specific processing demands that can affect time perception.

Rhythmic tapping is a particular instance of temporal production. Here, subjects first produce repet-

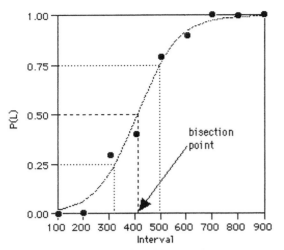

Fig. 1. Plotting the proportion of intervals judged to be long as a function of interval duration. Standard short and long durations were 100 and 900 ms, respectively. The bisection interval is the duration classified as 'long' on 50% of trials. The difference limen corresponds to half the difference classified as long on 75% of trials and that classified as 'long' on 25% of trials. The Weber fraction corresponds to the difference limen divided by the bisection point.

itive movements in pace with a metronome at a constant rate (i.e., synchronization phase) and then without a metronome (i.e., continuation phase). According to Wing and Kristofferson (1973a,b) two processes are involved in timing intervals with such a procedure: a 'timekeeper' and an 'implementation system'. More interestingly, from the total variability of the inter-response intervals it is possible to obtain separate estimates of the two processes. Collyer and Church (1998) have recently provided a complete description of this task and of the many inferences that can be drawn by a thorough analysis of the data that it generates.

In *temporal reproduction* the subject is presented with a stimulus duration and must reproduce that duration. The peak-interval procedure is a further instance of temporal reproduction which is particularly useful to study long interval timing (typically from 10 to 120 s). It has been widely used with animals and only recently it has been modified to test humans (Rakitin, Gibbon, Penney et al., 1998). Subjects first receive a series of fixed intervals in which they are rewarded for making responses (e.g., pressing a lever) after a criterion period following the appear-

ance of a signal. Then, fixed intervals are presented along with 'peak' or probe trials, in which there is no reward and the signal is left on for a relatively long period of time. In peak trials, subjects show increase in responding rate up to about the criterion time with a decline thereafter reflecting the appreciation that there will be no reward in that trial. Averaged over trials, responses are near the rewarded criterion time with a variance that is proportional to the peak time. This has been called the scalar property and has been the founding block of the Scalar Timing Theory, probably the most influential information processing approach to interval timing (Gibbon, Church and Meck, 1984).

Models for duration processing

Duration processing is at a higher hierarchic level than discrimination of simultaneity, and perception of temporal order. In order to be understood it needs to be decomposed in many constituent parts. There are two levels of modeling duration processing: establishing the basic components of the system and specifying how the different components are distributed and implemented in the nervous system.

Cognitive modeling

The scalar timing theory

The components that are necessary for prospective duration processing have been extensively studied by Gibbon and coworkers (Gibbon, 1977) in the framework of the Scalar Timing Theory. This is essentially a mathematical model describing the formal properties of the cognitive processes that operates when an organism is confronted with a timing task. A detailed description of the model is beyond the purpose of this chapter. The interested reader can find a detailed description of it in several papers (e.g., Gibbon, 1986; Gibbon and Church, 1990; Gibbon et al., 1984). The model (see Fig. 2) posits three stages (clock, memory, and decision). The clock consists of a pacemaker, a clock, and an accumulator. The pacemaker emits pulses at some (high) rate. A switch, which is controlled by a timing signal, gates the pulses in the accumulator. As long as the animal is timing the stimulus, pulses accumulate and are counted by the accumulator. At the memory stage,

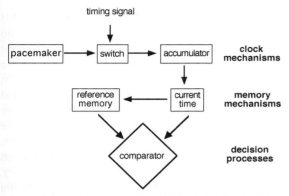

timing signal

Fig. 2. Schematic representation of the Scalar Timing Theory model for duration processing (modified from Gibbon et al., 1984).

working memory is loaded with the output of the accumulator and serves as a buffer for the current temporal information while reference memory stores past temporal information. At the decision stage, a comparator computes the difference between the elapsing interval and the remembered duration, divides this difference by the value of the standard, and tests whether the quotient exceeds the decision threshold.

As formulated, the Scalar Theory of Timing is based on two stringent psychophysical relationships (Allan, 1998): that the Psychophysical Law for time has an exponent of 1.0, i.e., subjective time is linear to real time and that there is also a linear relationship between standard deviation of subjective time and mean subjective time (Weber's Law).

Both these relationships have received firm empirical support in the human literature. However, a few authors have reported data that deviated from these linearities: Eisler (1976, 1996) has repeatedly argued that the Psychophysical Law for time is a power function with an exponent of 0.9. This would imply that either the pulses of the clock have a decelerating rate or the working memory is leaking as a function of time. With regard to Weber's Law, data inconsistent with a linear function have been generated, after extended practice, by Kristofferson (1980, 1984). Yet, these findings have never been replicated.

Examples of small but systematic deviation from linearities have also been reported in human experiments by Collyer, Broadbent and Church (1992, 1994) and by Collyer and Church (1998). They

concluded that nonlinearities point to an underlying discreteness in timing and that a multiple-oscillator model of timing can provide a better qualitative fit to the observed data rather than a model that hypothesizes a single oscillator for different time-ranges (Church and Broadbent, 1991).

The multiple time scale model

The Multiple Time Scale (MTS) model has been recently introduced by Staddon and Higa (1999) to explain animal timing without resorting to the concept of an internal pacemaker. It is based on the assumption that salient events produce a temporary state in the organism that varies with time. Values of that state constitute a stimulus that controls when responding will occur. The model substitutes the perceptual representation of time by the pacemaker–accumulator system in the Scalar Timing Theory with a series of cascaded habituation units. However, the model has not been elaborated in detail to specify any assumption about memory storage and retrieval, or decision rules (Church, 1999). Its main appeal is that, even with a small number of units, it can accommodate interval timing to cover any specified range (Staddon, Higa and Chelaru, 1999). So far, there have been no attempts to apply the MTS model to human timing.

Neural network modeling

Cognitive models of time are mainly concerned about providing a coherent quantitative account of timing experiments by specifying the modular structure of the system. Neural models are more concerned about explaining how time might be represented in the nervous system. There are two opposite views of how this might be achieved: (1) taking advantage of an endogenous oscillatory mechanism (single or distributed clock models); or (2) building upon a few basic neuronal properties to obtain separate representations for each specific duration (the hourglass model).

Single clock counter model

Clock counter models derive their appeal from the fact that they can easily accommodate the need to integrate timing in both perception and production domains, across different modalities, and along a wide range of duration. One such model has been proposed by Treisman (1963). He maintained that the pacemaker is composed of two parts: an oscillator and a calibration unit. The oscillator would produce an output at a constant frequency. The calibration unit would scale the base frequency as a function of task demands. This could provide flexible timing information.

Periodic timing mechanisms, like those subserving circadian rhythms, have been associated to the existence of the suprachiasmatic nucleus of the brainstem (Rusak and Zucker, 1979). Not surprisingly, there has been a quest for a similar mechanism regulating interval timing, with little success. No single brain structure is uniquely associated with a severe selective disruption of timing. Human timing is therefore more likely to result from the interaction of a number of structures, each with its specific role.

Distributed models

Pacemaker models do not need to assume that there is a single oscillator. Neural network models can encode intervals over a range of duration based either on harmonically related oscillators (Church and Broadbent, 1990) or on a population of oscillators distributed around a mean frequency (Miall, 1996).

Oscillatory activity around the frequency of 40 Hz (the gamma band) have been recorded both all over the cortex (Basar, Gönder and Ungan, 1975) and from neurons firing in synchrony with cells located both within a functional column and in spatially separated columns (Gray, König, Engel and Singer, 1989; Gray and Singer, 1989).

Although the functional significance of these oscillations is far from clear, it has be hypothesized (Stryker, 1989) that it might be the basis for linking ('binding') perceptual information (i.e., associating all the information belonging to a single object). According to Basar-Eroglu et al. (1996) the gamma band is a universal "building block" of brain activity in different species. Several pieces of evidence have been collected to support this notion. In an experiment by Joliot, Ribary and Llinas (1994) human subjects were instructed to attend to either single or paired auditory stimuli while their brain magnetic activity was recorded. Results showed that with paired stimuli, at shorter intervals (<12 ms) only one 40-Hz response, to the first stimulus, was

observed. With longer intervals a second 40-Hz response appeared, which coincided with the subject's perception of a second clearly distinct auditory stimulus. The authors argued that activity near 40 Hz represents a neurophysiological correlate to the temporal processing of auditory stimuli and that it might relate not only to primary sensory processing but also to the temporal binding underlying cognition. In the same line, Lisman and Idiart (1995) have modeled a neural network based on properties of known brain oscillations and have suggested a specific role for these oscillation in short-term memory functions. According to this model, each memory is stored in a different high-frequency ('40 Hz') subcycle of a low-frequency oscillation. Memory patterns could repeat on each low-frequency (5 to 12 Hz) oscillation, so that approximately seven high-frequency subcycles might be nested in a low-frequency oscillation (which could explain the well-known 7 ± 2 upper limit of human short-term memory). Consequently, the authors also argued that brain oscillations might be a timing mechanism for controlling the serial processing of short-term memories.

More recently, Rodriguez, George, Lachaux et al. (1999) demonstrated that in humans face perception (but not perception of inverted pictures of faces) elicits phase synchrony in the gamma range. This activity spans over a large region of the cortex (including the left parieto-occipital and fronto-temporal regions) and along all the duration of the task (i.e., from stimulus presentation to motor response), suggesting that it also subserves large-scale cognitive integration.

One might well hypothesize that the same distributed oscillatory activity is taken as input (the pacemaker) for the different functions that are attributed to the internal clock.

The interval measure (hourglass) models

According to these models, the representation of time may be distributed across a set of interval timers, each with its own processing cycle. Buonomano and Merzenich (1995) have provided an elegant demonstration that an artificial neural network, by taking advantage of a few well characterized neuronal properties, can transform temporal information into a spatial code in a self-organizing manner. In this kind of networks, if stimulus 'A' then 'B' is presented, 'A' will produce a change in the network state as a result of time-dependent neuronal properties and stimulus 'B' will then produce a pattern of activity that codes for 'B preceded by A' rather than simply 'B'. Furthermore, without changing any model parameter, the network is able to perform complex temporal pattern discrimination and to generalize to similar temporal patterns. This kind of networks can contribute to processing of information in the hundreds of millisecond range, which is mostly relevant in tasks such as speech recognition, frequency discrimination, music perception, and motion processing.

Support for the spatial transform model came from a study using hippocampal long-term potentiation (Buonomano, Hickmott and Merzenich, 1997). Ivry (1996) has hypothesized that a similar arrangement with different 'hourglass' neurons, coding for different intervals might occur in the cerebellar cortex. Within this view, cerebellar neurons might be tuned to particular intervals and linked to specific input and output systems. This could allow maximal flexibility to the behavior, which would not be constrained by any fundamental frequency. There is some evidence from temporal coupling experiments that motor control might be accomplished by separate timing mechanism. Helmuth and Ivry (1996) asked subjects to tap with the right or the left hand, or both hands. The variability of tapping with either hand was greatly reduced when the two hands tapped together in comparison to when either hands tapped alone. The authors argued that this could be explained by assuming that two independent timing signals were generated, one for each hand, and that these signals were averaged by a common output gate.

Extending this notion, this model assumes multiple timing mechanisms, each associated with a specific input or output system. Such an arrangement might be appropriate for timing durations within the range of hundreds of milliseconds. However, intrinsic neuronal properties, such as paired pulse facilitation and slow inhibitory postsynaptic potential have a too short time course to explain duration processing in the range of seconds or minutes. Furthermore, the notion of timing mechanisms linked to specific input and output systems is at variance with studies investigating generalization of interval learning (Nagarajan, Blake, Wright et al., 1998; Wright,

Buonomano, Mahncke and Merzenich, 1997). The results of these studies demonstrate that although learning cannot generalize across untrained temporal intervals, it can generalize across modalities (e.g., from the somatosensory to the auditory system) and to untrained locations (e.g., contralateral skin location or untrained sound frequencies).

Indeed, neither internal clock models nor interval measure models would predict such characteristics of learning. Recent models of temporal processing on the scale of tens to hundred of millisecond incorporate mechanisms of short-term synaptic plasticity (Buonomano, 2000). However, it might prove difficult to extend the applicability of this model to longer intervals.

Neural systems involved in temporal processing

In the last two decades several investigators have tried to identify the neural systems involved in temporal processing, taking advantage of different methods. The development of sophisticated behavioral paradigms, new data analysis techniques, and theoretical models have helped very much the progress in this field. In the following sections we will review the results obtained by studying the effects of drugs, of brain lesions and degenerative disorders, and those obtained by neuroimaging studies.

Neurotransmitters involved in timing

Studies investigating the role of neurotransmitters in timing behavior are especially concerned with intervals in the range of seconds-to-minute. Results were mainly obtained in rats and have been recently reviewed and interpreted by Meck (1996) in the framework of the Scalar Theory of Timing. The most important findings concern dopamine and acetylcholine.

It has been repeatedly demonstrated that neuroleptic drugs such as haloperidol decrease clock speed, while the administration of dopamine agonists such as methamphetamine can increase it (Maricq, Roberts and Church, 1981). This has been shown training rats to discriminate criterion intervals and administering the antagonist (or agonist) drug before testing interval discrimination. The results are a change in accuracy (bias), without any change in the variability of performance. In other words, when the internal clock speeds up (as for the effect of a dopaminergic drug) subjects overestimate time. On the contrary, when the internal clock decreases its speed (as due to haloperidol), they underestimate it. Neuroleptic affinity for dopamine D_2 receptors predicts the drug effect on timing (Meck, 1986), thus suggesting that these receptors play a major role in determining the rate of temporal integration for time perception.

These findings provide a strong argument for claiming a pivotal role of the basal ganglia in timing.

Interestingly, manipulations of the cholinergic system also affects timing but in a different way (Meck, 1983; Meck and Church, 1987). Drugs acting on the central cholinergic system affect both accuracy and precision (variability) of timing and cause gradual but persistent shifts of the psychophysical functions related to time. So, training time discrimination under physostigmine (a drug that increases the effective levels of acetylcholine) causes a decrease of remembered durations of reinforced times and decreased variability that persists beyond the termination of the treatment. An opposite persistent shift of the psychophysical function along with increased variability is observed with atropine (an anticholinergic drug). The magnitude of these shifts are dose dependent and proportional to the interval being timed, indicating a change in memory storage speed, i.e., the rate at which temporal information is transferred to reference memory. There is no definite evidence about whether the M_1 or the M_2 muscarinic receptor subtype mediates the behavioral effects on duration discrimination produced by cholinergic drugs (Meck, 1996). Acetylcholine is a neurotransmitter diffusely distributed over the cerebral cortex, including the Frontal Cortex and the Meynert's Nucleus Basalis Magnocellularis. Interestingly, lesions of both these structures in rats (Meck, Church, Wenk and Olton, 1987) cause an effect which is similar to the administration of atropine (overestimation of the expected time of reinforcement) but opposite to that caused by lesions in the hippocampal system.

The effects of brain lesion and degenerative brain disorders on temporal processing

Studies on humans with brain damage have demonstrated that several regions of the brain are involved in temporal information processing. Besides the prefrontal cortex, we can list the hippocampus, the cerebellum, the basal ganglia, the parietal cortex. I will briefly review the available data to determine if there is any empirical support to dissociate the role played by each brain area.

Hippocampus
Richards (1973) studied temporal reproduction by H.M., a famous and intensively studied patient who became severely amnesic after a bilateral removal of the medial temporal cortex (Milner, 1962; Milner, Corkin and Teuber, 1968; Scoville and Milner, 1964). H.M. was asked to reproduce intervals ranging from 1 to 300 seconds. Results showed that he was as accurate as normal subjects only for intervals less than 20 seconds. After this interval, he consistently underestimated the passage of time, so that, by extrapolation, it was possible to determine H.M.'s equivalence for 1 hour as 3 minutes.

The same observation was not replicated by Shaw and Aggleton (1994) in a group of three postencephalitic amnesic subjects who supposedly had an extensive damage to the temporal lobe, including the hippocampus. However, encephalitic patients may have also widespread damage to other cortical areas.

Hopkins, Kesner and Goldstein (1995a) have investigated the role of the hippocampus in temporal processing in a study involving a group of subjects with hypoxic brain injury. These patients had profound anterograde amnesia. Although hypoxia may also cause a more diffuse brain damage, according to MRI data they had bilateral damage to the hippocampus, but no damage to the entorhinal cortex, parahippocampal cortex, or temporal cortex. They also showed no sign of prefrontal cortex dysfunction. Hypoxic patients, patients with damage to the prefrontal cortex, and normal controls were administered with a temporal discrimination task. The task required to remember the duration of either 1 or 3 s across a delay varying from 1 to 20 s. Results demonstrated that hypoxic subjects were impaired relative to control and prefrontal patients for all delays. Prefrontal damaged subjects showed no deficit. A couple of control tasks were also carried out to demonstrate that the impairment could not be attributed to defective object memory or to inability to estimate time intervals of 1 and 3 s.

Animal data (Jackson, Kesner and Amann, 1998) also demonstrated that extensive hippocampal lesions impaired rats' ability of remembering duration of exposure of an object at short and 10-s delays, while they had no difficulty in discriminating between durations.

In conclusion, the hippocampus seems to be involved in consolidating the representation of temporal intervals and maintaining it across a delay. Over- or under-estimation may depend on the integrity of areas involved in comparing ongoing intervals with those stored in the reference memory.

The cerebellum

The cerebellum traditionally has been viewed as a neural structure devoted to motor control. However, recent neurobehavioral and functional neuroimaging studies support the view that the cerebellum may be involved in nonmotor functions, including attention (Akshoomoff and Courchesne, 1992; Allen, 1997; Courchesne, Townsend, Akshoomoff et al., 1994), working memory (Klingberg, Kawashima and Roland, 1996), semantic association (Petersen, Fox, Posner et al., 1989), verbal learning (Andreasen, O'Leary, Arndt et al., 1995), memory (Appollonio et al., 1993), and problem solving (Kim, Ugurbil and Strick, 1994).

It has also been shown that patients with lesion in the cerebellum are impaired in both motor and perceptual tasks requiring accurate timing in the range of hundreds of milliseconds (Ivry and Keele, 1988, 1989; Nichelli, Alway and Grafman, 1996). Nichelli et al. (1996) examined a group of patients with cerebellar degeneration using a time bisection paradigm. Results showed impaired patients' performance both on 100–900 ms and on 100–600 ms time bisection condition. With longer intervals the pattern of performance also showed a precision deficit which pointed to defective sustained attention and/or decision processes. A recent study (Casini and Ivry, 1999) has examined patients with either prefrontal or cerebellar lesions on temporal and nontemporal perceptual tasks under two levels of attentional load. Trials involved a comparison between a standard tone and a subsequent one that varied in frequency, duration, or both. Results showed that patients with frontal lobe lesions were significantly impaired on both tasks whereas the variability of cerebellar patients

increased in the duration task only. This dissociation suggests that frontal patients' deficits on temporal processing of intervals in the hundreds of millisecond range can be related to the attentional demands of such tasks. Cerebellar patients on the contrary seem to have a more specific problem related to timing, at least in timing intervals in this time range.

Based both on animal and human data it has been proposed (Clarke, Ivry, Grinband et al., 1996) that the cerebellum is capable of representing temporal information ranging from a few milliseconds to an upper bound of a couple of seconds. With longer intervals either performing the task is shifted to different neural structures or there is a qualitative change in our capabilities to represent temporal information.

A different view has been proposed by Malapani and Gibbon (1998). These authors trained patients with focal lesions of the cerebellum to reproduce durations of 8, 12, and 21 s. Results demonstrated that cerebellar lesions do not affect accuracy. However, they showed increased variability proportional to the target time by patients with lateral cerebellar lesions. This finding was interpreted in the framework of the Scalar Timing Theory as suggesting a role of cerebellum in storage and retrieval of temporal memories or at the point of response decision (see Fig. 1).

The basal ganglia

Based on animal studies on the role of neurotransmitters, it has been hypothesized that basal ganglia play a key role in controlling the internal clock (Meck, 1986, 1996). In humans, Parkinson disease (PD) has served as the model of basal ganglia dysfunction. Several studies have demonstrated that PD patients are impaired in discriminating brief durations (Artieda, Pastor, Lacruz and Obeso, 1992; Harrington, Haaland and Knight, 1998; Hellström, Lang, Portin and Rinne, 1997; Rammsayer and Classen, 1997). Pastor, Artieda, Jahanshahi and Obeso (1992) examined a group of 44 unmedicated PD patients on a test of verbal time estimation and in several time reproduction tasks. Patients with PD underestimated the duration of a time interval in the verbal time estimation task and showed overproduction of time intervals when required to reproduce a short time sample. Absolute errors were greater in the reproduction of longer time intervals in both controls and

PD patients, but especially in the latter. Results led the authors to suggest that the 'internal clock', is abnormally slow in PD. However, in a subsequent study, Malapani, Rakitin, Levy et al. (1998) examined PD patients when brain dopamine transmission was impaired (OFF state) and when dopamine transmission was reestablished, at the time of maximal clinical benefit following administration of levodopa + apomorphine (ON state). Patients reproduced target times of 8 and 21 s trained in blocked trials with the peak interval procedure, which were veridical in the ON state, comparable to normative performance by healthy young and aged controls. In the OFF state, temporal reproduction was impaired in both accuracy and precision (variance). The 8-s signal was reproduced as longer and the 21-s signal was reproduced as shorter than they actually were. These results did not represent a typical defective clock pattern (which would have caused a precision but not an accuracy effect). On the contrary, accuracy distortions implicate a dysfunctional temporal memory representation either on storage or retrieval, which might overshadow distortions in timekeeping processes.

Note that several studies (Blanchet, Marie, Dauvillier et al., 2000; Brown and Marsden, 1988; Dujardin, Degreef, Rogelet et al., 1999; Fournet, Moreaud, Roulin et al., 2000) have demonstrated that PD patients have attention and/or working memory deficit, which might play a role in determining their performance on temporal tasks.

In summary, both animal and human data provide a strong support to the hypothesis of a role of the basal ganglia in timing. Yet, it is far from clear whether basal ganglia damage impairs internal clock mechanisms or nontemporal processes supporting timing.

The prefrontal cortex

Memory for time is usually divided into three major components: (1) temporal order, which refers to memory for the sequential occurrence of events; (2) duration, i.e., memory for intervals between events; (3) time perspective, which refers to using memory for anticipating future events. A number of lesion studies, both on humans and animals, have suggested that the prefrontal cortex is involved somehow in mediating all these aspects of memory for time. I will briefly review the experimental evidence for each of these aspects.

Temporal order. A number of clinical studies have pointed out that patients with damage to the prefrontal cortex cannot (1) remember the order in which information has been experienced, (2) plan and create a complex set of movements, (3) program a temporally ordered set of activities.

For example it has been demonstrated (Butters, Kasniak, Glisky and Elinger, 1994; Lewinsohn, Zieler, Libet et al., 1972; McAndrews and Milner, 1991; Milner, 1971; Milner, McAndrews and Leonard, 1990) that frontal lobe lesions are associated with a selective deficit in recency judgements, i.e., in discriminating more from less recent event. Patients with frontal lobe damage, and to some extent, temporal-lobe-resected patients with hippocampal damage, are also impaired in judging the frequency of occurrence of abstract designs (Milner, Petrides and Smith, 1985) and of words (Jurado, Junque, Pujol et al., 1997) within a list. In addition, frontal-lobe-damaged patients are impaired in self-ordering a sequence of stimuli presented one at a time (Petrides and Milner, 1982), in reproducing the correct order of a list of 15 words, and at arranging in chronological order 15 factual events that occurred in the past (Shimamura, Janowsky and Squire, 1990). They are also very poor in reproducing a series of ordered facial movements, even when they are very good in remembering the components of the sequence (Kolb and Milner, 1981).

Previous research indicates that patients who become amnesic after hippocampal lesion may also have some difficulties in processing temporal-order information. To better understand the relative role of hippocampal and frontal systems in temporal-order memory, Kesner, Hopkins and Fineman (1994) devised several memory tasks by which they could test both order and item memory with the same stimuli. In all these tasks the study phase contained a sequence of stimuli, and the test phase consisted in presenting two stimuli. In case of item memory testing, only one of the two stimuli was presented in the study phase and subjects were requested to select it. For testing order recognition both stimuli were presented in the study phase and subjects were requested to indicate which of the two stimuli occurred earlier in the study sequence. Subjects with hypoxic brain injury resulting in hippocampal damage were markedly impaired both in item and in order memory (Hopkins et al., 1995a). On the contrary, prefrontal-cortex-damaged subjects showed no deficit on item recognition but they had severe deficit on all order recognition tests relative to controls (Kesner et al., 1994).

In a further series of experiments, Hopkins, Kesner and Goldstein (1995b) demonstrated that prefrontal-cortex-damaged patients, differently from hippocampal patients, were impaired in remembering temporal order for all temporal distances and could not take advantage of knowledge-based information to memorize the temporal order of specific sequences of events.

Memory for temporal order has been also studied in rats using an eight-arm radial maze (Chiba, Kesner and Gibson, 1997; Chiba, Kesner and Reynolds, 1994). In these experiments, during the study phase, rats are allowed to visit each of eight arms in a predetermined order. Then, in the test phase they are required to choose which of two arms they had visited earlier in the study phase. The arms selected as test arms varied according to temporal distance (i.e., to the number of arms occurring between them). Results demonstrate that after medial prefrontal lesions, rats are impaired in remembering the temporal order of the arm to be visited. Such impairment is observed for all temporal distances and both for new sequences and for well learned constant sequences (Chiba et al., 1997). On the contrary, rats with dorsal hippocampal lesions are only impaired on new sequences for temporal distances of 0 and 2, but they perform well on new sequences for the longest temporal distances of 4 and 6, and with constant sequences, for all temporal distances. As a consequence of these studies it appears that the hippocampus is sensitive to interference in a data-based memory system, while the medial prefrontal cortex mediates temporal order within a knowledge-based memory system (Kesner, 1998).

In monkeys, dorsolateral prefrontal cortex lesions result in deficits in temporal ordering of events comparable to what has been described in rats (Brody and Pribram, 1978; Pinto-Hamuy and Linck, 1964). In one experiment Pribram and Tubbs (1967) demonstrated that monkeys with frontal lesions could not perform a 5-s right–left delayed alternation task (order deficit) but were unimpaired with a 15-s delay between a right–left couplet of responses. In this latter situation it was hypothesized that animals could chunk each couplet as an item of information and did not need to remember the order of a right–left couple of response.

Duration. The experimental evidence that links prefrontal damage in humans to impaired duration processing is derived from a couple of studies. In one study (Nichelli, Clark, Hollnagel and Grafman, 1995), subjects classified test intervals as more similar to either a short or a long interval that had been presented at the start of a test block. Results demonstrated that patients with focal frontal lesions were impaired at discriminating intervals in both the millisecond (100–900 ms) and second (8–32 s) range. However, they were unimpaired in a similar task involving linear length discrimination.

With a different procedure (Parametric Estimation by Sequential Testing, PEST), Mangels, Ivry and Shimizu (1998) demonstrated that prefrontal damage impairs temporal discrimination in the seconds range (around a standard of 4 s), but not in the millisecond (around 400 ms) range.

Using the same PEST procedure to test temporal discrimination in the millisecond range (around standards of 300 and 600 ms) Harrington, Haaland and Hermanowicz (1998) found that right hemisphere patients were significantly impaired. Lesion overlays of patients with right hemisphere lesions and impaired timing included patients who had lesions in the prefrontal cortex or in the inferior parietal cortex.

Indirect evidence of the role of the prefrontal cortex in duration processing can also be derived from a study investigating time production and reproduction in patients with the Korsakoff syndrome (Shaw and Aggleton, 1994). Characteristically, these patients show amnesia and frontal-lobe-associated cognitive deficits. Results demonstrated that Korsakoff's patients were impaired both in time production and in two time reproduction tasks. Moreover, the number of errors in all three tasks was positively correlated with errors in Cognitive Estimation, a frontal-lobe-mediated task in which the subject is asked to estimate the length, the price, or other characteristics of different objects.

Despite difficulty in directly comparing analyses of cognitive functions in man and rodents, a number of experiments in rats have provided important insights in the basic mechanisms involved in temporal discrimination. Olton (1989) trained rats to press a lever after a fixed stimulus interval. At the end of the interval, the food was delivered and the stimulus was shut off. Then, he measured rats' ability to predict the time of food arrival with probe trials during which no food was delivered for any response. The response rate of normal rats, summed over many probe trials with no reinforcement, had a distinct peak that occurred at the time when food was normally delivered during the fixed interval training. After a bilateral lesion of the frontal cortex rats continued to perform the general aspects of the task appropriately but produced a rightward shift of the peak time in trials with no reinforcement, indicating that they expected reinforcement at a time later than when reinforcement was provided on food trials. This finding was interpreted in the framework of the Scalar Timing Theory (Gibbon et al., 1984) as indicating interference with the reference memory mechanism (that stores information about the actual time of reinforcement during food trials). A similar rightward shift (overestimation) of the expected time was obtained after lesions of the nucleus basalis magnocellularis, an area of the forebrain that has significant projections to the frontal cortex. On the contrary lesions in the hippocampal system produce an opposite leftward shift, indicating underestimation of the expected reinforcement (Meck, Church and Olton, 1984; Meck et al., 1987).

Time perspective. It has been suggested that the prefrontal cortex mediates memory for future events, which refers to remembering to do things (prospective memory) as contrasted with retrospective memory, which refers to remembering of information and past events (Della Barba, 1993). A number of studies have emphasized that performing a planned action at a given time requires monitoring the interval between formulation of intention and execution of the action (Ceci and Bronfenbrenner, 1985; Harris, 1984; Winograd, 1988). It has been suggested that time monitoring procedures can be viewed as repeated test–wait cycles until the test indicates that it is time for the action (Harris and Wilkins, 1982; Miller,

Galanter and Pribram, 1960). Test–wait cycles are more frequent early in the waiting period, in order to synchronize one's psychological clock with the passage of real time (Ceci, Baker and Bronfenbrenner, 1988). Once synchronization is accomplished, time monitoring becomes less frequent. However, as the target time approaches, observations of the clock increase in frequency by an amount that is related to the accuracy of responding (Harris, 1982). Studies of normal subjects (Einstein and McDaniel, 1990; Kvavilashvili and Ellis, 1996) have identified three subtypes of prospective memory (i.e., time-, event-, and activity-based). A time-based prospective memory task involves remembering to perform an action at a specific time while an event-based prospective memory task requires the subject to perform an action when an external cue appears (e.g., remembering to say something to someone when you meet him or her). Activity-based prospective memory also involves an external cue but it does not require the interruption of the ongoing activity (e.g., switching off the oven after cooking). A number of studies have demonstrated that traumatic brain injury affects prospective memory. More recently, Shum, Valentine and Cutmore (1999) demonstrated that it affects all the three subtypes of prospective memory. Shallice and Burgess (1991) investigated the ability to carry out a variety of prospective memory tasks in three patients who had sustained traumatic injuries involving the frontal lobe. Two of them were unimpaired in a number of tests of frontal functions. However, they all showed severe deficits in a number of open-ended multiple subgoal tasks, typically spending too long on each individual task. It was argued that lack of motivation, retrospective memory deficits, or general cognitive impairments could not account for their impairment, which was rather at the level that bridges cognitive and memory abilities with motivational requirements. According to this theoretical framework, prospective memory is a function of the Supervisory Attentional System (Norman and Shallice, 1986) which involves the articulation of a goal and the formulation of the plan to achieve the goal. A different view is envisaged by Grafman's Managerial Knowledge Unit (MKU) representational model (Grafman, 1989, 1995). MKUs are composed by series of events, actions or ideas that are linked together to form a knowledge unit (e.g., a schema) which are typically stored in the prefrontal cortex. MKUs have a typical duration of activation (e.g., leaving the house in the morning involves taking a shower, eating breakfast, and similar events, each with its typical duration). Prospective memory would be obtained by the possibility of binding representations either based on the duration of an event or on its boundaries.

The role of the prefrontal cortex in prospective memory is supported by a study (McDaniel, Glisky, Rubin et al., 1999) comparing older adults divided into four groups on the basis of their scores on two composite measures: one assessing frontal lobe function and the other assessing medial temporal lobe function. Results demonstrated that subjects who performed well on the frontal tasks have better prospective memory than subjects who performed relatively poorly on the same task. There was no significant difference in prospective memory performance attributable to medial temporal functioning.

More direct evidence about the role of the frontal lobes in prospective memory was provided by a PET study by Okuda, Fujii, Yamadori et al. (1998). These authors administered a task requiring subjects to retain and remember a planned action while performing an ongoing routine activity. Activations related with prospective memory were identified in the right dorsolateral and ventrolateral prefrontal cortices as well as in midline medial frontal lobe.

In conclusion, there is consistent agreement that prospective memory involves the prefrontal cortex. However, there have been no studies that have tried to disentangle the neural substrate of event- or activity-based from time-based prospective memory.

Neuroimaging studies

Functional neuroimaging studies have produced further evidence of the neural substrate of temporal processing. Jueptner, Rijntjes, Weiller et al. (1995) used positron emission tomography (PET) to localize a cerebellar timing function in humans. They asked subjects to compare a test interval with a standard interval. Results indicated that the superior part of the vermis, the adjacent cerebellar hemispheres, the basal ganglia, and the cingulate cortex are involved in discriminating auditory intervals lasting less than 1 s. Rao, Harrington, Haaland et al. (1997) demonstrated that tapping at a constant internal rhythm

(target interval 300 and 600 ms) produced activation of both a medial premotor system (including caudal SMA, left putamen, and ventrolateral thalamus) and the right inferior frontal gyrus.

Onoe, Komori, Onoe et al. (2001) performed PET studies with rhesus monkeys during time discrimination tasks of intervals ranging from 400 to 1500 ms. Changes in rCBF that covaried significantly with the durations of the target being perceived by animals were found in the dorsolateral prefrontal cortex, the posterior part of the inferior parietal cortex, basal ganglia, and posterior cingulate cortex. Furthermore, a loss of neuronal function in the dorsolateral prefrontal cortex caused by a local application of bicuculline resulted in the selective reduction of performance in time discrimination tasks. The results indicate that a neural network composed of the posterior inferior parietal cortex and the dorsolateral prefrontal cortex plays a crucial role in the temporal monitoring process in time perception.

Maquet, Lejune, Pouthas et al. (1996) studied temporal processing with a temporal generalization task in which the subjects had to judge whether the duration of the illumination of a green LED was equal to or different from that of a previously presented standard. A significant increase in regional cerebral blood flow (rCBF) was observed in the right prefrontal cortex, right inferior parietal lobule, anterior cingulate cortex, vermis, and a region corresponding to the left fusiform gyrus. A similar increase was also obtained by an intensity generalization task in which the judgment concerned the intensity of the LED. However, in a follow-up study Lejune, Maquet, Bonnet et al. (1997), combined activation associated with temporal generalization and synchronization tasks to highlight a temporal processing network including, again, right prefrontal, inferior parietal, anterior cingulate cortex, left putamen, and left cerebellar hemisphere.

Using functional magnetic resonance imaging, Basso, Nichelli, Grafman et al. (1997) investigated the neural substrate of a temporal production task. Subjects were asked to judge if a digit probe belonged to a string of digits presented immediately before but to provide their response only after 1.5 s had elapsed. This time estimation condition, compared with control working memory and motor tasks, was associated with increased activity in the middle occipital gyri, in the right inferior parietal lobe, and bilaterally in the dorsolateral prefrontal cortex.

In a subsequent study, Basso, Nichelli, Pugliese et al. (1999) investigated the functional magnetic resonance correlates of the time bisection paradigm. Despite striking task differences with the previous study, results revealed increased activity in many areas that were also associated with the temporal production task. Activation related with the temporal bisection included the right inferior parietal lobe, the right middle frontal gyrus, the right superior frontal gyrus, and the left putamen.

In conclusion neuroimaging data seem to indicate that processing temporal durations is based on a complex circuit including at least the right inferior parietal lobule, and the dorsolateral prefrontal cortex. Although the role of each brain area is far from established, we hypothesize that stimuli can initiate phase-locked oscillatory activities, which serve as basis for feature binding and duration perception. In the case of a visual stimulus the oscillatory activity might take place in extrastriate visual areas. This constitutes the 'pacemaker' of the internal clock. Pulses emitted from the pacemaker need to be accumulated and counted in brain structures capable of serving multiple sensory streams. We claim that the right inferior parietal cortex plays a pivotal role in such timing processes and that it can adapt to information coming from multiple sources. Incoming intervals need then to be compared with reference intervals. We surmise that bilateral prefrontal cortex performs this operation.

Conclusion

Converging evidence indicates that the prefrontal cortex plays a pivotal role in different temporal processing tasks (e.g., temporal discrimination, temporal production, temporal-order memory, time-based prospective memory). Indeed, a 'time factor' is part of some of the most influential theories on frontal lobe functioning. According to Fuster (1985b) the prefrontal cortex is principally involved in representing the "*temporal structure of behavior*", which manifests itself in short-term memory and in short-term motor sets (Fuster, 1995). Fuster has also argued that, in order to encode temporal aspects of behav-

iors, the prefrontal cortex mediates "*cross-temporal contingencies*" (Fuster, 1985a). Cross-temporal contingencies can be interpreted as associative relationships between events that are related to each other because they are parts of a set of actions that have a common goal.

Goldman-Rakic (1987) has demonstrated prefrontal cortical neurons firing only during the delay between the presentation of a stimulus set to memorize and the presentation of a probe for response. According to Goldman-Rakic (1995), the prefrontal cortex serves as a working memory that temporarily "keeps active" a representation of a stimulus until a response is provided. Even if the meaning of 'keeping active' is unclear in this context, it is noteworthy that prefrontal neurons can code stimulus location and not just a memory of the object itself. In this regard the function of the prefrontal cortex is viewed as that of maintaining information and knowledge across temporal intervals.

Grafman (1989) has proposed a different model of prefrontal cortex function. According to his view, cortical function is better described in terms of stored representation than in terms of abstract operations. There appears to be an evolutionary trend from cognitive architectures in which knowledge units represent a single feature of a stimulus event to units that represent a series of events. Examples of the former include edge detectors in the visual system. More complex units include objects, lexical items, syntactic frames, and cognitive macrostructures like plans, scripts, schemas, and mental models, which are subserved by the so-called Managerial Knowledge Units (MKUs). The latter are typically stored in the prefrontal cortex.

In this framework, event order constraints and duration are an inherent property of representations. The more complex the knowledge unit, the longer the duration of its activation time and the more critical the order of occurrence of its components. This theory has no need to hypothesize the existence of an internal clock. Different regions of the cerebral cortex would be activated by temporal tasks depending on the time range of the trial and the number and complexity of events composing the stimulus presentation. As a consequence, the purported role of prefrontal cortex in temporal tasks might depend both on the time range that is examined, the format of the stimulus presentation, and the subject's memory strategy.

Acknowledgements

This work was completed during a short stay at the Cognitive Neuroscience Section of the National Institutes of Neurological Disorders and Stroke in Bethesda, MD (USA). I want to thank Jordan Grafman for his friendly and warm support and for providing access to the facilities that allowed me to prepare the manuscript. I would also like to thank Annalena Venneri for her helpful comments.

References

Akshoomoff NA, Courchesne E: A new role for the cerebellum in cognitive operations. Behavioral Neuroscience: 106; 731–738, 1992.

Allan LG: The perception of time. Perception and Psychophysics: 26; 340–354, 1979.

Allan LG: The influence of the scalar timing model on human timing research. Behavioural Processes: 44; 101–117, 1998.

Allen G: Attentional activation of the cerebellum independent of motor involvement. Science: 275; 1940–1943, 1997.

Andreasen NC, O'Leary DS, Arndt S, Cizadlo T, Hurtig R, Rezai K et al.: Short-term and long-term verbal memory: a positron emission tomography study. Proceedings of the National Academy of Sciences of the United States of America: 92; 5111–5115, 1995.

Appollonio IM, Grafman J, Schwartz V, Massaquoi S, Hallet M: Memory in patients with cerebellar degeneration. Neurology, 43; 1535–1544, 1993.

Artieda J, Pastor MA, Lacruz F, Obeso JA: Temporal discrimination is abnormal in Parkinson's disease. Brain: 115 (Part 1); 199–210, 1992.

Basar E: Brain Function and Oscillations. II. Integrative Brain Function. Neurophysiology and Cognitive Processes. Berlin: Springer, 1999.

Basar E, Gönder A, Ungan P: Dynamics of brain rhythmic and evoked potentials. II. Studies in the auditory pathway, reticular formation, and hippocampus during waking stage. Biological Cybernetics: 25; 41–48, 1975.

Basar-Eroglu C, Strüber D, Schürman M, Stadler M, Basar E: Gamma banda responses in the brain: A short review of psychophysical correlates and functional significance. Journal of Psychophysiology: 24; 101–112, 1996.

Basso G, Nichelli P, Grafman J, Warton CM, Peterson M: The brain structures involved in temporal production: a fMRI study. Neuroimage: 5; S77, 1997.

Basso G, Nichelli P, Pugliese M, Pietrini P, Salat D, Grafman J: Functional MRI exploration of the time interval bisection paradigm. Neuroimage: 9; S400, 1999.

Beling I: Über das Zeitgedächtnis der Bienen. Zeitschrift für vergleichende Physiologies: 9; 259–338, 1929.

Blanchet S, Marie RM, Dauvillier F, Landeau B, Benali K, Eustache F et al.: Cognitive processes involved in delayed non-matching-to-sample performance in Parkinson's disease. European Journal of Neurology: 7; 473–483, 2000.

Block RA, George EJ, Reed MA: A watched pot sometimes boils. A study of duration experience. Acta Psychologica: 46; 81–94, 1980.

Brody BA, Pribram KH: The role of the frontal and parietal cortex in cognitive processing: test of spatial and sequence functions. Brain: 101; 603–633, 1978.

Brown RG, Marsden CD: Internal versus external cues and the control of attention in Parkinson's disease. Brain: 111; 323–345, 1988.

Brown SW: Time perception and attention: The effects of prospective versus retrospective paradigms and task demands on perceived duration. Perception and Psychophysics: 38; 115–124, 1985.

Buonomano D, Hickmott P, Merzenich M: Context-sensitive synaptic plasticity and temporal-to-spatial transformations in hippocampal slices. Proceedings of the National Academy of Sciences of the United States of America: 94; 10403–10408, 1997.

Buonomano DV: Decoding temporal information: A model based on short-term synaptic plasticity. Journal of Neuroscience: 20; 1129–1141, 2000.

Buonomano DV, Merzenich MM: Temporal information transformed into spatial code by a neural network with realistic proprieties. Science: 267; 1028–1030, 1995.

Butters MA, Kasniak AW, Glisky EL, Elinger PJ, Shacter DL: Recency discrimination deficits in frontal lobe patients. Neuropsychology: 8; 343–353, 1994.

Cahoon D, Edmonds EM: The watched pot still won't boil: Expectancy as a variable in estimating the passage of time. Bulletin of the Psychonomic Society: 16; 115–116, 1980.

Campbell J: Human vs. animal time. In Pastor MA, Artieda J (Eds), Time, Internal Clocks and Movement. Amsterdam: Elsevier, pp. 115–126, 1996.

Casini L, Ivry RB: Effects of divided attention on temporal processing in patients with lesions of the cerebellum or frontal lobe. Neuropsychology: 13; 10–21, 1999.

Casini L, Macar F: Effects of attention manipulation on judgments of duration and intensity in the visual modality. Memory and Cognition: 25; 812–818, 1997.

Ceci SJ, Baker JG, Bronfenbrenner U: Prospective remembering, temporal calibration, and context. In Gruneberg MM, Morris PE, Sykes RN (Eds), Practical Aspects of Memory: Current Research and Issues. Chichester: Wiley, pp. 360–365, 1988.

Ceci SJ, Bronfenbrenner U: 'Don't forget to take the cupcakes out of the oven': prospective memory, strategic time monitoring and context. Child Development: 56; 152–164, 1985.

Chiba AA, Kesner RP, Gibson CJ: Memory for temporal order on new and familial spatial location sequences: Role of the medial prefrontal cortex. Learning and Memory: 4; 311–317, 1997.

Chiba AA, Kesner RP, Reynolds AM: Memory for temporal

location as a function of temporal lag in rats: Role of hippocampus and medial prefrontal cortex. Behavioral and Neural Biology: 61; 123–131, 1994.

Church RM: Evaluation of quantitative theories of timing. Journal of Experimental Analysis of Behavior: 71; 253–256, 1999.

Church RM, Broadbent HA: Alternative representation of time, number, and rate. Cognition: 37; 55–81, 1990.

Church RM, Broadbent HA: A connectionist model of timing. In Commons ML, Grossberg S, Staddon JE (Eds), Neural Network Models of Conditioning and Action. Hillsdale, NJ: Lawrence Erlbaum Associates, pp. 225–240, 1991.

Clarke S, Ivry R, Grinband J, Roberts S, Shimizu N: Exploring the domain of the cerebellar timing system. In Pastor MA, Artieda J (Eds), Time, Internal Clocks and Movement. Amsterdam: Elsevier, pp. 257–280, 1996.

Collyer CE, Broadbent HA, Church RM: Categorical time production: Evidence for a discrete timing in motor control. Perception and Psychophysics: 51; 134–144, 1992.

Collyer CE, Broadbent HA, Church RM: Preferred rate of repetitive tapping and categorical time production. Perception and Psychophysics: 55; 443–453, 1994.

Collyer CE, Church RM: Interresponse intervals in continuation tapping. In Rosembaum DA, Collyer CE (Eds), Timing of Behavior: Neural, Psychological, and Computational Perspectives. Cambridge, MA: MIT Press, pp. 64–87, 1998.

Courchesne E, Townsend J, Akshoomoff NA, Saitoh O, Yeung-Courchesne R, Lincoln AJ et al.: Impairment in shifting attention in autistic and cerebellar patients. Behavioral Neurosciences: 108; 848–865, 1994.

Della Barba G: Prospective memory: a 'new' memory system. In Boller F, Grafman J (Eds), Handbook of Neuropsychology, Vol. 8. Amsterdam: Elsevier, Amsterdam, pp. 239–251, 1993.

Dujardin K, Degreef JF, Rogelet P, Defebvre L, Destee A: Impairment of the supervisory attentional system in early untreated patients with Parkinson's disease. Journal of Neurology: 246; 783–788, 1999.

Einstein GO, McDaniel MA: Normal aging and prospective memory. Journal of Experimental Psychology: Learning Memory and Cognition: 16; 717–726, 1990.

Eisler H: Experiments on subjective duration 1868–1975: A collection of power function experiments. Psychological Bulletin: 83; 1154–1171, 1976.

Eisler H: Time perception from a psychophysicist's perspective. In Helfrich H (Ed), Time and Mind. Seattle, WA: Hogrefe and Huber, pp. 65–86, 1996.

Fortin C, Breton R: Temporal interval production and processing in working memory. Perception and Psychophysics: 57; 203–215, 1995.

Fortin C, Rousseau R: Time estimation as an index of processing demand in memory search. Perception and Psychophysics: 42; 377–382, 1987.

Fortin C, Rousseau R, Bourque P, Kirouac E: Time estimation and concurrent nontemporal processing: specific interference from short-term-memory demands. Perception and Psychophysics: 53; 536–548, 1993.

Fournet N, Moreaud O, Roulin JL, Naegele B, Pellat J: Working memory functioning in medicated Parkinson's disease patients and the effect of withdrawal of dopaminergic medication. Neuropsychology: 14; 247–253, 2000.

Fraisse P: Perception and estimation of time. Annual Review of Psychology: 35; 1–36, 1984.

Fuster JM: The prefrontal cortex, mediator of cross temporal contingencies. Human Neurobiology: 4; 169–179, 1985a.

Fuster JM: Temporal organization of behavior. Human Neurobiology: 4; 57–60, 1985b.

Fuster JM: Memory in the Cerebral Cortex. Cambridge, MA: MIT Press, 1995.

Gallistel CR: The Organization of Learning. Cambridge, MA: MIT Press, 1990.

Gibbon J: Scalar expectancy theory and Weber's law in animal timing. Psychological Review: 84; 278–325, 1977.

Gibbon J: The structure of subjective time: How time flies. The Psychology of Learning and Motivation, Vol. 20. New York: Academic Press, pp. 105–135, 1986.

Gibbon J, Church RM: Representation of time. Cognition: 37; 23–54, 1990.

Gibbon J, Church RM, Meck WH: Scalar timing in memory. In Gibbon J, Allan L (Eds), Annals of the New York Academy of Sciences: Timing and Time Perception, Vol. 423. The New York: New York Academy of Sciences, pp. 52–77, 1984.

Goldman-Rakic PS: Circuitry of primate prefrontal cortex and regulation of behavior by representational memory. In Plum F (Ed), Handbook of Physiology — The Nervous System. Washington, DC: American Physiological Society, pp. 373–417, 1987.

Goldman-Rakic PS: Architecture of the prefrontal cortex and the Central Executive. Annals of the New York Academy of Sciences: 769; 71–83, 1995.

Grafman J: Plans, actions, and mental sets: Managerial Knowledge Units in the frontal lobes. In Perecman E (Ed), Integrating Theory and Practice in Clinical Neuropsychology. Hillsdale, NJ: Lawrence Erlbaum Associates, pp. 93–138, 1989.

Grafman J: Similarities and distinctions among current models of prefrontal cortical functions. In Grafman J, Holyoak KJ, Boller F (Eds), Structure and Functions of the Human Prefrontal Cortex, Vol. 769. New York: The New York Academy of Sciences, pp. 337–368, 1995.

Gray CM, König P, Engel AK, Singer W: Oscillatory responses in cat visual cortex exhibit inter-columnar synchronization which reflects global stimulus properties. Nature: 338; 334–337, 1989.

Gray CM, Singer W: Stimulus-specific neuronal oscillation in orientation columns of cat visual-cortex. Proceedings of the National Academy of Sciences of the United States of America: 86; 1698–1702, 1989.

Harrington DL, Haaland KY, Hermanowicz N: Temporal processing in the basal ganglia. Neuropsychology: 12; 3–12, 1998.

Harrington DL, Haaland KY, Knight RT: Cortical networks underlying mechanisms of time perception. Journal of Neuroscience: 18; 1085–1095, 1998.

Harris JE: External memory aids. In Neisser U (Ed), Memory Observed: Remembering in Natural Contexts. San Francisco, CA: W.H. Freeman, pp. 337–342, 1982.

Harris JE: Remembering to do things: a forgotten topic. In Harris JE, Morris PE (Eds), Everyday Memory: Actions and Absent-mindedness. London: Academic Press, pp. 71–92, 1984.

Harris JE, Wilkins AJ: Remembering to do things. Human Learning: 1; 123–136, 1982.

Hellström Å, Lang H, Portin R, Rinne J: Tone duration discrimination in Parkinson's disease. Neuropsychologia: 35; 737–740, 1997.

Helmuth LL, Ivry RB: When two hands are better than one: reduced timing variability during bimanual movements. Journal of Experimental Psychology: Human Perception and Performance: 22; 278–293, 1996.

Hicks RE, Miller GW, Kinsbourne M: Prospective and retrospective judgments of time as a function of amount of information processed. American Journal of Psychology: 89; 719–730, 1976.

Hirsh IJ: Auditory perception of temporal order. Journal of the Acoustical Society of America: 31; 758–767, 1959.

Hirsh JL, Sherrick CEJ: Perceived order in different sense modalities. Journal of Experimental Psychology: 62; 423–432, 1961.

Hopkins RO, Kesner RP, Goldstein M: Item and order recognition memory in subjects with hypoxic brain injury. Brain and Cognition: 27; 180–201, 1995a.

Hopkins RO, Kesner RP, Goldstein M: Memory for novel and familiar spatial and linguistic temporal distance information in hypoxic subjects. Journal of International Neuropsychological Society: 1; 454–468, 1995b.

Ivry R: The representation of temporal information in perception and motor control. Current Opinion in Neurobiology: 6; 851–857, 1996.

Ivry R, Keele SW: Dissociation of the lateral and medial cerebellum in movement timing and movement execution. Experimental Brain Research: 73; 167–180, 1988.

Ivry RL, Keele SW: Timing functions of the cerebellum. Journal of Cognitive Neuroscience: 1; 136–152, 1989.

Jackson PA, Kesner RP, Amann K: Memory for duration: Role of hippocampus and medial prefrontal cortex. Neurobiology of Learning and Memory: 70; 328–348, 1998.

Joliot M, Ribary U, Llinas R: Human oscillatory brain activity near 40 Hz coexist with cognitive temporal binding. Proceedings of the National Academy of Sciences of the United States of America: 91; 11748–11751, 1994.

Jueptner M, Rijntjes M, Weiller C, Faiss JH, Timmann D, Mueller SP et al.: Localization of a cerebellar timing process using PET. Neurology: 45; 1540–1545, 1995.

Jurado MA, Junque C, Pujol J, Oliver B, Vendrell P: Impaired estimation of word frequency in frontal lobe patients. Neuropsychologia: 35; 635–641, 1997.

Kesner RP: Neural mediation of memory for time: Role of the hippocampus and medial prefrontal cortex. Psychonomic Bulletin and Review: 5; 585–596, 1998.

Kesner RP, Hopkins RO, Fineman B: Item and order dissociation in humans with prefrontal cortex damage. Neuropsychologia: 32; 881–891, 1994.

Kim SG, Ugurbil K, Strick PL: Activation of a cerebellar output nucleus during cognitive processing. Science: 265; 949–951, 1994.

Klingberg T, Kawashima R, Roland PE: Activation of multi-modal cortical areas underlies short-term memory. European Journal of Neuroscience: 8; 1965–1971, 1996.

Kolb B, Milner B: Performance of complex arm and facial movements after focal brain lesions. Neuropsychologia: 19; 491–504, 1981.

Kristofferson AB: A quantal step in duration discrimination. Perception and Psychophysics: 27; 300–306, 1980.

Kristofferson AB: Quantal and deterministic timing in human duration discrimination. In Gibbon J, Allan L (Eds), Timing and Time Perception, Vol. 423. New York: New York Academy of Sciences, pp. 3–15, 1984.

Kvavilashvili L, Ellis J: Varieties of intention: Some distinctions and classifications. In Brandimonte M, Einstein GO, McDaniel MA (Eds), Prospective Memory: Theory and Applications. Mahwah, NJ: Lawrence Erlbaum, pp. 23–62, 1996.

Lejune H, Maquet P, Bonnet M, Casini L, Ferrara A, Macar F et al.: The basic pattern of activation in motor and sensory temporal tasks: positron emission tomography data. Neuroscience Letters: 235; 21–24, 1997.

Lewinsohn PM, Zieler JL, Libet J, Eyeberg S, Nielson G: Short-term memory: A comparison between frontal and non-frontal right- and left-hemisphere brain-damaged patients. Journal of Comparative and Physiological Psychology: 81; 248–255, 1972.

Lisman JE, Idiart MAP: Storage of 7 ± 2 short-term memories in oscillatory subcycles. Science: 267; 1512–1515, 1995.

Macar F: Neural bases of internal timers: A brief overview. Cahiers de Psychologie Cognitive-Current Psychology of Cognition: 17; 847–865, 1998.

Malapani C, Gibbon J: Cerebellar dysfunction of temporal processing in the seconds range in humans. Neuroreport: 17; 3907–3912, 1998.

Malapani C, Rakitin B, Levy R, Meck WH, Deweer B, Dubois B et al.: Coupled temporal memories in Parkinson's disease: A dopamine related dysfunction. Journal of Cognitive Neuroscience: 10; 316–331, 1998.

Mangels JA, Ivry RB, Shimizu N: Dissociable contributions of the prefrontal and neocerebellar cortex to time perception. Cognitive Brain Research: 7; 15–39, 1998.

Maquet P, Lejune H, Pouthas V, Bonnet M, Casini L, Macar F et al.: Brain activation induced by estimation of duration: A PET study. Neuroimage: 3; 119–126, 1996.

Maricq AV, Roberts S, Church RM: Methamphetamine and time estimation. Journal of Experimental Psychology: Animal Behavior Processes: 7; 18–30, 1981.

McAndrews MP, Milner B: The frontal cortex and memory for temporal order. Neuropsychologia: 29; 849–859, 1991.

McDaniel MA, Glisky EL, Rubin SR, Guynn MJ, Routhieaux BC: Prospective memory: a neuropsychological study. Neuropsychology: 13; 103–110, 1999.

Meck WH: Selective adjustment of the speed of internal clock and memory processes. Journal of Experimental Psychology: Animal Behaviour Processes: 9; 171–201, 1983.

Meck WH: Affinity for the dopamine D_2 receptor predicts neuroleptic potency in decreasing the speed of an internal clock.

Pharmacology Biochemistry and Behavior: 25; 1185–1189, 1986.

Meck WH: Neuropharmacology of timing and time perception. Cognitive Brain Research: 3; 227–242, 1996.

Meck WH, Church RM: Cholinergic modulation of the content of temporal memory. Behavioral Neuroscience: 101; 457–464, 1987.

Meck WH, Church RM, Olton DS: Hippocampus, time, and memory. Behavioral Neuroscience: 98; 3–22, 1984.

Meck WH, Church RM, Wenk GL, Olton DD: Nucleus basalis magnocellularis and medial septal area lesions differently impair temporal memory. Journal of Neuroscience: 7; 3505–3511, 1987.

Miall, C: Models of neural timing. In Pastor MA, Artieda J (Eds), Time, Internal Clock, and Movement. New York: Elsevier, pp. 69–94, 1996.

Miller GA, Galanter E, Pribram KH: Plans and the Structure of Behavior. New York: Holt, Rinehart and Winston, 1960.

Milner B: Les troubles de la mémoire accompagnant des lésions hippocampiques bilatérales. In Passuant P (Ed), Physiologie de l'hippocampe. Paris: Centre National de la Recherche Scientifique, pp. 257–272, 1962.

Milner B: Interhemispheric differences in the localization of psychological processes in man. British Medical Bulletin: 27; 272–277, 1971.

Milner B, Corkin S, Teuber H-L: Further analysis of the hippocampal amnesic syndrome: 14-year follow-up study of H.M. Neuropsychologia: 6; 215–234, 1968.

Milner B, McAndrews MP, Leonard G: Frontal lobes and memory for the temporal order of recent events. Cold Spring Harbor Symposia in Quantitative Biology: 55; 987–994, 1990.

Milner B, Petrides M, Smith ML: Frontal lobes and the temporal organization of memory. Human Neurobiology: 4; 137–142, 1985.

Nagarajan SS, Blake DT, Wright BA, Byl N, Merzenich MM: Practice-related improvements in somatosensory interval discrimination are temporally specific but generalize across skin location, hemisphere, and modality. Journal of Neuroscience: 18; 1559–1570, 1998.

Nichelli P, Alway D, Grafman J: Perceptual timing in cerebellar degeneration. Neuropsychologia: 34; 863–871, 1996.

Nichelli P, Clark K, Hollnagel C, Grafman J: Duration processing after frontal lobe lesions. In Grafman J, Holyoak KJ, Boller F (Eds), Structure and Function of the Human Prefrontal Cortex. Annals of the New York Academy of Sciences, Vol. 769. New York: New York Academy of Sciences, pp. 183–190, 1995.

Norman DA, Shallice T: Attention to action: Willed and automatic control of behavior. In Davidson RJ, Schwarz GE, Shapiro D (Eds), Consciousness and Self-Regulation: Advances in Research and Theory. New York: Plenum Press, pp. 1–18, 1986.

Okuda J, Fujii T, Yamadori A, Kawashima R, Tsukiura T, Fukatsu R et al.: Participation of the prefrontal cortices in prospective memory: evidence from a PET study in humans. Neuroscience Letters: 253; 127–130, 1998.

Olton DS: Frontal cortex, timing and memory. Neuropsychologia: 27; 121–130, 1989.

Onoe H, Komori M, Onoe K, Takechi H, Tsukada H, Watanabe Y: Cortical networks recruited for time perception: a monkey positron emission tomography (PET) study. Neuroimage: 13; 37–45, 2001.

Pastor MA, Artieda J, Jahanshahi M, Obeso JA: Time estimation and reproduction is abnormal in Parkinson's disease. Brain: 115; 211–225, 1992.

Pastore R, Farrington S: Measuring the difference limen for identification of order of onset for complex auditory stimuli. Perception and Psychophysics: 58; 510–526, 1996.

Pastore R, Harris L, Kaplan J: Temporal order identification: some parameters dependencies. Journal of the Acoustical Society of America: 71; 430–436, 1982.

Petersen SE, Fox PT, Posner MI, Mintun M, Raichle E: Positron emission tomographic studies of the processing of single words. Journal of Cognitive Neuroscience: 1; 153–170, 1989.

Petrides M, Milner B: Deficit on subject-ordered task after frontal- and temporal-lobe lesions in man. Neuropsychologia: 20; 249–262, 1982.

Piéron H: The Sensations. New Haven, CT: Yale University Press, 1952.

Pinto-Hamuy T, Linck P: The effect of frontal lesions on performance of sequential tasks by monkeys. Transactions of the American Neurological Association: 89; 243–244, 1964.

Pöppel E: Time perception. In Held R, Leibowitz W, Teuber H-L (Eds), Handbook of Sensory Physiology, Vol. 8. Heidelberg: Springer, pp. 713–729, 1978.

Pöppel E, Schill K, von Steinbüchel N: Sensory integration within temporally neutral system states: A hypothesis. Naturwissenschaften: 77; 89–91, 1990.

Pribram KH, Tubbs WE: Short-term memory, parsing, and the primate frontal cortex. Science: 156; 1765–1767, 1967.

Rakitin BC, Gibbon J, Penney TB, Malapani C, Hinton SC, Meck WH: Scalar Expectancy Theory and peak-interval timing in humans. Journal of Experimental Psychology: Animal Behavior Processes: 24; 15–33, 1998.

Rammsayer TH, Classen W: Impaired temporal duration discrimination in Parkinson's disease: Temporal processing of brief durations as an indicator of degeneration of dopaminergic neurons in the basal ganglia. International Journal of Neuroscience: 91; 45–55, 1997.

Rao SM, Harrington DL, Haaland KY, Bobholz JA, Cox RW, Binder JR: Distributed neural systems underlying the timing of movements. Journal of Neuroscience: 17; 5528–5535, 1997.

Richards W: Time reproductions by H.M. Acta Psychologica: 37; 279–282, 1973.

Rodriguez E, George N, Lachaux J-P, Martinerie J, Renault B, Varela F: Perception's shadow: long-distance synchronization of human brain activity. Nature: 397; 430–433, 1999.

Rusak B, Zucker I: Neural regulation of circadian rhythms. Physiological Review: 59; 449–526, 1979.

Scoville WB, Milner B: Loss of recent memory after bilateral hippocampal lesions. Journal of Neurology Neurosurgery and Psychiatry: 20; 11–21, 1964.

Shallice T, Burgess P: Deficits in strategy application following frontal lobe damage in man. Brain: 114; 727–741, 1991.

Shaw C, Aggleton JP: The ability of amnesic subjects to estimate time intervals. Neuropsychologia: 32; 857–873, 1994.

Shimamura AP, Janowsky JS, Squire LR: Memory for temporal order of events in patients with frontal lobe lesions and amnesic patients. Neuropsychologia: 28; 803–813, 1990.

Shum D, Valentine M, Cutmore T: Performance of individuals with severe long-term traumatic brain injury on time-, event-, and activity-based prospective memory tasks. Journal of Clinical and Experimental Neuropsychology: 21; 49–58, 1999.

Staddon JER, Higa JJ: Time and memory: Towards a pacemaker-free theory of interval timing. Journal of the Experimental Analysis of Behavior: 71; 215–251, 1999.

Staddon JER, Higa JJ, Chelaru IM: Time, trace, memory. Journal of Experimental Analysis of Behavior: 71; 293–301, 1999.

Stryker MP: Is grandmother an oscillation? Nature: 338; 297–298, 1989.

Treisman M: Temporal discrimination and the indifference interval: Implications for a model of the 'internal clock'. Psychological Monographs: 77; 1–31, 1963.

Wearden J, Lejune H: Across the great divide: Animal psychology and time in humans. Time and Society: 2; 87–106, 1993.

Wing AM, Kristofferson AB: Response delays and the timing of discrete motor responses. Perception and Psychophysics: 14; 5–12, 1973a.

Wing AM, Kristofferson AB: The timing of interresponse intervals. Perception and Psychophysics: 13; 455–460, 1973b.

Winograd E: Some observations on prospective remembering. In Gruneberg MM, Morris PE, Sykes RN (Eds), Practical Aspects of Memory: Current Research and Issues. Chichester: Wiley, pp. 349–353, 1988.

Wittmann M: Time perception and temporal processing levels of the brain. Chronobiology International: 16; 17–32, 1999.

Wright BA, Buonomano D, Mahncke H, Merzenich M: Learning and generalization of auditory temporal interval discrimination. Journal of Neuroscience: 17; 3956–3963, 1997.

Zakay D: On prospective time estimation, temporal relevance and temporal uncertainty. In Macar F, Pouthas V, Friedman WJ (Eds), Time, Action and Cognition. Dordrecht: Kluwer, pp. 109–117, 1992.

Zakay D, Nitzan D, Glicksohn J: The influence of task difficulty and external tempo on subjective time estimation. Perception and Psychophysics: 34; 451–456, 1983.

Handbook of Neuropsychology, 2nd Edition, Vol. 7
J. Grafman (Ed)

CHAPTER 10

Neural network models of prefrontal cortex and cognitive control

Jonathan D. Cohen *

*Department of Psychology, Center for the Study of Brain, Mind and Behavior, Princeton University, Green Hall,
Princeton, NJ 08544, USA*

Introduction

Human cognitive function is characterized by two equally remarkable features: the ability to make and carry out long-term plans, while at the same time preserving the capacity to flexibly and productively adapt to novel circumstances. These abilities have long suggested that the human brain is endowed with sophisticated mechanisms for executive function, or cognitive control. Understanding these mechanisms, however, has been an elusive scientific objective — both at the psychological and neural levels. Numerous theories at both levels of analysis have suggested that there are specialized systems that support this ability. This chapter will focus on one such system: the prefrontal cortex (PFC). However, it is critically important to first establish the appropriate context for such a consideration: it is highly unlikely that any one structure of the brain is solely responsible for executive function. The term 'executive function' itself conjures the image of a powerful gray-haired individual sitting at the top of the corporate hierarchy, in full control of all the affairs of a company. However, even this image of a corporate executive and her charge overestimates the scope of her role, failing to capture the complex reality and distributed nature of control in an actual company. In fact, one might view the CEO of a company more accurately

as a 'court of last resort'. While it is true that the CEO can exert influence just about anywhere in the company, it is certainly not the case that every decision and every element of control rests in her hands. Such 'micromanagement' would, in fact, be a disastrous corporate policy, as no one person has the time or the knowledge necessary to perform such a ubiquitous role. Rather, control is distributed, exercised at local levels by individuals with expertise in the relevant areas. The CEO is engaged only when problems arise that exceed local expertise, when conflicts occur between local experts, or when plans with wide-ranging or long-term consequences must be evaluated and put into effect.

In this chapter, we will consider the role that the PFC plays in control, which, in some respects, resembles that of the CEO of a company. As suggested by the review by Miller and Asaad, 2002, this volume), the PFC sits at the top of the processing hierarchy in the brain, receiving inputs and sending projections to virtually every other major processing area. Furthermore, neurobiological studies in non-human primates and neuropsychological and neuroimaging studies in humans provide strong support for the view that the PFC represents high level rules, or goals that guide behavioral performance, especially when this involves newly formed, or weak associations between stimuli and responses (see Miller and Cohen, 2001 for a review) — that is, when more specialized or dedicated systems within the brain cannot adequately support performance. In this chapter, we will consider the specific mecha-

* Tel.: +1 (609) 258-2696; Fax: +1 (609) 258-2549;
E-mail: jdc@princeton.edu

nisms by which the PFC carries out this function. However, once again, it is important to keep in mind that the PFC is not the only structure in the brain that can exhibit control. Indeed, the same basic mechanisms of control — or ones very similar — almost certainly operate locally in other structures. Thus, we should not think of these as necessarily unique to the PFC, nor should we think about the PFC as the only structure with the capacity for control. That said, there is a preponderance of evidence to support the idea that the PFC is highly specialized for its role in control. Indeed, this is the area of the brain that is most highly expanded in humans relative to other species, damage to this area is associated with impairments in characteristically human cognitive functions that are directly dependent upon control (such as planning, or the ability to respond adaptively in novel circumstances), and disturbances of this structure have been consistently implicated in neuropsychiatric diseases such as schizophrenia that also appear to be uniquely human. For these reasons, this chapter focuses on the PFC as a way of understanding the neural mechanisms that underlie cognitive control.

The idea that the PFC plays a critical role in cognitive control is not new. The very first neurologists, neuroscientists and neuropsychologists quickly recognized the importance of the frontal lobes in the control of behavior and higher cognitive function (e.g., Bianchi, 1922; Kraeplin, 1950; Luria, 1969). More recently, scientists have elaborated these ideas within the framework of modern cognitive psychological theory. For example, Shallice (1982) hypothesized that the PFC functions as a 'supervisory attentional system' (SAS), responsible for goal-oriented behavior, by intervening when automatic processes fail. This is most likely in novel task situations, when appropriate automatic behaviors are not available or are in conflict with one another, or when the task demands that these be overridden in favor of less habitual, but task-appropriate responses (e.g., naming the color of a Stroop stimulus). Similarly, Grafman (1994) has proposed that the PFC is responsible for representing action schemata or event markers, that serve to organize elements of behavior into complexes that have the appropriate temporal structure and are elicited under the appropriate environmental circumstances for specified tasks.

While these theories provide a good qualitative account of the role of the PFC in cognition, and in some cases relate this to specific computational mechanisms (e.g., Kimberg and Farah, 1993; Shallice and Burgess, 1996), they do not provide insights into how such mechanisms are implemented neurally. Recently, investigators have begun to use neural network models to better understand how the PFC may carry out its functions. Such models attempt to simulate the behavioral performance of human subjects (or animals) in cognitive tasks using neurally plausible processing mechanisms (e.g., the spread of activity among simple processing units along weighted connections), without necessarily incorporating all of the biophysical details that may be involved in real neuronal function. The goal of this effort is to identify principles of neural function that are most relevant to behavior. Using this approach, Braver, Cohen and Servan-Schreiber (1995), Cohen and Servan-Schreiber (1992), Dehaene and Changeux (1989, 1992) and Levine and Prueitt (1989) and have all described models of prefrontal function, and used these to simulate the performance of normal and frontally damaged patients in tasks that are sensitive to PFC damage, such as the Wisconsin Card Sort, Stroop task and others. These models capture many of the essential findings regarding behaviors associated with prefrontal function. Furthermore, they help isolate and identify fundamental principles of neural function that define the unique contribution that the PFC makes to cognitive control. At the same time, the limitations of these models reveal gaps in our understanding of the mechanisms that underlie PFC function and cognitive control more generally. These issues are the focus of this chapter.

Guided activation

All neural network models that address the function of the PFC simulate this as the activation of a set of units that can be thought of as representing the task-relevant 'rule'; that is, units whose activation leads to the production of a response other than the one most strongly associated with the stimulus. However, in most models, the PFC units themselves are not responsible for carrying out performance; rather, they influence the activity of other units whose responsibility this is. This is clearly illustrated by a model of the Stroop task developed by Cohen, Dunbar and McClelland (1990).

This model (see Fig. 1) is made up of five sets of units required for carrying out each of the two sub-tasks (color naming and word reading): two sets of input units representing each of the two types of stimulus features (e.g., the colors red and green, and the visual/orthographic features associated with the words RED and GREEN); a set of output units representing each potential response (e.g., the words 'red' and 'green' [1]); and two sets of intermediate units that provide a pathway between each set of input units and the output units. Thus, the connections between these various units define two processing pathways, one for reading the word and another for naming the color of the stimulus. Connections between units of different sets are excitatory (information flow), while connections within each set are inhibitory (competition for representation).[2] The connections along the word reading pathway are of greater strength, as a consequence of more extensive and more consistent training than for the naming of colors. The result is that when an incongruent Stroop stimulus is presented (such as the word 'GREEN' printed in red ink), information flowing along the word pathway dominates the competition at the response level, and the model responds to the word (see Fig. 1A). This captures the fact, noted earlier, that in the absence of instructions subjects read the word (i.e., say 'green'). Critical to our interests, however, is the fact that when subjects are instructed to do so, they can name the color (i.e., say 'red' to the very same stimulus).

In this model of the Stroop effect, the ability to elicit the weaker response requires the addition of a set of units (context units in Fig. 1), each of which is connected to intermediate units in one of the two

processing pathways. Activating one of these units biases processing in favor of the corresponding pathway by providing additional input to (i.e., 'priming') the intermediate units along that pathway, placing them within a more sensitive range of their processing function. In the case of the color pathway, this allows them to more effectively compete with and prevail over activity flowing along the stronger word pathway (see Fig. 1B). This biasing effect corresponds precisely to the role of top-down attentional control in the biased competition model proposed by Desimone and Duncan (1995). In effect, they establish the mapping between stimuli and responses required to perform the task. We assume that the task demand units in our model correspond to the function of neurons within the PFC. This model and closely related ones have been used to simulate the performance of both normal subjects and patients with frontal damage in a wide range of tasks that tap cognitive functions commonly associated with PFC function, such as working memory, attention, behavioral inhibition, planning and problem solving (e.g., Braver et al., 1995; Cohen and Servan-Schreiber, 1992; Dehaene and Changeux, 1992; Mozer, 1991; Phaf, Van der Heiden and Hudson, 1990; O'Reilly, Braver and Cohen, 1999; O'Reilly and Munakata, 2000).

Modulatory vs. transmissive function of the PFC

The Stroop model brings several features of the proposed function of the PFC into clear focus. First, it emphasizes the view that the role of the PFC in control is modulatory rather than transmissive. Much as the CEO of a company issues instructions that influence the actions of others, who actually carry them out, the PFC influences the flow of activity along pathways elsewhere in the brain that actually perform the task. For example, in the model, activating the color unit does not in itself transmit information about a particular response (red or green). Rather, it simply insures that activity flowing along the color naming pathway will have a greater influence over the response than activity flowing along the word pathway. As a more concrete analogy, the function of the PFC can be likened to that of a switch operator in a system of railroad tracks: we can think of the brain as a set of tracks (pathways) connecting various origins (e.g., stimuli) to destinations (responses).

---

[1] In the model depicted in Fig. 1, single processing units are used to represent combinations of features that make up each word. In reality, of course, it is assumed that these are represented by distributed patterns over multiple units. Simulations using such distributed representations exhibit all of the same behaviors with regard to attention and cognitive control as the simpler model depicted here, which was chosen for clarity of presentation.

[2] While this was not the precise arrangement in the original implementation of the model (Cohen et al., 1990), subsequent implementations of the model have confirmed that these assumptions produce similar results (e.g., Cohen and Huston, 1994).

A. No Control

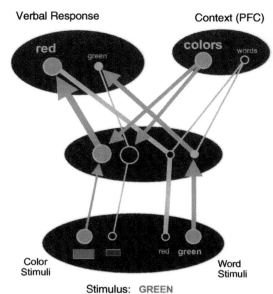

B. Control

Fig. 1. Model of the Stroop task (adapted from Cohen et al., 1990). Circles represent processing unit. Heavier lines connecting units designate stronger connection weights (e.g., along the word reading vs. the color naming pathway). More intensely colored units and connections designate flow of activity. Larger units designate increased sensitivity. (A) Response of the system to an incongruent Stroop stimulus in the absence of top-down control, showing greater flow of activity along the stronger word pathway, and a response to the word. (B) Activating the appropriate unit in the context layer provides top-down control, sensitizing units in the color naming pathway, and allowing the system to produce a color naming response to the incongruent stimulus.

The goal is to get the trains (activity carrying information) at each origin to their proper destination as efficiently as possible, avoiding any collisions. When the track is clear (i.e., a train can get from its origin to destination without risk of running into any others), then no intervention is needed (i.e., the behavior can be carried out automatically, and will not rely on the PFC). However, if two trains must cross the same bit of track, then some coordination is needed. The trains must be re-timed, re-directed, or in some cases one must be cancelled.[3] Thus, on this view, the role of the PFC is to establish the proper flow of activity along the pathways required to perform a particular task (or set of tasks), by biasing units along the appropriate pathways in more

posterior and/or subcortical areas. Representations within the PFC function as rules, goals, or schemata, insofar as the biasing effects of their activity set up the appropriate relationship between a stimulus (or category of stimuli) and an associated response (or set of responses), by guiding the flow of activity in other parts of the brain. Note that this function is not necessarily restricted to mappings from stimuli to responses, but applies equally well to mappings involving internal states (e.g., thoughts, memories, emotions, etc.), either as 'origins', 'destinations', or both. Note also that this function is not necessarily unique to the PFC. There may also be more local 'switch operators', responsive to regular and regionally specific patterns of traffic. However, at points of contact between different regions, or when new traffic patterns must be implemented, higher level control is required. This is presumably the contribution made by the PFC.

This distinction between modulation vs. transmission is consistent with the classic pattern of neu-

---

[3] This view is very similar to the multiple resources view that has been espoused in the literature on attention (cf. Navon and Gopher, 1979;). Here, we argue that the role of the PFC is to manage those circumstances under which multiple demands are placed on the same resources.

ropsychological deficits associated with frontal lobe damage. The components of a complex behavior are usually left intact, however the subject is not able to coordinate them in a task-appropriate way (for example, a patient who, when preparing coffee, first stirred and then added cream; Duncan, Emslie, Williams et al., 1996; Levine, Stuss, Milberg et al., 1998; Shallice, 1982). The notion that the function of the PFC is primarily modulatory also makes some interesting and testable predictions. For example, it predicts that in neuroimaging studies, while it should be possible to find circumstances that activate more posterior areas without activation of the PFC, it should be much less common to activate the PFC without their associated posterior structures. In other words, while it should be possible for transmission to occur without the need for modulation (e.g., word reading in the Stroop task), it does not make sense to have modulation in the absence of transmission.

Active maintenance in the service of control

The Stroop model also illustrates another critical feature of theories about the role of the PFC in executive function: the importance of active representation as a mechanism of control. For a representation to have a biasing influence, it must be activated. For this influence to endure (e.g., over the course of performing a task), it must be actively maintained. For example, to continue color naming, the activity of the color unit must maintained, or else the system will 'resort' to word reading, or at least be subject to undue interference from that pathway. This feature of the model provides a point of contact between theories of PFC function and cognitive psychological constructs such as controlled vs. automatic processing and working memory. The Stroop model was developed in an effort to specify the mechanisms underlying the relationship between controlled vs. automatic processing. The model suggested that this is in fact a continuum, defined by the relative strength of the pathway supporting a task-relevant process (i.e., the weight of the connection along that pathway) compared to those carrying competing information. Thus, weaker pathways (such as for color naming) rely more on top down support than stronger ones (word reading), especially when confronted with conflicting information from a stronger

competing pathway (cf. Simulation 5 of Cohen et al., 1990).

In most neural network models of the PFC, the activity of units used to simulate the PFC provide the necessary biasing influence to support the flow of activity along a desired pathway, comparable to the effects of the task demand units in the Stroop model. Accordingly, an increase in the demand for control requires greater or more enduring activation of the corresponding units in the PFC. This concurs with accumulating evidence from the neuroimaging literature that tasks thought to rely on controlled processing consistently engage the PFC (Baker, Rogers, Owen et al., 1996; Banich, Milham, Atchley et al., 2001; Cohen, Forman, Braver et al., 1994a; Cohen, Perlstein, Braver et al., 1997; Frith, Friston, Liddle and Frackowiak, 1991; MacDonald, Cohen, Stenger and Carter, 2000; Smith and Jonides, 1999). It is also consistent with both behavioral and neuroimaging findings concerning the effects of practice on automaticity and the involvement of the PFC. The Stroop model provides a mechanistic account of the long-standing observation that as a task becomes more practiced, its reliance on control is reduced; that is, it becomes automatic. In the model, this is because practice strengthens the connections along the corresponding pathway, and simulations capture detailed quantitative effects of practice both on measures of performance (e.g., power law improvements in speed of response) and concurrent changes in the reliance on control (e.g., Stroop interference). Reduced reliance on control means that less activity is needed in the task demand units to perform the task, or protect it from interference. From a neural perspective, as a pathway is repeatedly selected by the PFC bias signals, activity-dependent plasticity mechanisms can strengthen the pathway. Over time, these circuits can function independently of the PFC, and performance of the task becomes more automatic. This concurs with neuroimaging and neurophysiological observations that consistent practice on a task reduces the amount of PFC activity observed (Karni, Meyer, Jezzard et al., 1995; Knight, 1984, 1997; Petersen, van Mier, Fiez and Raichle, 1998; Shadmehr and Holcomb, 1997; Yamaguchi and Knight, 1991), and that PFC damage can impair new learning but spare performance on well-practiced tasks (Rushworth, Nixon, Eacott and Passingham, 1997).

This perspective also provides an interesting interpretation of the relationship between PFC function and working memory. Traditional theories of working memory have distinguished between storage and executive components (Baddeley, 1986), with the former responsible for maintaining information online (i.e., in an activated state), and the latter responsible for manipulating this information. Neuropsychological interpretations of this theory have placed the storage component in posterior structures, such as the parietal and temporal cortices (e.g., Gathercole, 1994), while the executive component has been almost universally associated with the PFC. However, as reviewed in Miller and Asaad (2002, this volume), neurophysiological studies have presented a very different view, consistently implicating the PFC in active maintenance (i.e., the storage function of working memory). Indeed, strong arguments have been made that this storage function of the PFC is necessary for working memory-dependent behaviors, and that the information stored within the PFC can be quite specific and detailed (e.g., Barone and Joseph, 1989; Funahashi, Bruce and Goldman-Rakic, 1989; Funahashi, Inoue and Kubota, 1997; Wilson, O Scalaidhe and Goldman-Rakic, 1993). This view seems to contrast with the more traditional neuropsychological view that the PFC subserves the executive but not the maintenance functions of working memory. Neural network models offer a possible resolution of this dilemma. They suggest that executive control involves the active maintenance of a particular type of information: representations that can influence processing needed for task performance. Thus, on this view, the PFC is responsible for active maintenance in the service of executive control. This view concurs with cognitive psychological theories based on production system architectures (e.g., ACT*; Anderson, 1983), which similarly posit that behaviors involving executive function require the active maintenance of goal representations in working memory.

This approach helps unify the role that the PFC plays in the variety of cognitive functions with which it has been associated, such as selective attention, behavioral inhibition, and executive function in working memory. These can all be seen as varying reflections, in behavior, of the operation of a single underlying mechanism of cognitive control: the biasing effects of representations in the PFC on processing in pathways responsible for task performance. As suggested by the biased competition model of Desimone and Duncan (1995), selective attention and behavioral inhibition may be two sides of the same coin: attention is the effect of biasing competition in favor of task-relevant information, and inhibition is the consequence that this has for the irrelevant information. Note that according to this view, inhibition occurs due to local competition between conflicting representations (e.g., between the two responses in the Stroop model), rather than centrally by the PFC. The 'binding' function of selective attention (e.g., Gelade and Treisman, 1980) can also be explained by such a mechanism, if we think of PFC representations in this case as selecting the desired combination of stimulus features (over the other competing combinations) to be mapped onto the response. Finally, we can think of the executive function of working memory as the active maintenance of PFC representations needed to bias competition among posterior pathways in favor of those that support performance of the desired task.

As we have emphasized repeatedly, it is important to recognize that the PFC may not be the only mechanism for cognitive control. Not only may similar mechanisms operate in other parts of the brain, but there may be other types of mechanisms critical to cognitive control. For example, the hippocampal complex is thought to house a system responsible for the rapid formation of new associations. This can be thought of as providing a form of 'weighted-based' control, which allows arbitrary representations to be rapidly bound together, and thereby influence processing in other structures. A full discussion of this 'binding' mechanism of control, and its relationship to theories of the PFC is beyond the scope of this chapter, but can be found elsewhere (e.g., Cohen and O'Reilly, 1996; O'Reilly and Munakata, 2000; O'Reilly et al., 1999). It is important to note here, however, that a weight-based 'binding' mechanism alone cannot account for the full flexibility of behavior, which requires an activity-based 'biasing' mechanism of the sort that we have ascribed to the PFC.

Other issues

The human capacity for cognitive control is notable for the wide range of tasks that it can support, and how quickly and flexibly it can adapt to the demands of new tasks. Insofar as this relies on representations within the PFC, it raises questions about the nature of these representations, and how they are dynamically updated in the service of control. Because of the importance of such issues to a complete theory of PFC function and cognitive control, several of these are reviewed below.

Representational capacity

If control occurs by the influence that activity patterns within the PFC have in guiding the flow of activity in other parts of the brain, then the PFC must house representations suitable for each task that relies on control. That is, the PFC must have units whose connectivity to other structures insures that the proper biasing effects will occur. For example, the Stroop model described above assumes that there are units corresponding to each of the two mappings (words and colors onto verbal responses) relevant to the two tasks. Neurophysiological studies reviewed in Miller and Asaad (2002, this volume) suggest that neurons within the PFC may fulfill such a function, exhibiting patterns of activity consistent with a role in establishing task-appropriate mappings (e.g., between the shape or color of a stimulus and the direction of an eye movement). Indeed, as the task shifts (e.g., the response to a particular stimulus), so too does the unit activated in the PFC (Rainer, Asaad and Miller, 1998). Recent neuroimaging studies provide convergent support, indicating that PFC activity occurs when behavior relies upon explicit knowledge about rules (e.g., in categorizing stimuli — Smith, Jonides, Marshuetz and Koeppe, 1998) or arbitrarily determined conjunctions of stimulus features (Prabhakaran, Narayanan, Zhao and Gabrieli, 2000).

However, the tremendous range of tasks of which people are capable, and that can be assumed to rely on cognitive control, raise important questions about the capacity of the PFC to support the necessary scope of representations. The large size of the PFC (approximately 30% of the cortical mass), coupled with its anatomic connectivity, suggest that it can support a wide number and range of mappings between stimuli, internal representations (such as memories and emotions), and responses, and therefore a

wide range of tasks. However, there is a potentially limitless number of such tasks that a person can be asked to perform (e.g., place your left thumb on your nose the next time you eat peanut butter). If we assume that each requires a separate, dedicated unit (or set of units) within the PFC, we quickly confront the problem of a combinatorial explosion. This demands at least one of two possible solutions: either the PFC harbors a set of representations that constitute a powerful task vocabulary, such that combinations of these representations can implement any task of which a person is capable; or, ongoing plasticity establishes new representations as they are needed. A combination of these factors is likely to provide the answer. The fact that the PFC undergoes the most extended period of development of any brain structure, which is paralleled by the development of cognitive control (both of which do not fully mature until late in adolescence), suggests that the PFC may have a sufficiently rich base of experience upon which to develop a sophisticated vocabulary of representations. At the same time, evidence has begun to accrue that PFC exhibits ongoing plasticity, with representations getting 'tuned' or new ones emerging as new mappings are required (Bichot, Schall and Thompson, 1996; Schultz and Dickinson, 2000). As yet, however, the mechanisms that govern either the initial development of representations within the PFC or their 'tuning', are largely unknown either at the neurobiological or computational levels (e.g., what learning algorithms apply). This could involve the modification of existing synapses, the formation of new ones (either within the PFC itself or elsewhere, such as the hippocampal complex), or even the recruitment of entirely new neurons (Gould, Reeves, Graziano and Gross, 1999). In any case, it seems assured that this will involve a combination of associative and reward-based learning rules that operate on connections among PFC units and between these and the units in posterior pathways that they influence.

Organizational scheme

A question that is closely related to the nature of representations within the PFC is how they are organized. Most areas of the brain show clear forms of topographic organization, according to principles that closely relate to their function. This suggests that information within the PFC may also be meaningfully organized, and that understanding the principles of this organization will provide insights into PFC function. Various schemes have been proposed. For example, one possibility is that the PFC is organized by function, with different regions carrying out qualitatively different operations. One longstanding view is that orbitofrontal areas are associated with behavioral inhibition, while ventrolateral and dorsal regions are associated with memory or attentional functions (Dias, Robbins and Roberts, 1997; Fuster, 1989; Goldman-Rakic, 1987). Another more recent suggestion is that ventral regions support maintenance of information (memory) while dorsal regions are responsible for the manipulation of such information (Owen, Evans and Petrides, 1996; Petrides, 1996). Such distinctions have heuristic appeal. However, neural network models suggest another intriguing possibility.

If different regions of the PFC represent different types of information, then perhaps variations in the biasing influence that they produce can account for apparent dissociations of function. For example, activity of orbital PFC is most frequently observed in tasks involving social, emotional, or appetitive stimuli (Hecaen and Albert, 1978; O'Doherty, Rolls, Francis et al., 2000; Price, 1999; Stuss and Benson, 1986; Swedo, Shapiro, Grady et al., 1989), while more dorsal regions are activated in tasks involving more 'cognitive' dimensions of stimuli, such as their form, location or sequential order. Social and appetitive stimuli are typically associated with greater response strength asymmetries than cognitive ones (for example, the relative strength of the urge to sneak a finger-full of icing off a recently baked cake and the more restrained but socially appropriate response of commenting on how nice it smells is likely to be much greater than the difference between word reading and color naming). Thus, the impression that orbital PFC subserves an inhibitory function may be explained by the fact that it is critical for biasing task-relevant processes (e.g. socially acceptable behaviors) against stronger competing alternatives (e.g., appetitive urges) than is typically so in the cognitive domain. Similarly, apparent functional dissociations between maintenance (ventral) vs. manipulation (dorsal) may reduce to the

differing representational demands that tasks have for sequential (dorsal) vs. non-sequential (ventral) information.

A variety of other schemes have also been proposed for the representational organization of the PFC. These include featural dimensions, sensory vs. motor, sequential vs. non-sequential, representational abstractness or complexity, and temporal extent over which representations must be maintained (e.g., Barone and Joseph, 1989; Casey, Forman, Franzen et al., 2001; Wilson et al., 1993), some of which have been subject to explicit computational modeling (e.g., O'Reilly, Noelle, Braver and Cohen, in press). While existing neural network models do not provide deep insights into which, if any, of these, is most likely to be correct, they do suggest important insights into this issue. First, they suggest that it is unlikely that different types of information will be represented in a modular, or discretely localized form. This is because the role ascribed to the PFC requires that it represent relationships across content domains as much as within them. While laboratory tasks, such as the Stroop task, demonstrate that subjects are capable of isolating different dimensions of a stimulus for processing, real world behavior more frequently requires that we recognize and respond to relationships across such dimensions.

Second, they suggest that learning will play an important role in the formation of representations in the PFC (cf. the capacity issue discussed above); this may have an important influence on representational organization. This has been suggested by neural network modeling efforts, which illustrate that training circumstances can have a profound influence on the extent to which representations that develop within the PFC are uni- or multi-modal. For example, Braver and Cohen (1995) constructed a network with two pathways for task performance, each of which had projections into a 'memory' layer used to represent the PFC. These projections were partially overlapping, so that some units within the PFC layer received projections from both pathways, while others received projections from only one of the pathways. The model was then trained on a simple delayed-response task, in which it had to learn the proper pattern of sustained activity within the PFC layer to associate stimuli with their corresponding responses after a delay. The model was trained in two ways: in a blocked condition, it was trained first on stimuli in one of the pathways, and only after it had learned this was it trained on stimuli in the other pathway. This resulted in a clear segregation of units within the PFC layer, with distinct pools representing stimuli in each of the two pathways. In a mixed condition, training was interleaved for stimuli in the two pathways. This resulted in the preferential selection of PFC units that received projections from both pathways; that is, there was a high incidence of 'multimodal' representations that developed to perform the task. These findings are consistent with empirical results, in which studies using blocked training have produced evidence of segregation in the representation of different types of information in the PFC (e.g., Wilson et al., 1993), while in studies using mixed training designs, or explicitly requiring the conjoint use of different types of information, multimodal units are observed within the PFC (Fuster, Bauer and Jervey, 1982; Rao, Rainer and Miller, 1997; White and Wise, 1999). These results suggest that not only may plasticity be important for providing task-relevant representations in the PFC, but also that the specific nature of experience may play a critical role in defining the organization of representations within the PFC.

Mechanisms of active maintenance

Virtually all modern theories of PFC function emphasize its capacity for active maintenance. Remarkably, however, there has been a dearth of empirical research on the mechanisms that are responsible for sustained activity. There are a number of theoretical possibilities for how this might occur, that can be roughly divided into two classes: *cellular* and *circuit-based*. Cellular models propose unit-bistability as the basis of sustained activity, which is dependent on the biophysical properties of individual cells. The transitions between states are triggered by inputs to the PFC, but maintained via the activation of specific voltage-dependent conductances (e.g., NMDA — Wang, 1999). Circuit-based models, on the other hand, propose that the recirculation of activity through closed (or 'recurrent') loops of interconnected neurons sustain activity. These loops could be intrinsic to the PFC (Melchitzky, Sesack, Pucak and Lewis, 1998; Pucak,

Levitt, Lund and Lewis, 1996) or they might involve other structures (such as the cortex–striatal–globus pallidus–thalamus–cortex loop; Alexander, Delong and Strick, 1986). Whether comprised of connections that are local to the PFC or involving other structures, systems of recurrently connected units — referred to in computational terms as attractor systems (see Fig. 2A) — have been subject to detailed analysis (e.g., Ermentrout and Cowan, 1979; Hopfield, 1982; Hopfield and Tank, 1986), and have begun to figure prominently in theoretical work on the mechanisms underlying PFC function (e.g., Braver et al., 1995; Dehaene and Changeux, 1989; Zipser, Kehoe, Littlewort and Fuster, 1993).

A better understanding of the mechanisms underlying active maintenance may also provide insight into one of the most striking and perplexing properties of cognitive control: its severely limited capacity. This has long been recognized in cognitive psychology (Broadbent, 1958; Posner and Snyder, 1975; Shiffrin and Schneider, 1977), and is painfully apparent to anyone who has tried to talk on the phone and read E-mail at the same time. We must be careful to distinguish this form of capacity — which has to do with how many representations can be maintained *at the same time* — from the notion of representational capacity referred to above, which has to do with the number of different representations that are *possible*. The resource limitation of cognitive control has played an explanatory role in many important models of human cognition (e.g., Cowan, 1998; Engle, Kane and Tuholski, 1999; Just and Carpenter, 1992; Posner and Snyder, 1975; Shiffrin and Schneider, 1977). However, to date, no theory has provided an explanation of the limitation *itself*. It is possible that this limitation reflects some inherent biological constraint, such as the energetic requirements of actively maintaining representations in the PFC. More likely, however, it reflects some fundamental, and potentially interesting computational properties of the system. For example, there may be some inherent limit in the number of representations that can be concurrently maintained within an attractor network (cf. Usher and Cohen, 1999). Alternatively, there may be an inescapable trade-off between the breadth of PFC extrinsic connectivity (required to effect control across the rest of the system) and the number of representations that can be

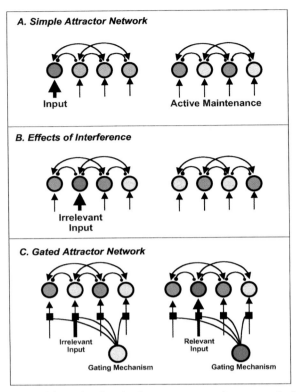

Fig. 2. Schematic of a simple attractor network, composed of four recurrently connected units. Excitatory connections are shown with arrows and inhibitory connections as lines terminating with filled circles. (A) External input to the leftmost unit excites the third unit from the left with which it shares an excitatory connection, and suppresses the other units. Recirculation of activity across mutually excitatory connections between the first and third units allows them to sustain their activity, and suppress the other units. This represents one attractor state of the system. (B) External input to the second unit allows this to compete with the first and third units, suppressing them, and recruiting the fourth unit, thus driving the system into another attractor state. (C) The addition of a gating unit can prevent the influence of irrelevant input, by modulating the input to the attractor network (modulatory connections shown by lines terminating in a filled square). When the gating unit is inactive no input reaches the attractor network (leftmost figure), whereas when the gating unit is active they can (rightmost figure).

active at the same time (without the entire system erupting in activity). In any event, resource limitations are a sine qua non of cognitive control, and therefore provide an important benchmark for theories that seek to explain its underlying mechanisms.

Dynamics of control

So far, we have considered only static forms of control, in which a task involves a single goal (e.g., name the color of the stimulus in the Stroop task), and for which a single representation within the PFC (e.g., the color unit) is sufficient to guide the flow of activity required for performance. In reality, however, control is highly dynamic. People move from one task to the next, and new goals replace old ones as they are fulfilled or more pressing ones arise. People also adaptively adjust the degree of control they exert as circumstances demand, for example becoming more cautious or allocating greater attention as a task becomes more difficult. Neural network models can account for these dynamic features of control — such as the sequencing of goals, or graded adjustments in the focus of attention — by assuming that there are mechanisms for appropriately updating or adjusting the strength of representations within the PFC, and monitoring the results. Indeed, such mechanisms are no doubt fundamental to the role that the PFC plays in higher cognitive functions such as decision making, strategic planning, and problem solving. The specific nature of these mechanisms are presently an important focus of research, as basic questions about their operation come into focus.

Updating of representations within the PFC

The mechanisms responsible for updating representations within the PFC must be able to satisfy two conflicting demands: on the one hand, they must be responsive to changes in the environment that demand the activation of new goals. Otherwise, behaviors associated with the currently activated goal will rigidly persist even when a new goal becomes more important or perhaps imperative; that is, failure to update will cause perseveration. On the other hand, the system must not be so responsive to the environment that every little change elicits an update in PFC activity. Otherwise, the system will fail to persist in achieving a desired goal; that is, it will be too distractible. Neurophysiological studies suggest that PFC representations are selectively responsive to task-relevant stimuli (Rainer et al., 1998), yet they are robust to interference from distractors (Miller, Erickson and Desimone, 1996). Furthermore, two hallmarks of damage to the PFC are increased distractiblity (inappropriate updating) and perseveration (inadequate updating). These observations point out a problem for attractor network models of the PFC: their state is strongly determined by their inputs. Thus, presentation of a new input will quickly drive such a system toward a different attractor state (Fig. 2B). This may be advantageous if the input signals the need for a shift in control (i.e., an updating of the representation in the PFC). However, if this is not the case, the input will disrupt control, and derail behavior. Although such networks can be configured to display resistance to disruption, this impairs their ability to be updated when it is appropriate. These observations suggest the operation of a mechanism that appropriately updates the state of the PFC in response to task demands.

Cohen, Braver and O'Reilly (1996) have proposed that dopamine may play such a role. This hypothesis integrates anatomical, physiological and behavior findings (Chiodo and Berger, 1986; Hernandez-Lopez, Bargas, Surmeier et al., 1997; Lewis, Hayes, Lund and Oeth, 1992; Luciana, Depue, Arbisi and Leon, 1992; Penit-Soria, Audinat and Crepel, 1987; Sawaguchi and Goldman-Rakic, 1991; Schultz, 1986 Schultz, Apicella and Ljungberg, 1993; Williams and Goldman-Rakic, 1993) with prior theoretical work regarding the modulatory functions of dopamine (Servan-Schreiber, 1990; Cohen and Servan-Schreiber, 1992), to suggest that dopamine release may influence the responsivity of units in the PFC to their afferent input, 'gating' access to and thereby regulating the updating of representations in the PFC (see Fig. 2C). Timing is a critical feature of such a gating mechanism: the signal to gate input must be rapid, and timed to appropriately coincide with conditions under which an update in control is adaptive. This is consistent with recent studies indicating that dopamine release — once thought to be slow and non-specific — has a phasic component with timing characteristics consistent with its proposed role in gating. Specifically, Schultz (1986; Schultz et al., 1993) have found that transient, stimulus-locked dopamine activity occurs in response to stimuli that are themselves not predicted, but that predict a later meaningful event. This is precisely the timing required for a gating signal responsible for updating goal representations. For example, imagine that you are walking to work,

when out of the corner of your eye you notice a $20 bill lying on the ground. This unpredicted stimulus predicts reward, but only if you update your current goal and bend down to pick up the bill.

The hypothesis that dopamine implements a gating function, controlling access to the PFC, may also address a critical question: How does the system learn what stimuli should elicit such a signal, and update the representations in the PFC. DA has long been associated with reward-based learning, and the pattern of release observed by Schultz has recently been interpreted by Montague and colleagues as playing a critical role in reinforcement learning and decision making behavior (Egelman, Person and Montague, 1998; Montague, Dayan and Sejnowski, 1996). In a series of sophisticated computational modeling and empirical studies, they have shown that phasic release of dopamine could function as a widely distributed error signal, driving the learning of temporal predictors of reinforcement. Intriguingly, the parameter used in these models to simulate the learning effects of DA is formally equivalent to the parameter that has been used in models simulating its modulatory effects on unit responsivity in the PFC (i.e., the gating function).

The dual and concurrent influences of DA on gating and learning suggest the intriguing possibility that phasic DA activity may be sufficient to produce cortical associations between the information being gated, and a triggering of the gating signal in the future. That is, if the system learns *while* it gates, then perhaps it can learn *when* to gate. Recent computational modeling studies establish the plausibility of this hypothesis. In a simulation model of a simple working memory task, Braver and Cohen (2001) have shown that the coincident effects of DA on gating and reinforcement learning are sufficient for the model to discover which stimuli must be represented in working memory, and when they must be updated to successfully perform the task. This ability to self-organize averts the problem of a theoretical regress regarding control (i.e., a 'homunculus'), by allowing the system to learn on its own what signals should produce an updating of the contents of the PFC and when this should occur.

One issue that remains relatively unexplored is the capacity to pursue a sequence of subgoals without disrupting a superordinate goal (in the example above, bending down to pick up the $20 bill does not waylay the superordinate plan to get to work). This suggests that it is possible to update some representations within the PFC (those associated with subgoals) while preserving others (those associated with superordinate goals). The mechanism for updating described above does not address this need: when an updating signal occurs, all representations within the PFC layer are subject to updating. Yet, the ability for selective updating is fundamental to virtually all higher cognitive faculties, such as search processes underlying problem solving (Newell and Simon, 1972), and the ability to generate and understand the nested structure of natural languages. One approach is to assume that updating can take a more graded form. Computational models exploring this possibility have met with some success in simulating behaviors (and their disturbance with frontal damage) typically thought to reflect hierarchical goal structures (Botvinick and Plaut, under review). Another interesting approach hypothesizes that the basal ganglia play a role in regulating which representations are updated within the PFC (Frank, Loughry and O'Reilly, 2000). This work builds on the longstanding observation of circuitry that connects the PFC with the basal ganglia through a series of parallel loops, with each loop involving different regions of the PFC (Alexander et al., 1986). Frank et al. (2000) constructed a neural network model that incorporates this architecture, and assumes that the gating effects of dopamine are actually exerted at the level of the basal ganglia. This allows different regions of the PFC to be updated at different times. They configured such a model to successfully perform a hierarchical working memory task, in which some representations had to be updated while others were maintained. A challenge for this model is now to be able to learn to perform such a task on its own, perhaps using the reinforcement learning mechanisms discussed above.

Monitoring and the allocation of control
The foregoing discussion has focused on the need to appropriately update representations in the PFC as new goals arise. However, as noted earlier, people also show a facility for adapting the *degree* of control they allocate to a task as the demands that it places on performance change. For example, you pay closer attention to the road on a dark and rainy night than on a bright sunny day. Strategic adjustments in control have also been observed in the laboratory under a variety of experimental conditions: subjects slow down on the trial following an error, and increase the focus of attention when confronted with distractors (e.g., Botvinick, Nystrom, Fissell et al., 1999; Gratton, Coles and Donchin, 1992; Laming, 1968; Logan, Zbrodoff and Fostey, 1983; Tzelgov, Henik and Berger, 1992). Such adjustments are adaptive in view of the well-recognized capacity limits on cognitive control discussed above. If the capacity to allocate control is limited, then it is best to allocate only the amount needed to perform a task acceptably, and reserve the remainder for tasks that meet other needs. Thus, while driving on a rainy night requires your full attention, it might be better spent talking to a companion when driving conditions permit.

These phenomena can be accommodated by assuming that the degree of control allocated to a task corresponds to the level of activity of its representation in the PFC. This is clearly illustrated by the Stroop model. To achieve comparable levels of performance, color naming demands significantly greater activation of the color unit than word reading does of the word unit (cf. Simulation 5 of Cohen et al., 1990). If we assume that competition occurs among representations in the PFC, as it does in other parts of the brain, then as a PFC representation becomes more activated it will compete more with others in the PFC. Consequently, as one task increases its demands for control (i.e., requiring a more strongly activated PFC representation), the less will be available for others.

Presumably, the system has adapted to give each task the minimum amount of control (i.e., level of PFC activity) needed to support acceptable performance. However, this may vary across circumstances (e.g. from neutral to incongruent trials in the Stroop task), and the capacity to make strategic allocations of control depends on the ability to monitor performance, and determine when control is needed and when it can be diminished or withdrawn. Despite the obvious nature of this inference, and its early recognition in cognitive psychology (Kahneman, 1973), astoundingly little research has addressed the mechanisms responsible for such monitoring, at either the psychological or neurobiological levels. Recently, however, a series of human electrophysiological, neuroimaging, and modeling studies have begun to address this topic. These suggest that certain regions of the anterior cingulate cortex, a midline frontal structure, may play a critical role in monitoring processing. This was first suggested by the discovery of the error-related negativity (ERN), a scalp-recorded event related potential (ERP) that is closely associated with the commission of errors and appears to emanate from the anterior cingulate cortex (Falkenstein, Hohnsbein, Hoormann and Blanke, 1991; Gehring, Goss, Coles et al., 1993).

More recently, functional neuroimaging studies have confirmed error-related activation of the anterior cingulate cortex (Carter, Braver, Barch et al., 1998). Interestingly, such activity is also observed under conditions of response conflict, even when the subject has not committed any errors (e.g., Botvinick et al., 1999; Carter, Macdonald, Botvinick et al., 2000). Modeling work has demonstrated that such conflict is closely associated with errors (Botvinick, Braver, Carter et al., 2001; Yeung, Botvinick and Cohen, under review). In these models, conflict is operationalized as the computational energy of the network. The energy of a neural network corresponds to the compatibility of activated representations, and the stability of its current state (Hopfield, 1982). Higher energy (or conflict) is associated with processing uncertainty, and therefore a higher likelihood of errors. Thus, monitoring conflict can be used to mark the impending likelihood of errors, without requiring the system to wait for an actual error to occur. More generally, conflict can be used as an index of the need for control. As we have seen above, a primary function of control is to prevent or reduce conflict (e.g., to prevent interference of the word while naming the color of a Stroop stimulus). Thus, the presence of conflict serves as a indication of the need for the additional allocation of control (e.g., control is needed when naming the color of an incongruent Stroop stimulus; see Fig. 1). Modeling work has shown that conflict monitoring can account for the diverse circumstances under which activity has been observed in certain areas of anterior cingulate cortex in neuroimaging studies (Botvinick et al., 2001), as well as many features of the ERN (Yeung et al., under review). Furthermore, this work

has shown that by coupling conflict monitoring to adjustments in the allocation of control (e.g., allowing the presence of conflict to amplify the pattern of activity within the PFC; see Fig. 3), the model can simulate empirical data concerning adjustments in control that human subjects make in response to conflict (Botvinick et al., 2001).

The distinct roles of the PFC in the execution of cognitive control and anterior cingulate cortex in conflict monitoring are illustrated in a recent fMRI study by MacDonald et al. (2000), using an instructed version of the Stroop task (see Fig. 4). In each trial, subjects were given a cue indicating whether they were to name the color or read the word in the subsequent display. The cue was followed by a delay of several seconds, and then either a congruent or an incongruent Stroop stimulus was displayed. Fig. 4 shows that, during the delay, increasing activity was observed within a region of dorsolateral PFC, greater for color naming (the more demanding task) than word reading. There was no differential activation observed within the anterior cingulate cortex during this period. In contrast, strong activation was observed in the anterior cingulate cortex during the period of stimulus presentation and responding, and this was significantly greater for incongruent than congruent stimuli. There was no differential response within the PFC during this period.

These findings provide strong support for several of the hypotheses we have discussed: that the demands for control are associated with activity in the PFC; that tasks demanding greater control elicit stronger activity within the PFC; and that regions of the anterior cingulate cortex respond selectively to conflict in processing. These findings also corroborate earlier indications (Botvinick et al., 1999; Carter et al., 2000) that while the anterior cingulate cortex responds to conflict in processing, it does not appear to play a direct role in the allocation of control. If this is so, it raises the intriguing question of how information about conflict is translated into adjustments of control. One possibility is that this is mediated by brainstem neuromodulatory systems, such as the locus coeruleus (see Cohen, Botvinick and Carter, 2000). This structure is responsible for almost all of the norepinephrine released in the brain, and has long been recognized as playing a central role in arousal. However, recent studies of this system (Aston-Jones,

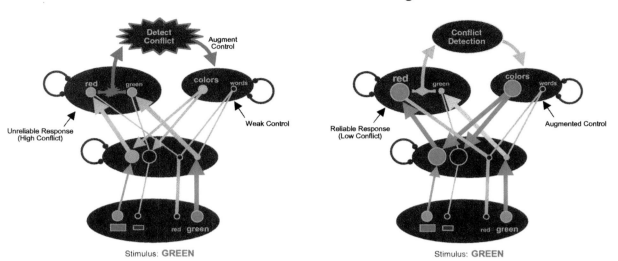

Fig. 3. The operation of a conflict detection mechanism. (A) Weak control (e.g., due to comparable activity of word and color units in context layer), in the presence of an incongruent Stroop stimulus, leads to co-activation of competing units in the output layer of the network. Because units within each layer are mutually inhibitory (designated by looped connections with filled circles), this produces conflict, activating the conflict detection mechanism. This produces a signal to augment control in the context layer. (B) The effects of augmenting control (increasing the activity of the color unit in the context layer) is to reduce conflict in the output layer.

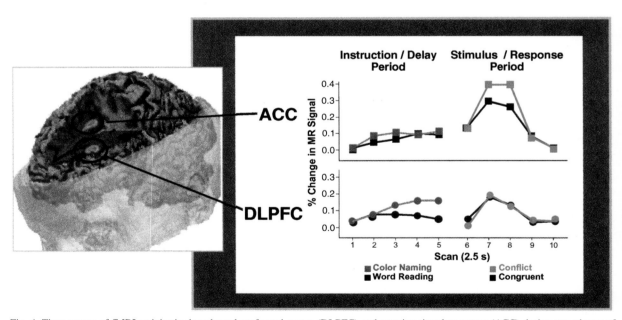

Fig. 4. Time course of fMRI activity in dorsolateral prefrontal cortex (DLPFC) and anterior cingulate cortex (ACC) during two phases of a trial in the instructed Stroop task used by MacDonald et al. (2000). During the instruction and preparatory period, there is significantly greater activation of DLPFC for color naming than word reading, but no difference in ACC. During the stimulus and response phase, there is greater activation of ACC for conflict than congruent stimuli, but no difference between these for DLPFC. (Taken from Miller and Cohen, 2001.)

Rajkowski, Kubiak and Alexinsky, 1994; Usher, Cohen, Servan-Schreiber et al., 1999) have suggested that the locus coeruleus may play a more specific role in modulating control (e.g., task preparedness, and the focus of attention), and there is now evidence of an important anatomic projection from anterior cingulate cortex to locus coeruleus (Rajnowski, Lu, Zhu et al., 2000). The functional significance of such a link, and its role in cognitive control, remains largely conjectural. However, this possibility echoes similar conjectures regarding of the role of dopamine in cognitive control (as discussed above), suggesting that these brainstem systems — long thought to play non-specific roles in mentation — may in fact be centrally involved in the fine-grained regulation of cognitive control.

Prospective control and planning

Perhaps the most distinguishing characteristic of human cognition is its ability to plan for the future. That is, when a worthwhile goal cannot be productively pursued at the moment, we can chose to do so later, at some more appropriate time. Such deferral can span remarkable durations of time, such as plans to pursue a career, or for retirement. The mechanisms of active maintenance that we have discussed so far do not provide a satisfying account of this ability. For example, when I plan in the morning to go to the grocery store on the way home from work, it seems unlikely that I actively maintain a representation of this goal in the PFC throughout the day. Rather, it seems to rely on some other type of storage, with re-activation of the goal at the appropriate time. This may involve interactions between the PFC and other brain systems capable of rapid learning, such as the hippocampal complex (cf. Cohen and O'Reilly, 1996). Thus, it is possible that the hippocampal complex rapidly encodes an association between the desired goal representation(s) within the PFC (e.g., go to the grocery store) and features of the circumstance under which the goal should be evoked (e.g., setting sun, clock striking 5). Then, as these circumstances arise, the appropriate representation within the PFC is associatively activated, guiding subsequent performance in accordance with the goal (e.g., turning right at the light rather than the habitual left toward home). However, the detailed nature of such interactions, and their relationship to the dopaminergic gating and learning mechanisms described above, remain to be fully specified.

Summary and conclusion

One of the great mysteries of the brain is how purposeful, goal-directed behavior emerges from the millions of relatively simple processing units that are its basic computational elements. This chapter summarized work over the past decade using neural network models, which suggests that this capacity stems from the biasing influence that activity of units within the PFC has on processing in other parts of the brain, guiding the flow of activity along pathways that can support the performance of an intended task. This function is supported by the extensive connections that the PFC has to the rest of the brain, its ability to actively maintain and appropriately update representations, and the ability of its circuitry to be modified by experience.

The neural network models we have considered provide a framework within which to specify explicit hypotheses about the mechanisms underlying the role of the PFC in cognitive control. We have reviewed a number of these, some of which have begun to take explicit form in simulation models. We have also provided a sampling of the many questions that remain about these mechanisms and the functioning of the PFC. Most importantly, perhaps, the models we have considered illustrate how neurally plausible mechanisms can exhibit the properties of self-organization and self-regulation required to account for cognitive control, without recourse to unexplained sources of intelligence (e.g., an 'homunculus'). Even if the particular mechanisms described turn out to be incorrect, at the very least they provide examples of how the use of neural network models, and the effort to be mechanistically explicit, can provide valuable guidance in this conceptually demanding pursuit. The human brain is arguably the most complex device in the known universe, and its capacity for higher cognitive function remains one of its deepest mysteries. It seems highly unlikely that significant progress in understanding how it operates can be made without an ever tighter coupling of behavioral and neurobiological experiments with detailed computational analysis and modeling.

Acknowledgements

The author would like to acknowledge the generous support of the NIMH and NARSAD, as well as B.J. Casey, Todd Braver, David Noelle and Randy O'Reilly, whose collaborations have had a profound influence over, and represent a collective source of knowledge and theoretical ideas presented in this review.

References

Alexander GE, Delong MR, Strick PL: Parallel organization of functionally segregated circuits linking basal ganglia and cortex. Annual Review of Neuroscience: 9; 357–381, 1986.

Allport DA, Tipper SP, Chmiel NRJ: Perceptual integration and postcategorical filtering. In Posner MI, Marin OCM (Eds), Attention and Performance, Vol. XI. Hillsdale, NJ: Lawrence Erlbaum, 1985.

Anderson JR: The Architecture of Cognition. Cambridge, MA: Harvard University Press, 1983.

Aston-Jones G, Rajkowski J, Kubiak P, Alexinsky T: Locus coeruleus neurons in monkey are selectively activated by attended cues in a vigilance task. Journal of Neuroscience: 14(7); 4467–4480, 1994.

Baddeley A: Working Memory. Oxford: Clarendon Press, 1986.

Baker SC, Rogers RD, Owen AM, Frith CD, Dolan R et al.: Neural systems engaged by planning: a PET study of the Tower of London Task. Neuropsychologia: 34; 515–526, 1996.

Banich MT, Milham MP, Atchley R, Cohen NJ, Webb A et al.: Prefrontal regions play a predominant role in imposing an attentional 'set': evidence from fMRI. Cognitive Brain Research, in press, 2001.

Barone P, Joseph JP: Prefrontal cortex and spatial sequencing in macaque monkey. Experimental Brain Research: 78; 447–464, 1989.

Bianchi L: The Mechanism of the Brain and the Function of the Frontal Lobes. Edinburgh: Livingstone, 1922.

Bichot NP, Schall JD, Thompson KG: Visual feature selectivity in frontal eye fields induced by experience in mature macaques. Nature: 381; 697–699, 1996.

Botvinick MM, Braver TS, Carter CS, Barch DM, Cohen JD: Conflict monitoring and cognitive control. Psychological Review, in press, 2001.

Botvinick M, Nystrom LE, Fissell K, Carter CS, Cohen JD: Conflict monitoring versus selection-for-action in anterior cingulate cortex. Nature: 402; 179–181, 1999.

Botvinick MM, Plaut DC: Doing without schema hierarchies: A recurrent connectionist approach to routine sequential action and its pathologies, under review.

Braver TS, Cohen JD: Organization of working memory representations depends upon the nature of training. Proceedings of the Cognitive Neuroscience Society: 2; 95, 1995.

Braver TS, Cohen JD: On the control of control: The role of dopamine in regulating prefrontal function and working mem-

ory. In Monsell S, Driver J (Eds), Attention and Performance XVIII; Control of Cognitive Processes, in press, 2001.

Braver TS, Cohen JD, Servan-Schreiber D: A computational model of prefrontal cortex function. In Touretzky DS, Tesauro G, Leen TK (Eds), Advances in Neural Information Processing Systems. Cambridge, MA: MIT Press, pp. 141–148, 1995.

Broadbent DE: Perception and Communication. London: Pergamon, 1958.

Carter CS, Braver TS, Barch DM, Botvinick MM, Noll D, Cohen JD: Anterior cingulate cortex, error detection, and the online monitoring of performance. Science: 280; 747–749, 1998.

Carter CS, Macdonald AM, Botvinick M, Ross LL, Stenger VA et al.: Parsing executive processes: strategic vs. evaluative functions of the anterior cingulate cortex. Proceedings of the National Academy of Sciences of the United States of America: 97; 1944–1948, 2000.

Casey BJ, Forman SD, Franzen P, Berkowitz A, Braver TS et al.: Sensitivity of prefrontal cortex to changes in target probability: a functional MRI study. Human Brain Mapping, in press, 2001.

Chiodo L, Berger T: Interactions between dopamine and amino-acid induced excitation and inhibition in the striatum. Brain Research: 375; 198–203, 1986.

Cohen AH, Holmes PJ, Rand RH: The nature of coupling between segmental oscillators of the lamprey spinal generator for locomotion: A model. Journal of Mathematical Biology: 13; 345–369, 1982.

Cohen AH, Rand RH, Holmes PJ: Systems of coupled oscillators as models of central pattern generators. In Cohen AH, Rossignol S, Grillner S (Eds), Neural Control of Rhythmic Movements in Vertebrates. New York: Wiley, pp. 333–367, 1988.

Cohen JD: Special issue: functional topography of prefrontal cortex. Neuroimage: 11; 378–379, 2000.

Cohen JD, Barch DM, Carter CS, Servan-Schreiber D: Schizophrenic deficits in the processing of context: converging evidence from three theoretically motivated cognitive tasks. Journal of Abnormal Psychology: 108; 120–133, 1999.

Cohen JD, Botvinick MM, Carter CS: Anterior cingulate and prefrontal cortex: who's in control? Nature Neuroscience: 3(5); 421–423, 2000.

Cohen JD, Braver TS, O'Reilly RC: A computational approach to prefrontal cortex, cognitive control, and schizophrenia: Recent developments and current challenges. Philosophical Transactions of the Royal Society of London Series B: Biological Sciences: 351; 1515–1527, 1996.

Cohen JD, Dunbar K, McClelland JL: On the control of automatic processes: a parallel distributed processing account of the Stroop effect. Psychological Reviews: 97; 332–361, 1990.

Cohen JD, Forman SD, Braver TS, Casey BJ, Servan-Schreiber D, Noll DC: Activation of prefrontal cortex in a non-spatial working memory task with functional MRI. Human Brain Mapping: 1; 293–304, 1994a.

Cohen JD, Huston TA: Progress in the use of interactive models for understanding attention and performance. In Umiltá C, Moccovitch M (Eds), Attention and Performance, Vol. XV. Cambridge, MA: MIT Press, 1994.

Cohen JD, O'Reilly RC: A preliminary theory of the interactions between the prefrontal cortex and hippocampus that contribute to planning and prospective memory. In Brandimonte M, Einstein G, McDaniel M (Eds), Prospective Memory: Theory and Applications. Hillsdale, NJ: Erlbaum, 1996.

Cohen JD, Perlstein WM, Braver TS, Nystrom LE, Noll DC et al.: Temporal dynamics of brain activation during a working memory task. Nature: 386; 604–608, 1997.

Cohen JD, Romero RD, Farah MJ, Servan-Schreiber D: Mechanisms of spatial attention: the relation of macrostructure to microstructure in parietal neglect. Journal of Cognitive Neuroscience: 6; 377–387, 1994b.

Cohen JD, Servan-Schreiber D: Context, cortex and dopamine: a connectionist approach to behavior and biology in schizophrenia. Psychological Reviews: 99; 45–77, 1992.

Cohen JD, Servan-Schreiber D: A theory of dopamine function and its role in cognitive deficits in schizophrenia. Schizophrenia Bulletin: 19; 85–104, 1993.

Cohen JD, Servan-Schreiber D, McClelland JL: A parallel distributed processing approach to automaticity. American Journal of Psychology: 105; 239–269, 1992.

Cowan N: Evolving conceptions of memory storage, selective attention, and their mutual constraints within the human information processing system. Psychological Bulletin: 104; 163–191, 1998.

Dehaene S, Changeux JP: A simple model of prefrontal cortex function in delayed-response tasks. Journal of Cognitive Neuroscience: 1; 244–261, 1989.

Dehaene S, Changeux JP: The Wisconsin card sorting test: theoretical analysis and modeling in a neuronal network. Cerebral Cortex: 1; 62–79, 1992.

Desimone R, Duncan J: Neural mechanisms of selective visual attention. Annual Review of Neuroscience: 18; 193–222, 1995.

Dias R, Robbins TW, Roberts AC: Dissociable forms of inhibitory control within prefrontal cortex with an analog of the Wisconsin Card Sort Test: restriction to novel situations and independence from 'on-line' processing. Journal of Neuroscience: 17; 9285–9297, 1997.

Duncan J, Emslie H, Williams P, Johnson R, Freer C: Intelligence and the frontal lobe: the organization of goal-directed behavior. Cognitive Psychology: 30; 257–303, 1996.

Egelman DM, Person C, Montague PR: A computational role of dopamine delivery in human decision-making. Journal of Cognitive Neuroscience: 10; 623–630, 1998.

Engle RW, Kane M, Tuholski S: Individual differences in working memory capacity and what they tell us about controlled attention, general fluid intelligence, and functions of the prefrontal cortex. In Miyake A, Shah P (Eds), Mechanisms of Active Maintenance and Executive Control. New York: Cambridge University Press, 1999.

Ermentrout GB, Cowan JD: Temporal oscillations in neuronal nets. Journal of Mathematical Biology: 7(3); 265–280, 1979.

Falkenstein M, Hohnsbein J, Hoormann J, Blanke L: Effects of crossmodal divided attention on late ERP components. II. Error processing in choice reaction tasks. Electroencephalographic and Clinical Neurophysiology: 78; 447–455, 1991.

Frank MJ, Loughry B, O'Reilly RC: Interactions between frontal cortex and basal ganglia in working memory: A computational model. Cognitive, Affective and Behavioral Neuroscience: 1; 137–160, 2000.

Frith CD, Friston K, Liddle PF, Frackowiak RSJ: Willed action and the prefrontal cortex in man: a study with PET. Proceedings of the Royal Society of London Series B: 244; 241–246, 1991.

Funahashi S, Bruce CJ, Goldman-Rakic PS: Mnemonic coding of visual space in the monkey's dorsolateral prefrontal cortex. Journal of Neurophysiology: 61; 331–349, 1989.

Funahashi S, Inoue M, Kubota K: Delay-period activity in the primate prefrontal cortex encoding multiple spatial positions and their order of presentation. Behavioural Brain Research: 84; 203–223, 1997.

Fuster JM: The Prefrontal Cortex, Vol 2. New York: Raven, 1989.

Fuster JM, Bauer RH, Jervey JP: Cellular discharge in the dorsolateral prefrontal cortex of the monkey in cognitive tasks. Experimental Neurology: 77; 679–694, 1982.

Gathercole SE: Neuropsychology and working memory: a review. Neuropsychology: 8; 494–505, 1994.

Gehring WJ, Goss B, Coles MGH, Meyer DE, Donchin E: A neural system for error detection and compensation. Psychological Science: 4; 385–390, 1993.

Gelade G, Treisman A: A feature-integration theory of attention. Cognitive Psychology: 12; 97–136, 1980.

Goldman-Rakic PS: Circuitry of primate prefrontal cortex and regulation of behavior by representational memory. In Plum F (Ed), Handbook of Physiology: The Nervous System. Bethesda, MD: American Physiological Society, pp. 373–417, 1987.

Gould E, Reeves AJ, Graziano MS, Gross CG: Neurogenesis in the neocortex of adult primates. Science: 286; 548–552, 1999.

Grafman J: Alternative frameworks for the conceptualization of prefrontal functions. In Boller F, Grafman J (Ed), Handbook of Neuropsychology. Amsterdam: Elsevier, pp. 187, 1994.

Gratton G, Coles MGH, Donchin E: Optimizing the use of information: strategic control of activation of responses. Journal of Experimental Psychology: 121; 480–506, 1992.

Hecaen H, Albert ML: Human Neuropsychology. New York: Wiley, 1978.

Hernandez-Lopez S, Bargas J, Surmeier DJ, Reyes A, Galarraga E: D1 receptor activation enhances evoked discharge in neostriatal medium spiny neurons by modulating an L-type Ca^{2+} conductance. Journal of Neuroscience: 17; 3334–3342, 1997.

Hopfield JJ: Neural networks and physical systems with emergent collective computational abilities. Proceedings of the National Academy of Science of the United States of America: 79; 2554–2558, 1982.

Hopfield JJ, Tank DW: Computing with neural circuits: a model. Science: 233; 625–633, 1986.

Just MA, Carpenter PA: A capacity theory of comprehension: individual differences in working memory. Psychological Reviews: 99; 122–149, 1992.

Kahneman D: Attention and Effort. Englewood Cliffs, NJ: Prentice Hall, 1973.

Karni A, Meyer G, Jezzard P, Adams MM, Turner R, Ungerleider LG: Functional MRI evidence for adult motor cortex plasticity during motor skill learning. Nature: 377; 155–158, 1995.

Kimberg DY, Farah MJ: A unified account of cognitive impairments following frontal lobe damage: The role of working memory in complex organized behavior. Journal of Experimental Psychology: 122; 411–428, 1993.

Knight RT: Decreased response to novel stimuli after prefrontal lesions in man. Clinical Neurophysiology: 59; 9–20, 1984.

Knight RT: Distributed cortical network for visual attention. Journal of Cognitive Neuroscience: 9; 75–91, 1997.

Kraeplin E: Dementia praecox and paraphrenia. New York: International Universities Press, Inc., 1950.

Laming DRJ: Information Theory of Choice–Reaction Times. London: Academic, 1968.

Lewis DA, Hayes TL, Lund JS, Oeth KM: Dopamine and the neural circuitry of primate prefrontal cortex: Implications for schizophrenia research. Neuropsychopharmacology: 6; 127–134, 1992.

Levine B, Stuss DT, Milberg WP, Alexander MP, Schwartz M, Macdonald R: The effects of focal and diffuse brain damage on strategy application: evidence from focal lesions, traumatic brain injury and normal aging. Journal of the International Neuropsychological Society: 4; 247–264, 1998.

Levine DS, Prueitt PS: Modeling some effects of frontal lobe damage-novelty and perseveration. Neural Networks: 2; 103–116, 1989.

Logan GD, Zbrodoff NJ, Fostey ARW: Costs and benefits of strategy construction in a speeded discrimination task. Memory and Cognition: 11; 485–493, 1983.

Luciana M, Depue RA, Arbisi P, Leon A: Facilitation of working memory in humans by a D2 dopamine receptor agonist. Journal of Cognitive Neuroscience: 4; 58–68, 1992.

Luria AR: Frontal lobe syndromes. In Vinken PJ, Bruyn GW (Eds), Handbook of Clinical Neurology, Vol. 9. New York: Elsevier, pp. 725–757, 1969.

MacDonald AW, Cohen JD, Stenger VA, Carter CS: Dissociating the role of dorsolateral prefrontal cortex and anterior cingulate cortex in cognitive control. Science: 288; 1835–1838, 2000.

Melchitzky DS, Sesack SR, Pucak ML, Lewis DA: Synaptic targets of pyramidal neurons providing intrinsic horizontal connections in monkey prefrontal cortex. Journal of Comparative Neurology: 390; 211–224, 1998.

Miller EK, Asaad WF, The prefrontal cortex: conjunction and cognition. In Grafman J (Ed), The Frontal Lobes. Handbook of Neuropsychology, Vol 7, 2nd ed. Amsterdam: Elsevier, pp. 29–54, 2002.

Miller EK, Cohen, JD: An integrative theory of prefrontal cortex function. Annual Review of Neuroscience, in press, 2001.

Miller EK, Erickson CA, Desimone R: Neural mechanisms of visual working memory in prefrontal cortex of the macaque. Journal of Neuroscience: 16; 5154–5167, 1996.

Montague PR, Dayan P, Sejnowski TJ: A framework for mesencephalic dopamine systems based on predictive Hebbian learning. Journal of Neuroscience: 16; 1936–1947, 1996.

Mozer MC: The Perception of Multiple Objects: A Connectionist Approach. Cambridge, MA: MIT Press, 1991.

Navon D, Gopher D: On the economy of the human processing system. Psychological Review: 86; 214–255, 1979.

Newell A, Simon HA: Human Problem Solving. Englewood Cliffs, NJ: Prentice Hall, 1972.

O'Doherty J, Rolls ET, Francis S, Bowtell R, McGlone F et al.: Sensory-specific satiety-related olfactory activation of the human orbitofrontal cortex. NeuroReport: 11; 893–897, 2000.

O'Reilly RC, Braver TS, Cohen JD: A biologically-based computational model of working memory. In Miyake A, Shah P (Ed), Models of Working Memory: Mechanisms of Active Maintenance and Executive Control. New York: Cambridge University Press, 1999.

O'Reilly RC, Munakata Y: Computational Explorations in Cognitive Neuroscience: Understanding the Mind. Cambridge: MIT Press, 2000.

O'Reilly RC, Noelle DC, Braver TS, Cohen JD: Prefrontal cortex and dynamic categorization tasks: Representational organization and neuromodulatory control. Cortex: in press.

Owen AM, Evans AC, Petrides M: Evidence for a two-stage model of spatial working memory processing within the lateral frontal cortex: a positron emission tomography study. Cerebral Cortex: 6; 31–38, 1996.

Penit-Soria J, Audinat E, Crepel F: Excitation of rat prefrontal cortical neurons by dopamine: An in vitro electrophysiological study. Brain Research: 425; 263–274, 1987.

Petersen SE, van Mier H, Fiez JA, Raichle ME: The effects of practice on the functional anatomy of task performance. Proceedings of the National Academy of Sciences of the United States of America: 95; 853–860, 1998.

Petrides M: Specialized systems for the processing of mnemonic information within the primate frontal cortex. Philosophical Transactions of the Royal Society of London Series B: Biological Sciences: 351; 1455–1461, 1996.

Phaf RH, Van der Heiden AHC, Hudson PTW: SLAM: a connectionist model for attention in visual selection tasks. Cognitive Psychology: 22; 273–341, 1990.

Posner MI, Snyder CRR: Attention and cognitive control. In Solso RL (Ed), Information Processing and Cognition. Hillsdale, NJ: Erlbaum, 1975.

Prabhakaran V, Narayanan K, Zhao Z, Gabrieli JD: Integration of diverse information in working memory within the frontal lobe. Nature Neuroscience: 3; 85–90, 2000.

Price JL: Prefrontal cortical networks related to visceral function and mood. Annals of the New York Academy of Science: 877; 383–396, 1999.

Pucak ML, Levitt JB, Lund JS, Lewis DA: Patterns of intrinsic and associational circuitry in monkey prefrontal cortex. Journal of Comparative Neurology: 376; 614–630, 1996.

Rainer G, Asaad WF, Miller EK: Selective representation of relevant information by neurons in the primate prefrontal cortex. Nature: 393; 577–579, 1998.

Rajnowski J, Lu W, Zhu Y, Cohen JD, Aston-Jones G: Prominent projections from the anterior cingulate cortex to the locus coeruleus in rhesus monkey. Society for Neuroscience Abstracts: 26; 2230, 2000.

Rao SC, Rainer G, Miller EK: Integration of what and where in the primate prefrontal cortex. Science: 276; 821–824, 1997.

Rushworth MF, Nixon PD, Eacott MJ, Passingham RE: Ventral prefrontal cortex is not essential for working memory. Journal of Neuroscience: 17; 4829–4838, 1997.

Sawaguchi T, Goldman-Rakic PS: D1 dopamine receptors in prefrontal cortex: Involvement in working memory. Science: 251; 947–950, 1991.

Schultz W: Responses of midbrain dopamine neurons to behavioral trigger stimuli in the monkey. Journal of Neurophysiology: 56; 1439–1461, 1986.

Schultz W, Apicella P, Ljungberg T: Responses of monkey dopamine neurons to reward and conditioned stimuli during successive steps of learning a delayed response task. Journal of Neuroscience: 13; 900–913, 1993.

Schultz W, Dickinson A: Neuronal coding of prediction errors. Annual Review of Neuroscience: 23; 473–500, 2000.

Servan-Schreiber D: From physiology to behavior: Computational models of catecholamine modulation of information processing. Ph.D. Thesis: Carnegie Mellon University, 1990.

Shadmehr R, Holcomb H: Neural correlates of motor memory consolidation. Science: 277; 821–824, 1997.

Shallice T: Specific impairments of planning. Philosophical Transactions of the Royal Society of London Series B: Biological Sciences: 298; 199–209, 1982.

Shallice T, Burgess P: The domain of supervisory processes and temporal organization of behaviour. Philosophical Transactions of the Royal Society of London Series B: Biological Sciences: 351; 1405–1411, 1996.

Shiffrin RM, Schneider W: Controlled and automatic human information processing: II. Perceptual learning automaticity, attending and a general theory. Psychological Reviews: 84; 127–190, 1977.

Smith EE, Jonides J: Storage and executive processes in the frontal lobes. Science: 283; 1657–1661, 1999.

Smith EE, Jonides J, Marshuetz C, Koeppe RA: Components of verbal working memory: evidence from neuroimaging. Proceedings of the National Academy of Sciences of the United States of America: 95; 876–882, 1998.

Stuss DT, Benson DF: The Frontal Lobes. New York: Raven, 1986.

Swedo SE, Shapiro MB, Grady CL, Cheslow DL, Leonard HL et al.: Cerebral glucose metabolism in childhood-onset OCD. Archives of General Psychiatry: 46; 518–523, 1989.

Tzelgov J, Henik A, Berger J: Controlling Stroop effects by manipulating expectations for color words. Memory and Cognition: 20; 727–735, 1992.

Usher M, Cohen JD: Short term memory and selection process in a frontal-lobe model. In Heinke D, Humphries GW, Olsen A (Eds), Connectionist Models in Cognitive Neuroscience. Birmingham: Springer-Verlag, 1999.

Usher M, Cohen JD, Servan-Schreiber D, Rajkowski J, Aston-Jones G: The role of locus coeruleus in the regulation of cognitive performance. Science: 283; 549–554, 1999.

Wang XJ: Synaptic basis of cortical persistent activity: the importance of NMDA receptors to working memory. Journal of Neuroscience: 19; 9587–9603, 1999.

White IM, Wise SP: Rule-dependent neuronal activity in the prefrontal cortex. Experimental Brain Research: 126; 315–335, 1999.

Williams MS, Goldman-Rakic PS: Characterization of the dopaminergic innervation of the primate frontal cortex using a dopamine-specific antibody. Cerebral Cortex: 3; 199–222, 1993.

Wilson FAW, O Scalaidhe SP, Goldman-Rakic PS: Dissociation of object and spatial processing domains in primate prefrontal cortex. Science: 260; 1955–1958, 1993.

Yamaguchi S, Knight RT: Anterior and posterior association cortex contributions to the somatosensory P300. Journal of Neuroscience: 11; 2039–2054, 1991.

Yeung N, Botvinick MM, Cohen JD, under review: The neural basis of error detection: conflict monitoring and the error-related negativity, under review.

Zipser D, Kehoe B, Littlewort G, Fuster J: A spiking network model of short-term active memory. Journal of Neuroscience: 13; 3406–3420, 1993.

Subject Index *

---

* Underlined page numbers indicate in-depth treatment.